Bas Arts

THE POLITICAL INFLUENCE OF GLOBAL NGOs

Case Studies on the Climate and Biodiversity Conventions

International Books, 1998

ISBN 90 5727 012 9
Keywords: Environment/international affairs/NGOs/climate change/biodiversity

Copy editor: Annemarie Weitsel
Cover design: Marjo Starink
Cover photograph: Berlin Summit Action, Greenpeace/Vielmo
Desk Top Publishing: Hanneke Kossen
Printing: Drukkerij Haasbeek

International Books, A. Numankade 17, 3572 KP Utrecht, the Netherlands,
tel. +31 30 2731840, fax +31 30 2733614, e-mail: i-books@antenna.nl

Contents

List of Tables and Figures 9

List of Abbreviations and Acronyms 11

Preface 14

1 **General Introduction**
Subject and design of the study 16
1.1 Introduction 16
1.2 International environmental problems 17
1.3 International environmental policies 18
1.3.1 A short history 18
1.3.2 The United Nations Conference on Environment and Development (UNCED) 22
1.4 The political influence of NGOs 24
1.4.1 A short overview 24
1.4.2 NGOs and the UNCED 27
1.5 The design of the study 30
1.5.1 Iterative approach 30
1.5.2 Conceptual design 32
1.5.3 Technical design 37
1.6 Summary 46

2 **Theoretical Model**
A political arena perspective on NGOs in treaty formation and implementation 48
2.1 Introduction 48
2.2 The concept of 'NGO' 49
2.2.1 Definition 49
2.2.2 NGOs and the study of international relations 53
2.2.3 A political arena perspective 55
2.3 The concept of 'political influence' 57
2.3.1 Influence versus power 57
2.3.2 Agent versus structure 59
2.4 The concept of 'convention' 61
2.4.1 Definition 61
2.4.2 Conventions and regimes 63
2.4.3 Regimes as frames of reference 65
2.5 The context 67
2.5.1 The international system 67
2.5.2 Events and trends 69
2.6 The theoretical model 70
2.7 Summary 73

3 **Methodological Framework**
The EAC Method 74
3.1 Introduction 74
3.2 The assessment of political influence 75
3.2.1 Classical methodologies 75
3.2.2 Intensive process analysis 77
3.2.3 Adaptations 78
3.2.4 EAC Method 80
3.2.5 Extent of political influence 83
3.2.6 Discussion 87
3.3 The explanation of political influence 89
3.3.1 Explanatory factors 89
3.3.2 Comparative case-analysis 92
3.3.3 Respondents' views 94
3.4 The formulation of recommendations 95
3.4.1 The limits of prescriptive knowledge 95
3.4.2 Judgement criteria 95
3.4.3 Respondents' views 96
3.5 Summary 97

4 **NGOs and the UN Framework Convention on Climate Change** 98
4.1 Introduction 98
4.2 The UN Framework Convention on Climate Change (FCCC) 98
4.2.1 The issue of climate change 98
4.2.2 Contents of the FCCC 101
4.2.3 The policy cycle of the FCCC (1980-1997) 102
4.2.4 NGOs in the climate arena 108
4.3 NGOs in the formation phase (1990-1992) 110
4.3.1 The 'counterfactual' 110
4.3.2 Case 1: The objective of the FCCC 112
4.3.3 Case 2: Principles 117
4.3.4 Case 3: Target 121
4.3.5 Case 4: Provisions on implementation 128
4.4 NGOs in the implementation phase (1992-1995) 132
4.4.1 Case 5: AOSIS Protocol 133
4.4.2 Case 6: Joint implementation 138
4.4.3 Case 7: First review of national communications 144
4.4.4 Case 8: Berlin Mandate 148
4.5 An overview of the eight cases 153
4.6 Conclusions 155

5 **NGOs and the UNEP Convention on Biological Diversity** 157
5.1 Introduction 157
5.2 The UNEP Convention on Biological Diversity (CBD) 157
5.2.1 The issue of biological diversity 157
5.2.2 Contents of the CBD 160
5.2.3 The policy cycle of the CBD (1980-1997) 162
5.2.4 NGOs in the biodiversity arena 169
5.3 NGOs in the formation phase (1990-1992) 171
5.3.1 The 'counterfactual' 171
5.3.2 Case 9: The preamble 172
5.3.3 Case 10: Indigenous and local communities 177
5.3.4 Case 11: In-situ conservation 184
5.3.5 Case 12: Access to genetic resources 189
5.3.6 Case 13: Biosafety 196

5.4 NGOs in the implementation phase (1992-1995) 201
5.4.1 Case 14: Indigenous people 202
5.4.2 Case 15: Marine biodiversity 208
5.4.3 Case 16: Forests 211
5.4.4 Case 17: Biosafety 216
5.4.5 Case 18: Financial mechanism 220
5.5 An overview of the ten cases 227
5.6 Conclusions 228

6 The Explanation of Political Influence 229
6.1 Introduction 229
6.2 A hierarchy of cases 230
6.3 Explanations at the level of the political arena 234
6.3.1 NGOs and their interventions 234
6.3.2 States and other players 237
6.3.3 Policy issues 241
6.3.4 Negotiations and decision-making 242
6.4 Explanations at the level of the policy outcome 244
6.4.1 Regime elements 244
6.4.2 Domestic policy effects 246
6.4.3 Global effects 247
6.5 Explanations at the contextual level 248
6.5.1 UNCED and other contextual factors 248
6.5.2 Resources and regimes 249
6.5.3 Rules of the game, trends, events 253
6.6 The respondents' views 257
6.7 Enabling, constraining and neutral factors 261
6.8 Conclusions 266

7 Recommendations 267
7.1 Introduction 267
7.2 From explanatory factors to recommendations 268
7.2.1 Anticipation 268
7.2.2 Enabling factors 269
7.2.3 Constraining factors 277
7.2.4 Neutral factors 283
7.3 The respondents' views 286
7.4 A strategic model 291
7.5 Extrapolation 295
7.6 Conclusions 298

8 **Summary, General Conclusions and Discussion** 300
8.1 Introduction 300
8.2 Summary and debate 300
8.2.1 The political influence of global NGOs 300
8.2.2 The explanation of political influence 304
8.2.3 The increasing of political influence 307
8.3 Reflections on this study 310
8.3.1 Design 310
8.3.2 Theory 314
8.3.3 Methodology 317
8.3.4 Validity and reliability 318
8.4 Conclusions 320

Samenvatting 322

Bibliography 329

Appendix I: FCCC Documents 341

Appendix II: CBD Documents 346

Appendix III: Background of Respondents 351

Appendix IV: Curriculum Vitae 352

List of Tables and Figures

— Tables —

2.1 A classification of pressure groups 50
2.2 Activities of pressure and protest groups 52

3.1 The EAC Method 81
3.2 General format to assess political influence on a case-by-case basis 86
3.3 Possible explanatory factors of the NGOs' political influence 90
3.4 Format of the data-matrix to deduce explanatory factors of political influence 93

4.1 Activities of climate NGOs 109
4.2 The counterfactual of the FCCC as perceived by the respondents from the climate arena 111
4.3 Scores on political influence in case 1 (Objective FCCC) 116
4.4 Scores on political influence in case 2 (Principles FCCC) 121
4.5 Scores on political influence in case 3 (Target FCCC) 127
4.6 Scores on political influence in case 4 (Implementation FCCC) 131
4.7 Scores on political influence in case 5 (AOSIS Protocol) 138
4.8 Scores on political influence in case 6 (Joint Implementation) 144
4.9 Scores on political influence in case 7 (First review) 147
4.10 Scores on political influence in case 8 (Berlin Mandate) 152
4.11 Overview of the eight FCCC cases 154

5.1 The counterfactual of the CBD as perceived by the respondents from the biodiversity arena 171
5.2 Scores on political influence in case 9 (Preamble CBD) 177
5.3 Scores on political influence in case 10 (Indigenous people) 183
5.4 Scores on political influence in case 11 (In-situ conservation) 189
5.5 Scores on political influence in case 12 (Genetic resources) 195

5.6 Scores on political influence in case 13 (Biosafety) 201
5.7 Scores on political influence in case 14 (Indigenous people) 207
5.8 Scores on political influence in case 15 (Marine biodiversity) 211
5.9 Scores on political influence in case 16 (Forests) 215
5.10 Scores on political influence in case 17 (Biosafety) 220
5.11 Scores on political influence in case 18 (Financial mechanism) 225
5.12 Overview of the ten CBD cases 226

6.1 Hierarchy of cases based on the extent of the NGOs' political influence 232
6.2 Factors at the arena level I: NGOs and their interventions 236
6.3 Factors at the arena level II: states, other players, policy issues 238
6.4 Factors at the arena level III (policy phases, negotiations) and at the outcome level (regime elements, domestic effects) 243
6.5 Factors at the contextual level: Resource dominance and regimes 250
6.6 Trends and events in the international system in the period 1990-1995 255
6.7 Factors mentioned by respondents to explain the political influence of NGOs 258
6.8 Factors mentioned by respondents to explain the lack of political influence of NGOs 260
6.9 Enabling, constraining and neutral factors on the explanation of the NGOs' political influence 263

7.1 Factors to increase political influence mentioned by the respondents 288

8.1 FCCC and CBD cases and the extent of political influence of global NGOs 303
8.2 The most prominent explanatory factors of the NGOs' political influence 305

— Figures —

1.1 Framework of the study 34
2.1 The main concepts 49
2.2 A political arena model 71
3.1 Indicators of the EAC Method 82
6.1 Causal model on political influence 265
7.1 Strategic model for global NGOs 292

List of Abbreviations and Acronyms

AGBM	Ad Hoc Group on the Berlin Mandate
AOSIS	Alliance of Small Island States
BCAS	Bangladesh Centre for Advanced Studies
BINGO	Business Non-Governmental Organization
Bionet	Biodiversity Action Network
CAN	Climate Action Network
CBD	Convention on Biological Diversity
CEE	Central and Eastern Europe
CFCS	Chlorofluorocarbons
CH_4	Methane
CITES	Convention on International Trade in Endangered Species
CO_2	Carbon dioxide
COP	Conference of the Parties
CSCE	Conference on Security and Cooperation in Europe
CSD	Commission on Sustainable Development
CSE	Centre for Science and Environment
EAC	Ego-perception, Alter-perception, Causal analysis
EC	European Community (up to 1992)
Eco	An NGO journal published at intergovernmental, environmental negotiation sessions
ECOSOC	Economic and Social Council of the United Nations
EDF	Environmental Defense Fund
ELCI	Environment Liaison Centre International
EU	European Union (since 1992)
FAO	Food and Agriculture Organization
FCCC	Framework Convention on Climate Change

FIELD	Foundation for International Environmental Law and Development
FOEI	Friends of the Earth International
G7	Group of the Seven Main Industrialized Countries
G77	Group of Developing Countries
GATT	General Agreement on Tariffs and Trade
GEF	Global Environment Facility
GHG(s)	Greenhouse gas(es)
GRAIN	Genetic Resources Action International
ICCBD	Intergovernmental Committee on the Convention on Biological Diversity
IGO	Intergovernmental Organization
IIPA	Indian Institute of Public Administration
IILG	Interim International Liaison Group on Biological Diversity
ILO	International Labour Organization
INC	Intergovernmental Negotiation Committee
INGO	International Non-Governmental Organization
IO	International Organization
IPCC	Intergovernmental Panel on Climate Change
IPF	Intergovernmental Panel on Forests
IUCN	World Conservation Union
JI	Joint Implementation
JUSCANZ	Group of Japan, USA, Canada and New Zealand
LRTAP	Convention on Long Range Transboundary Air Pollution
NATO	North Atlantic Treaty Organization
NGO	Non-Governmental Organization
NRDC	Natural Resources Defense Council
O_3	Ozone
ODA	Official Development Assistance
OECD	Organization for Economic Cooperation and Development
OPEC	Organization of Petroleum Exporting Countries
PGR	Plant Genetic Resources
Rafi	Rural Advancement Foundation International

SBI	Subsidiary Body for Implementation
SBSTA	Subsidiary Body for Scientific and Technological Advice
SBSTTA	Subsidiary Body for Scientific, Technical, and Technological Advice
SEED	Solidarity, Equity, Ecology and Development
SNM	Stichting Natuur en Milieu (Dutch Society for Nature and Environment)
STAC	Scientific and Technological Advisory Committee
SWCC	Second World Climate Conference
TWN	Third World Network
UK	United Kingdom
UNCED	United Nations Conference on Environment and Development
UNCHE	United Nations Conference on the Human Environment
UNCTAD	United Nations Conference on Trade and Development
UNDP	United Nations Development Programme
UNEP	United Nations Environment Programme
UNESCO	United Nations Educational, Scientific and Cultural Organization
UNGA	United Nations General Assembly
USA	United States of America
WCED	World Commission on Environment and Development
WG	Working Group
WHO	World Health Organization
WIPO	World Intellectual Property Organization
WMO	World Meteorological Organization
WRI	World Resources Institute
WTO	World Trade Organization
WWF	World Wide Fund for Nature

Preface

The story of this book begins in Brazil, in Rio de Janeiro in June 1992. It might have been the 11th or 12th of June – I cannot recall the exact date – that the subject of this research project occurred to me for the first time. This happened in an awfully hot and crowded tent near the beautiful beaches of Rio. In this tent at the Global Forum, the non-governmental meeting organized parallel to the United Nations Conference on Environment and Development (UNCED), representatives of NGOs had gathered: 'yuppie' lobbyists, indigenous chiefs, environmental activists, punks and 'New Age' people. The aim was to discuss strategies for changing global society and politics in order to save the planet. But the atmosphere was a bit depressing. The Global Forum had, according to many, not been a success so far, since it turned out to be more of an alternative 'street-fair' than an attempt to seriously address UNCED matters, as some journalists reported with cynical joy. Also, there were rumours of bankruptcy of and corruption within the organization of the Global Forum (although these turned out to be false in the end). Last but not least, the venue for the Global Forum had put the activists in a position of isolation. While the UNCED took place in the Rio Centre some 25 miles outside town, the Global Forum was in Rio itself. Hence, demands, wishes and views from the Forum could neither easily nor quickly be handed over to the policy- and decision-makers gathered at the UNCED. Overall, NGO representatives felt unhappy and some sighed: 'What on earth are we doing in Rio?' They believed they could not have any impact on the UNCED outcome whatsoever. And then it crossed my mind, the question which in the end gave birth to this book: 'Could NGOs have influenced the outcome of the Earth Summit or not?'

So the people I need to thank first are all those NGO representatives who inspired me to write this thesis and who provided me with the necessary information, in particular those I met at meetings in Nassau (Bahamas), Geneva (Switzerland), Berlin and Bonn (Germany), Brussels (Belgium), and several times in the Netherlands. I want to explicitly mention two: Simone Lovera-Bilderbeek and Bill Hare. Without their assistance on several occasions, this book could not have been written. Looking at the other side of the negotiating table, I am grateful to all government representatives, diplomats and UN officials who agreed to be interviewed by me, although they had never met me (as the interviews were conducted by telephone). I was struck by their openness in talking to me about potentially sensitive matters.

Also, I owe a lot to my colleagues, friends and family. Without them this research project would definitely have failed (which is, incidentally, another example of 'influence'). Firstly, I would like to mention the very fruitful co-operation with my supervisors Prof. Klaas Bouwer, Prof. Nico Nelissen and Dr. Piet Verschuren. While Klaas focused on the details of my writing and Nico watched the headlines, Piet monitored my methodological approach. I particularly enjoyed our sessions in which we debated my work. Secondly, I would like to mention Leon Wecke and Prof. Gerrit Huizer. With their help I became involved in a project on sustainable development at the University of Nijmegen in the period 1989-1993, which – in retrospect – turned out to be the ideal preparation for this research project and book. Thirdly, I am grateful to my (former) colleagues at the Department of Environmental Policy Sciences, the Peace Research Centre and the Third World Centre. Their comments on my work were not *always* relevant (especially at the time when they were so jealous about my spending three weeks in the Bahamas), but generally they were a great help and joy to me. Fourthly, the staff members of our secretariat need to be mentioned, as they assisted me in finishing the manuscript. The same applies to Miriam Janssen, although her support – in the private sphere – very much transcended putting final touches to the manuscript. Fifthly, I am grateful to Monika Sie Dhian Ho and Bert-Jan Verbeek, who commented on my writings on international relation theories. They helped me, as a 'lost biologist', to feel more confident in a new academic field. Finally, I am extremely grateful to all my friends and family. They were a marvellous support in the period I wrote this book, mainly by *not* discussing it with me.

CHAPTER I

General Introduction
Subject and design of the study

— 1.1 Introduction —

The world faces a number of international environmental problems today, such as global warming and the depletion of the ozone layer, which require international environmental policies for their management and, eventually, their solution. Several countries and international organizations – such as the United Nations (UN) and European Union (EU) – are catalysts in drawing up such policies. However, Non-Governmental Organizations (NGOs) also play a role, and some academics even believe that it is difficult to overestimate the importance of NGOs for international environmental policies. The relevance of NGOs also became obvious in the context of the so-called UNCED process. The United Nations Conference on Environment and Development (UNCED) – or Earth Summit – was held in Rio de Janeiro, Brazil, in June 1992. Thousands of NGOs were involved, many of them environmental pressure groups. The question is whether this participation was effective in terms of political influence. In other words: did NGOs indeed have an impact on the outcome of Rio? Answers to this question differ, and, moreover, an assessment of NGO influence is quite a challenge from a scientific point of view. Therefore, this study focuses on this question by considering two cases: the Framework Convention on Climate Change (FCCC) and the Convention on Biological Diversity (CBD). Both treaties were signed by more than 150 countries at the Earth Summit.

In this chapter, section 2 briefly describes the nature of environmental problems worldwide and points up several trends. Section 3 focuses on the nature of international environmental policies. Different aspects of such policies are highlighted: international conferences, international law, action programmes, reports, and conflicts between the rich and poor countries. Section 4 deals with the political influence of NGOs at the global level. Several examples of (supposed) influence are considered, but the role of NGOs in the UNCED is emphasized. Section 5 focuses on the conceptual and technical design of this study: objectives, questions, strategies, techniques and materials are identified. Section 6, finally, gives a summary of this chapter.

— 1.2 International environmental problems —

Our planet is under siege. Assaults on the atmosphere – the greenhouse effect, the depletion of the ozone layer and increasing air pollution – pose a still unquantified threat to human life. The dumping of hazardous wastes, and land-based sources of pollution, present a similar threat to the oceans. On land we are destroying a tropical forest the size of Austria every year, and more than a hundred species of wild plants and animals are lost forever each day.

With these alarming words, Mostafa K. Tolba, former Executive Director of the United Nations Environment Programme (UNEP), opens *The World Environment 1972-1992* (Tolba and El-Kholy, 1992). This voluminous work reviews the trends in the global environment of the last two decades and identifies the issues that are at stake. It addresses the following international problems: air pollution; ozone depletion; climate change; availability of fresh water; coastal and marine degradation; land degradation; deforestation; habitat loss; loss of biological diversity; environmental hazards; toxic chemicals; hazardous wastes. All these issues affect, or may potentially affect, the quality of our physical environment and may therefore pose risks to human health or even to human survival. In consequence, these issues are defined as *environmental problems* (Boersema *et al.*, 1991; Bouwer, 1985, 1990; Bouwer and Leroy, 1995). Some are of a global nature (e.g. climate change), while others are regional (e.g. land degradation in Africa), or even local (e.g. coastal degradation). Whatever the geographical level, however, all these problems have environmental effects that go beyond the territories of individual states. Therefore, they are called *international* environmental problems (Caldwell, 1990).

According to the *World Environment 1972-1992*, the trends are not encouraging. To sum up some examples:

– The concentration of most air pollutants is still increasing, both in the rich countries and in the urban centres of the Third World. For example, the problem of acidification is still growing worse in many regions of the world despite the use of ever cleaner technologies in the West. Acidification is caused by emissions of sulphur and nitrogen oxides by industry and traffic amongst others. Lake and forest ecosystems are particularly vulnerable to acidification.

– The hole in the ozone layer over Antarctica is increasing in size, which poses a threat to human health (such as skin cancer due to an increase of UV-B radiation). However, the 1996 ban on the production of chlorofluorocarbons (CFCs) – one of the substances that causes this problem – is good news. Scientists expect the ozone layer to recover, albeit slowly.

– There is more and more evidence to suggest that human-induced climate change is indeed occurring, although some sceptical scientists insist that global warming is a chimera. If the climate does change, the consequences will be enormous. Both ecosystems and agricultural systems will alter fundamentally, and drought and famine in tropical regions may be aggravated.

– The regional seas and the oceans are also increasingly affected by humans. Examples are overfishing, oil releases from ships, and the dumping of hazardous or nuclear waste.

– The degradation of land – in particular of drylands – is a genuine threat. Agricultural lands deteriorate and may even desertificate. The poor of the world are the most seriously affected, all the more so since the loss of agricultural lands generally coincides with unequal ownership of and access to those lands on the one hand and population growth on the other.

– Deforestation is estimated to amount to an annual loss of approximately 17% of the forest cover in the tropical regions. In addition, boreal forests in Europe are vanishing.

– More and more species are becoming extinct, although biologists disagree on the exact number and rate. In addition, many species are threatened, for example 20% of all reptiles and 10% of all mammals.

It should be recognized, however, that these issues are not interpreted in the same way by everyone and that scientists draw different conclusions about the seriousness of the situation. Moreover, environmental problems are differently 'constructed' in various social contexts and eras (Liberatore, 1995). While a huge factory chimney is still a symbol of progress in several regions of the world, it is perceived as a threat to air quality and human health in others. In addition, optimistic or pessimistic views of human nature and human capacity lead to different pictures of the future (Van den Abbeele, 1985; Nelissen *et al.*, 1997). Some paint doom scenarios and warn against ecological disaster; others believe that environmental problems are mainly of a technical nature and can therefore be solved. Nevertheless, the fact that these problems pose risks to human society is generally accepted by most scientists and policy-makers.

— 1.3 International environmental policies —

1.3.1 A short history

The identification of international environmental problems has legitimized and intensified the formulation of international environmental policies, although it should be recognized that states may have motives other than environmental ones for contributing to such processes, such as security or economic interests (Junne, 1993). In this thesis, an international environmental policy is defined as any common and formal initiative taken by two or more states which aims at solving or softening an international environmental problem (Caldwell, 1990; Taylor, 1993; Van de Graaf and Hoppe, 1992). Such initiatives may be conferences, reports, declarations, action plans, binding treaties, funds, green taxes, or combinations of these. Nowadays, most policies are designed in the context of International Organizations (IOs) – such as the EU and UN – in collaboration with other players, such as business groups, scientists, and NGOs. International environmental *policies*, though, should be distinguished from international environmental *politics*, as the latter concept is wider in nature and scope. In this thesis, the term international environmental politics refers to the struggle between political players on the content of international environmental policies (Van de Graaf and Hoppe, 1992). Such a struggle generally takes the form of debate, agenda-setting, intergovernmental negotiations, diplomacy, and authoritative decision-making in international arenas. The input is different policy goals based on individual or collective interests of all the political actors involved. The output is a specific policy. In short, the term politics refers to both process and product, whereas the term policy refers only to the product itself.

It is generally assumed that the development of international environmental policies was initiated in 1972 (Caldwell, 1990; Caroll, 1988; Hurrell and Kingsbury, 1992). In that year, the United Nations Conference on the Human Environment (UNCHE) was held in Stockholm (Sweden). This meeting produced a declaration, an action plan, and a new, though small, UN body: the United Nations Environment Programme (UNEP). More significantly however, the UNCHE became the starting point for a large number of environmental initiatives at the global level. Today, there are some 150 multilateral legal instruments on the environment, two-thirds of which were established after 1972; UNEP was involved in nearly half of these latter cases (Rowlands and Greene, 1992). Such legal instruments concern the following areas (Bergesen and Parmann, 1993; Caldwell, 1990):

– nature conservation (for example: the Convention on International Trade in Endangered Species of Wild Flora and Fauna (CITES); the Ramsar Convention on Wetlands of International Importance; the Bonn Convention on the Conservation of Migratory Species of Wild Animals; the UNEP Convention on Biological Diversity);

– sea pollution (for example: the International Convention for the Prevention of Pollution from Ships, and the London Convention on the Prevention of Marine Pollution by Dumping of Wastes and Other Matter);

– nuclear and toxic waste (for example: the Basel Convention on the Control of Transboundary Movements of Hazardous Wastes);

– pollution of the atmosphere (for example: the Convention on Long-Range Transboundary Air Pollution (LRTAP); the Vienna Convention and Montreal Protocol for the Protection of the Ozone Layer; the United Nations Framework Convention on Climate Change).

Moreover, many international organizations and forums have put the issue of the environment on their agendas since the early 1970s. Various examples are: the General Agreement on Tariffs and Trade (GATT), the Group of Seven Industrialized Countries (G7), the Organization for Economic Cooperation and Development (OECD), the North Atlantic Treaty Organization (NATO), the Conference for Security and Cooperation in Europe (CSCE), the European Union (EU), the UN Conference on Trade and Development (UNCTAD), the UN Development Programme (UNDP), the World Bank, and others.

It would be misleading to state, however, that environmental policies 'started' in 1972 (McCormick, 1989). UN organizations had already been involved in the preservation of nature, the conservation of natural resources and the management of pollution before the 1970s. UNESCO established the International Union for the Protection of Nature in 1947 (this became the IUCN in 1956). The FAO, WHO, UNESCO, and ILO organized the UN Scientific Conference on the Conservation and Utilization of Resources in 1949. The so-called Biosphere Conference was held under UNESCO auspices in 1968. But even before World War II, international organizations had been involved in conservation policies, for example on migratory birds in Europe. Nonetheless, the focus and scale of activities changed after 1970. The emphasis shifted from nature (flora, fauna, resources) to the abiotic environment (air, water, soil), and policy-making was intensified.

Although the Stockholm conference marked the beginning of an era, public interest in environmental issues was subject to strong fluctuations (Arts and Mulleneers, 1995; Weiss *et al.*, 1994). The 1970s witnessed a sharp decrease of public interest due to the oil crises, economic stagnation and high unemployment rates. Moreover, the doom scenarios of many environmentalists, which had been so popular at the end of the 1960s, were increasingly criticized. Consequently, the environment issue slowly disappeared from the international agenda. It re-appeared unexpectedly in the mid-1980s, for several reasons. Firstly, the world economy was doing better in this period than in the 1970s. Secondly, new environmental issues had come up, such as the greenhouse effect and the ozone layer. Moreover, so-called shock events – the discovery of the hole in the ozone layer above Antarctica and the nuclear and industrial disasters in Chernobyl and Bhopal respectively – caused widespread concern. Thirdly, the easing of East-West tensions and the subsequent end of the Cold War created new opportunities in the environmental field (Opschoor, 1989b; Käkönen, 1994). Gorbachev's 'new thinking' included ecological aspects, and disarmament talks fed hopes of large sums of money being transferred to the environmental sector (peace dividend). What is more, several environmental movements played an important part in the civil protest which led to the fall of the Berlin Wall in 1989. Fourthly, the environment issue was taken more seriously in developing countries as well. Southern leaders were reluctant to deal with it in the 1970s, but they slowly accepted the issue in the 1980s. In other words, there was a growing global consensus on the need for dealing with environmental issues at that time.

Finally, new initiatives were undertaken by UN bodies. In 1982 UNEP reviewed the Stockholm Conference. This review was not encouraging, so new initiatives were launched. The World Commission on Environment and Development (WCED), headed by Gro Harlem Brundtland, was subsequently established and produced the famous report *Our Common Future* in 1987, which urges governments to pursue sustainable development, that is, a development style which covers economic, social, ecological and long-term considerations on a global scale (WCED, 1987). Since its publication, sustainable development has become one of the buzzwords in the international community. Furthermore, the Brundtland report foreshadowed the United Nations Conference on Environment and Development. When the UNCED was held in Rio de Janeiro in 1992, the environment issue was high on the international agenda. Since then it has abated somewhat, though not as dramatically as in the 1970s. Consequently, the environment is given due attention in international politics today, but the 'Rio hype' is over.

To conclude this section, it should be recognized that international environmental policies are of a different nature to national ones. Several differences can be distinguished here. First of all, the international community lacks a central authority which has top-down legislative and authoritative powers (Lieshout, 1993; Waltz, 1979). After all, the nation states of the world are sovereign, which formally implies territorial integrity for all states, no interference in internal affairs by others, and legal equality for all (Hocking and Smith, 1990). Therefore global policies are based on bottom-up co-operative arrangements – agreements, declarations, treaties – between states, in a more or less anarchic setting (Schermers, 1985a, 1985b). Consequently, common initiatives generally consist of broad, vague and weak compromises. Secondly, the environment issue is less salient in the international arena than in for example Western Europe. International politics generally focuses on security and economic issues (so-called 'high politics') and less on environmental ones ('low politics') (Koch *et al.*, 1994; Morgenthau, 1993). Consequently, agenda-setting on environmental issues is less obvious at global than at national level. Thirdly, international policies are generally based on negotiations aiming at unanimity, as well as on consensus decision-making. Therefore the law of the *lowest common denominator* operates (Hurrell and Kingsbury, 1992). This means that all parties implicitly have the right to veto any proposal, with the result that those who want least, impact the outcome most. Fourthly, the national implementation of international policies is difficult to enforce, given – again – the sovereignty of states (French, 1994; Liefferink *et al.*, 1993). Moreover, there are hardly any sanctions with which to punish deviant states. After all, the powers of international review bodies, tribunals and courts are in general limited. In short, compared to national ones, international policies are hard to design, implement and enforce.

Nonetheless, there are exceptions to all the problems mentioned above. For example, the meaning and effects of national sovereignty and international anarchy are relaxed by a great number of interdependencies among states, by international law and regimes as well as by international organizations (Keohane and Nye, 1989; Rittberger, 1993; Young, 1993). Some of the latter even have legislative powers (e.g. the European Commission). Besides, environmental issues do regularly appear high on the international agenda; qualified majority decision-making is sometimes possible; blockers of environmental negotiations have been overruled in several cases; and the possibility of using sanctions against deviant states – e.g. by trade measures – are provided for in some treaties (Carroll, 1988). Therefore international environmental policies may exceed the traditional constraints of the international system in a number of cases.

1.3.2 The UNCED

The United Nations Conference on Environment and Development – also referred to as the UNCED or Earth Summit – produced the Rio Declaration on Environment and Development, the Authoritative Statement on Forest Principles, Agenda 21, and the UN Commission on Sustainable Development (Johnson, 1993). In addition, more than 150 countries signed the UN Framework Convention on Climate Change (FCCC) and the UNEP Convention on Biological Diversity (CBD), although these two instruments were designed in distinct but parallel UN forums.

The Rio Declaration defines a number of legally non-binding principles, agreed upon by all nations, which should guide the design and implementation of international environmental and developmental policies. Important notions in this document are environmental management, human needs, national sovereignty, global responsibility, co-operation, precautionary approach, and prevention.

The Authoritative Statement on Forest Principles was a substitute for a binding forest treaty, which states had not been able to agree on before Rio. After all, forests are a very sensitive issue in global politics, in particular for Southern countries, which do not like to see their economic development hampered by conservation goals (Kolk, 1996). The non-binding Forest Principles are generally considered weak from an ecological point of view.

Agenda 21 is a non-binding action plan on sustainable development in the 21st century. It deals with topics such as poverty, consumption patterns, natural resources, forests, waste, health, oceans, decision-making, the role of NGOs, trade, and the implementation of measures. It covers all conceivable aspects of sustainable development, but many observers wonder whether states will be able and willing to implement all these good intentions. Given the lack of financial commitment by the industrialized countries at Rio, expectations do not run high. Finally, the UNCED established the UN Commission on Sustainable Development (CSD) to monitor the implementation of Agenda 21.

Many conflicts attend international environmental policies (Gupta *et al.*, 1993; Sprinz and Vaahtoranto, 1994; Van der Wurff, 1997). The different players involved generally defend different (national) interests: victims of environmental degradation will endorse other measures than those who cause it or those who are not affected by it. The same goes for those who have the technology and money to solve a problem and those who do not. International environmental policies are therefore characterized by intense struggles. The most visible one is the conflict between the developed and the developing countries (Lafferière, 1994; Miller, 1995). A North-South divide already became apparent during the preparations for the Stockholm Conference (1972). Contrary to the opinion of the Northern countries, the Southern countries were not convinced that the environment problem concerned them. Firstly, they considered it as a problem of luxury, as an effect of the Western, capitalist way of life. Secondly, they believed that it could not be a problem for developing countries yet, as these were simply too poor to be able to cause pollution on any significant scale. In addition, they feared that the environment could indeed become a problem for them, though not in a biophysical sense: the South feared that environmentalism could easily turn into protectionism and conditionalism. The first notion refers to the possibility that the rich countries would refuse Southern products access to their markets, legitimizing such protectionism with environmental arguments. The second notion concerns the fear of Southern countries that development aid would be tied to environmental conditions. Because of these worries, a common Third World position on international environmental policies has developed since the 1970s. It is based on six themes (Williams, 1993):

1. developing countries have the sovereign right to exploit their natural resources in order to accelerate their development process;
2. industrialized countries have to bear the major costs of international environmental policies, as the historical responsibility for the current problems lies with them;
3. environmental policy measures should not hinder the development process in the South;
4. the developing countries should have free access to modern, environmentally sound technology;
5. regular aid should not be tied to any environmental conditions;
6. additional resources should be transferred from the rich North to the South to enhance environmental protection there.

Due to this North-South controversy, international environmental policies increasingly addressed developmental aspects. This is also expressed in the names of the Stockholm and Rio conferences ('on the Human Environment' and 'on Environment and Development', respectively). We shall see that the North-South debate also had a major impact on the Climate and Biodiversity Conventions.

— 1.4 The political influence of NGOs —

1.4.1 A short overview

An NGO is generally defined as any organization which is not established by a government or group of governments (Feld and Jordan, 1983). Others refer to NGOs as non-profit pressure groups (Thompson-Feraru, 1974). The former definition is very broad – including for example churches, professional associations, terrorist groups and sports clubs – whereas the latter concerns groups that represent specific interests in political arenas. The latter definition is adopted within the context of this study, given the fact that it focuses on pressure groups – in particular environmental organizations – which try to influence international policies. The study of pressure groups, though, has merely focused on the national level (Van Noort *et al.*, 1987). Good examples are the studies by Wilson (1981) and Huberts (1988). The former deals with agricultural groups, unions, business organizations, voluntary associations, and their lobbyists in the American political system. Items such as interest definition, the backgrounds of such interests, the success of groups in terms of representation and influence, and the typical American nature of lobbying are analyzed. Huberts, on the other hand, assessed the influence of environmental protest and pressure groups on the design and construction of highways in the Netherlands. He concluded that in some cases these groups were influential.

In international relations literature the emphasis on pressure groups is less prominent. After all, the mainstream view focuses on nation states and their interactions (Lieshout, 1993; Morgenthau, 1993; Waltz, 1979). This state-biased perspective, though, has been challenged by many. For example Morss (1991, p. 55) writes:

> "The era in which the nations ruled the world is over. With the information revolution and the demise of the United States as the dominant world power, three groups have joined nations as important global players: transnational corporations, international organizations, and special interest groups."

Morss' third type of global players – special interests groups – refers to NGOs. Arguments to defend the proposition that these have become global players are the following: 1. their number has dramatically increased at the global level (from approximately 1,000 in 1950 to more than 15,000 now); 2. their constituencies have been tremendously broadened (for example the environmental movement is said to 'represent' millions of civilians worldwide); 3. their involvement in the UN system is formally recognized (Article 71 of the UN Charter); 4. they are believed to have impacted the content and course of international policies (Hocking and Smith, 1990; Van der Schot and Van de Veen, 1997). Points 1 and 2 are obvious, but points 3 and 4 need some clarification.

Article 71 of the UN Charter states that the Social and Economic Council (ECOSOC) may consult international NGOs (INGOs) which are concerned with matters within the ECOSOC's competence, amongst others the organization of UN conferences on special issues (Feld and Jordan, 1983; Jacobson, 1984). This clause on INGOs was added to the Charter due to pressure from mainly American NGOs and the US delegation while the UN system was established just after World War II (Woods, 1993). Since then, it has become standard practice that INGOs – and national NGOs at a later stage – may participate in UN conferences and meetings, at least if they cover the issues on the agenda and are formally accredited by UN staff. NGOs have observer status at such meetings, but they may generally make oral statements, and disseminate written position papers as well. In addition, they may have access to working groups in which negotiations between countries take place, but this is decided on a case-by-case basis by the officials. Countries may also add NGO representatives to their delegations at international meetings, but this is dependent on the willingness of individual countries and not common practice yet. Apart from this formal access, NGOs can lobby in the corridors of such UN meetings, and non-accredited NGOs may organize activities around the building where the negotiations take place. Such access raises the question whether NGOs can indeed influence intergovernmental policy-making. The book *Pressure Groups in the Global System* gives some examples with respect to that (Willets, 1982): the role of the Anti-Apartheid Movement in the economic boycott of South-Africa; the impact of Amnesty International on human rights issues; the influence of Friends of the Earth on nuclear policies; additionally, the effect of women's organizations on UN conferences, for example, are dealt with. The study concludes that on all these issues there have been spectacular successes for pressure groups (Willets, 1982, p. 195).

Similar claims are also made in the literature on international environmental policies. For example Caldwell (1990, p. 313) notes: "The importance of the NGOs in international environmental policy can hardly be overemphasized." Several examples of (supposed) influence are summed up below. Needless to say this selection is quite random and far from complete.

– European preservation groups – and not governments – were instrumental in the birth of one of the first intergovernmental bodies on nature protection established in 1913: the European Consultative Commission for the Protection of Nature. (Caldwell, 1990)

– The European Union set emission standards for 'clean' cars in 1991. Although these standards were a compromise among members of the EU and did not go as far as the NGOs' objectives, the environmental lobby in Brussels was – according to observers – quite effective. (Liefferink, 1995)
– More than 400 NGOs – mainly transnational and western ones – were officially represented at UNCHE in Stockholm (1972) and several of them were also involved in its preparation. In addition, parallel NGO activities were organized, such as the Environment Forum and the People's Forum. Although the effect of NGOs on the formal outcome of UNCHE is generally considered negligible in the relevant literature, the conference set a precedent with regard to the involvement of NGOs in global environmental policies. This was also due to the lobbying by NGOs, which did not only aim at environmental objectives, but at changing customs and procedures concerning NGO involvement in the UN system as well. (McCormick, 1989; Thomas, 1992; Thompson-Feraru, 1974)
– NGOs were key players in the final acceptance of the ivory trade ban under CITES (although this ban has recently been amended). They were the only ones in this arena who were able to truly link the local level (knowledge on the hunting of elephants and the (illegal) trade in ivory) to the global (knowledge on the diplomacy within the context of CITES). (Princen, 1994)
– The parties to the Antarctic Treaty adopted an Environmental Protocol in 1991. This protocol defines preservation objectives for the Antarctic region and, more importantly, issues a (temporary) ban on the mining of mineral resources. NGOs were instrumental in the adoption of this protocol by promoting the preservation of Antarctica in this arena for years. (Clark, 1994)
– NGOs were involved in the ozone layer issue since the beginning of the 1970s. They successfully campaigned against aerosols with CFCs and against McDonald's fast-food packaging, and lobbied for an immediate ban on the production of CFCs. In 1990 the countries which are parties to the Montreal Protocol adopted a phase-out of CFCs before the year 2000. (Rowlands, 1995)
– NGOs – particularly Friends of the Earth International and Greenpeace International – have lobbied for a total ban on whaling since the 1970s, and already did so at the UNCHE in Stockholm. A great deal of this lobbying was directed at the International Whaling Commission. Today a commercial moratorium is in place under the banner of this commission. (Burke, 1982; Rowlands and Greene, 1992)
– Greenpeace and others lobbied for the acceptance of the precautionary principle by the parties to the London Dumping Convention. It was accepted in 1990. Since then the parties are supposed to minimize the *production* of waste – which is destined to be dumped in seas and oceans – instead of minimizing the *effects* of waste on marine ecosystems. (Stairs and Taylor, 1992)
– More than 1,300 NGOs were accredited to the UNCED in Rio de Janeiro and some 7,500 gathered in the city at parallel conferences and meetings (Global

Forum, NGO Forum). Many of these organizations were also involved in the preparations for this conference and some even in the drafting of documents. (Finger, 1994)

It should be recognized, though, that the NGOs' influence is generally assumed in the examples above, not thoroughly assessed nor measured on the basis of standard methodologies. Besides, some reject the political influence of pressure groups in some of these areas, such as ozone (Rowlands, 1995). Furthermore, the NGOs' impact is not compared to the influence of other players who were also involved, for example business organizations, scientists, international organizations or individual countries. And the above examples are biased in the sense that NGOs had to face many failures in international environmental policies as well. For example, they have campaigned for a binding convention on the conservation and sustainable use of forests worldwide for years, without success (Humphreys, 1994; Kolk, 1996). Finally, it should be recognized that notions of political influence, power, impact, success, and effect – to mention the ones utilized above – are far from similar. Therefore conceptual delineations will be made below.

1.4.2 NGOs AND THE UNCED

According to many, the UNCED set a precedent with regard to NGO participation in international environmental policies (Doherty, 1994). Never before did so many NGOs participate in an UN conference, both as observers and – to some extent – as members of country delegations. Also it seemed as if they had never been so warmly welcomed by the authorities (Strong, 1992). Nonetheless, views differ among NGO representatives and academics as to whether the NGO participation in UNCED was really a success. Some believe it strengthened the movement, whereas others think the opposite. According to a representative of WWF (quoted in: Simms, 1993, p. 99): "The UNCED process has led to a strengthening and broadening of the role and capacities of NGOs in international relations." This view covers the general belief that the involvement of ever more NGOs from all regions of the world in the Earth Summit produced a rather strong and worldwide NGO network on environment and development. Such a network would enable NGOs to be better prepared to participate in international environmental policies in the future. Besides, non-governmental views from all continents were exchanged which made the definition of common messages and goals possible.

Chatterjee and Finger (1994, p. 104), however, come to a totally different conclusion, as they stress the contradictions among governments, business interests, NGOs, and grass-root organizations in Rio:

> "The UNCED process has divided, coopted, and weakened the green movement. On the one hand UNCED brought every possible NGO into the system of lobbying governments, while on the other hand it quietly promoted business to take over the solutions. NGOs are now trapped in a farce by which they have lent support to governments in return for small concessions on language, and thus legitimized the process of increased industrial development."

These authors claim that the Rio outcome mainly reflects government and business interests. After all, the ideology of economic growth and free trade is not questioned in any of the UNCED documents. Moreover, several critical issues – such as the role of transnational corporations and multilateral development banks in environmental destruction – were not seriously addressed either. Furthermore, the effective business lobby hijacked, according to these authors, the Rio summit, for example through the Business Council for Sustainable Development. The environmental movement, on the other hand, became divided. The UNCED machinery set up its own consultative mechanisms with NGOs through the Centre for Our Common Future, while existing NGO networks – such as the Environment Liaison Centre International – were neglected. Consequently, mainstream organizations became co-opted by the UNCED machinery, whereas grass-root organizations were more or less excluded. The same division applies to Northern and Southern NGOs: the former were dominant at UNCED and NGO meetings, the latter weak and under-represented.

Regarding the more direct impact of NGOs on the UNCED documents, views differ as well (Finger, 1994, p. 208):

> "It is difficult to assess the exact impact of the various environmental lobbying groups' activities on the negotiation documents. But analysis of Agenda 21 and of the Rio Declaration shows that NGOs influenced the wording only marginally. If NGO wording was adopted at all, it was incorporated into the texts in an almost token manner and in disregard of NGOs' underlying arguments."

An NGO representative is, on the other hand, more optimistic (quoted in: Arts, 1993, p. 129):

> "Many proposals of NGOs were carried into the documents through national delegations and new issues were introduced on the agenda due to NGO pressure. What is more, some NGO language was almost completely adopted by the negotiators, for example in the poverty chapter of Agenda 21."

Other sources confirm this latter view. For example Doherty (1994) and Enge and Malkenes (1993) refer to the Agenda 21 chapters on, respectively, major groups and consumption patterns as being written partly by NGOs.

A third element which provoked debate concerns the procedural effects of UNCED. Most observers believe that the conference has increased the formal opportunities for NGOs to participate in the UN system (Martens, 1993). The rules have become less strict. For example international NGOs which do not have official consultative status with the ECOSOC as well as national NGOs were given access to the UNCED process. Willets, on the other hand, claims that the possibilities for NGOs to participate in formal negotiations have probably decreased since Rio (Willets, 1996). After all, it was officially formulated that NGOs have no formal negotiation role at the first Preparatory Meeting of UNCED. This rule seems obvious, given the intergovernmental nature of UN negotiations, but it was never formulated as such before and, consequently, there was room for NGOs to intervene in the past. Now with the formal UNCED rule of 'no negotiation role for NGOs' the opportunities may have, according to Willets, decreased. Furthermore, the supposed impact of NGOs on the UNCED have caused the launching of anti-NGO campaigns behind the scenes, in particular by countries which do not like NGOs. Examples are some Arab and East-Asian countries. The former fear 'NGO dressing' by Islamic fundamentalists, whereas the latter try to block the criticism of human rights organizations of their policies. Therefore these nations try to prevent the view that the UNCED has set a precedent for future conferences and negotiations. Hence, the proposition that NGOs were able to upgrade their formal status thanks to the UNCED remains unproven.

It is obvious that these views yield different evaluations of UNCED (Doherty, 1994). Some NGO representatives and observers believe that it would be a waste of money and time for NGOs to continue to participate in such huge UN conferences and related meetings. The UNCED lesson is, according to them, clear: NGOs got frustrated, felt lost and were co-opted in this intergovernmental meeting. Others who have a more positive feeling about the outcome of Rio assume that the intervention of NGOs is continuously needed to maintain pressure on policy-makers. Therefore they try to learn lessons from the Earth Summit to apply to future events.

The different perceptions and evaluations with regard to the NGOs' success in Rio – either in terms of strengthening the movement, or their influence on documents or formal procedures – raise the question where those differences stem from. Of course experiences and world views of authors are not similar. For example, a liberal or a marxist will interpret the same event quite differently. In addition, diverse conceptualizations and assessments of success are employed here, however implicitly most of the time. Some assume that NGOs will be successful if (some of) their objectives are reflected in government documents; others if the number of NGOs attending a conference is large; others again if their message is disseminated through the press. Some authors, on the other hand, emphasize the NGOs' weakness, for example compared to the superpowers and transnational corporations which have, according to them, completely determined the outcome of the Earth Summit.

— 1.5 The design of the study —

1.5.1 Iterative approach

It was initially decided to conduct a study on the NGOs' impact on UNCED, especially since solid assessments of success are generally absent. Without such knowledge a valid evaluation is difficult to conduct. Furthermore, it was decided to interpret 'success' in terms of political influence, which is defined as the achievement of one's policy goals through one's own, intended, intervention in international environmental politics (Cox and Jacobson, 1973; Huberts, 1988, 1989; Huberts and Kleinnijenhuis, 1994). This focus on political influence is, given the disciplinary background of this study (policy sciences), no surprise. In addition, this topic has as yet hardly been covered by students of international relations. As Caldwell (1988, p. 20) notices: "The nature and extent of NGO influence on international environmental policy has not received comprehensive or detailed study." Therefore an assessment of political influence seems relevant not only from an NGO point of view, but from that of scientific literature as well. Consequently, this research aims at producing knowledge on the political influence of NGOs in order to contribute both to the scientific literature on this issue and to the evaluation of NGO participation in international environmental policies.

To pose a general objective is one thing, to put it in operation in terms of a coherent research design is another. Such a project is rather complex, as the translation of an objective into research questions which can be answered on the basis of the empirical data collected is less straightforward than it seems in advance. Therefore the design of this research project is given due attention below. It is based on the approach of Verschuren and Doorewaard (1995). According to them, the design of a research project consists of two groups of activities: conceptual and technical designing. The former refers to four elements: the formulation of the objective(s) of a study, the construction of a research framework, the formulation of research questions, and the delineation and elaboration of central concepts. The technical design consists of two elements: the research strategy and the research material.

Important to note is the iterative nature of the designing of a research project. Although it seems obvious that a researcher should follow a linear sequence of action – design a study first and then implement it – this approach is not always feasible. A researcher may be unable to formulate a fully transparent and consistent proposal behind his desk in advance of any field work, in particular where an explorative study is conducted. Consequently, many will start a first exploratory round of field work with a rather vague idea of the project in mind. Such a first confrontation with empirical data subsequently enables researchers to sharpen their ideas and reformulate their proposal in a much more coherent way. But Verschuren and Doorewaard go even further. They believe that the interplay between the design, implementation and report of a study should be a continuous process during the whole project period, something they label the *iterative* approach.[1] Such an approach will, according to them, make the design more consistent and creative. More consistent, because the different elements of a research project – for example the research questions, the theory, and the empirical data which are actually available – will be continuously adjusted against each other. More creative, because new ideas may be incorporated at any time.

The iterative approach is employed here as well. There were several reasons to do so. Apart from the general reasons mentioned above (consistency, creativeness), the complex nature of the object of this study urged the researcher to follow this line of thought. Due to the complexity of international research on a rather 'new' topic, it remained unclear for a long time which research questions, cases, NGOs, theoretical notions, and methodologies should be selected in order to complete the study. Besides, there was no suitable theoretical framework on NGOs in international environmental policies nor any suitable methodology to assess political influence at global level available in the literature. Therefore the conceptual as well as technical design of this research project slowly evolved during the project period. Not before a first small round of interviews had been conducted in the Netherlands in 1994 and an international meeting on the Biodiversity Convention had been attended by the researcher in the same year, could the design be more or less completed. But small adjustments were still made in the period 1995-1997.

---

1 The word *iterative* stems from the verb to *iterate* which means, according to the Oxford Dictionary, to say something again and again. In day-to-day English it even has the connotation of exaggerated or boring repetition. However, in social sciences the concept refers to a specific approach in analysis. The researcher should go forwards and backwards in different phases of analysis – problem definition, problem diagnosis, problem solution – in order to construct one consistent analysis and find satisfying solutions. This approach is inspired by the iterative thinking in mathematics, by which the outcome of one phase of analysis becomes the input of a next phase until there is (some) congruence in outcomes over several phases.

1.5.2 Conceptual design

As was outlined above, the conceptual design of a study consists of four elements: 1. the formulation of the objective(s); 2. the construction of a research framework; 3. the formulation of research questions; and 4. the delineation and elaboration of central concepts. These four elements will be dealt with in this subsection.

A Objectives

Although the objective was originally focused on an evaluation of the NGO participation in UNCED, the focus shifted towards a more knowledge-oriented study during the project period. It soon became obvious that the academic literature on the NGOs' political influence at global level was limited and that there were no trodden paths to an assessment of influence. In consequence, this study had to be of an explorative nature. Both a theoretical model and a methodology had to be designed, while field work had to take place at the same time. Therefore the focus shifted from an evaluation of NGO activities to: 1. the development of theory and methodology; and 2. the gathering of descriptive and explanatory knowledge on political influence. In addition, the object of study changed, something which will be legitimized below. The Earth Summit as object of study was replaced by two international environmental treaties, which were nonetheless formally linked to the UNCED process: the Convention on Biological Diversity (CBD) and the Framework Convention on Climate Change (FCCC). Furthermore, the period to be studied and the types of NGOs were restricted as well. The period now ranges from 1990 to 1995 and the NGOs involved are those non-profit pressure groups which intend to change climate and biodiversity policies at global level. Hence political, business and scientific organizations, as well as NGOs which limit their work to the national level, were excluded. These delineations were based on pragmatic and substantive grounds (see below). The shift from the UNCED to the FCCC and CBD also implied that an overall evaluation of the NGO participation in the UNCED process was no longer an option. Nonetheless, it was decided – given the background of the study – to maintain an evaluative objective, but now with regard to the FCCC and CBD.

This evaluative objective also underwent some change. Whereas it was meant to yield information with regard to a so-called *ex post* evaluation, in the end it turned out to be prescriptive. An *ex post* evaluation looks back at a process or a product and tries to assess whether a player involved did a good or bad job. A prescriptive evaluation, on the other hand, aims at formulating recommendations on how such a player can perform better in the future. It was decided to leave the *ex post* evaluation of the NGOs' participation in the FCCC and CBD to those NGOs themselves. After all, whether a specific amount of influence is subsequently perceived as a good or a bad performance is up to them, as there exists no objective criterium to judge this. More importantly, there is no information available to compare the outcome, such as similar knowledge on the NGOs' political influence in policy arenas other than the FCCC and CBD. Therefore one cannot decide whether NGOs did a better or worse job in the FCCC and CBD than in past performances. However, what can be done is look to the future while taking the FCCC and CBD as standards. Assuming that players can always perform better, it will be judged which lessons the NGOs can learn from the FCCC and CBD experiences. To achieve this objective, those factors which co-determined the success and failure of NGOs in the FCCC and CBD shall be identified and judged in the context of possible future actions of NGOs.

Having briefly outlined the history of the objectives of this study in the above, these can be formulated as follows:

1 The study aims at yielding descriptive as well as explanatory knowledge with regard to the political influence of global NGOs on the Climate and Biodiversity Conventions.
2 The study also aims at formulating recommendations for global NGOs to enhance their political influence in more or less similar policy arenas in the near future (prescriptive knowledge).

B *Research framework*

A research framework is, according to Verschuren and Doorewaard (1995, p. 40), a schematic reproduction of a study composed of its objectives and steps to achieve these. Such a scheme enables researchers to deduce more coherent questions from the objectives. Furthermore, it gives a handy overview of the study's structure to readers. Figure 1 represents the framework of this research project. The boxes on the left refer to the input of the study (theories, methodologies, empirical domain), those on the right to the outcome (knowledge). As the first objective in the latter subsection includes two kinds of knowledge, both descriptive and explanatory, two boxes of the research framework refer to it. These are, respectively, the boxes 'knowledge on political influence' and 'explanatory factors'. The second objective, on the other hand, does only refer to one kind of knowledge – prescriptive – and consequently to one box, 'recommendations for NGOs'.

The arrows in the framework should be read as follows. A vertical arrow means 'confrontation of' and a horizontal one 'leads to'. In short, a confrontation of theories and methodologies from literature should lead to the formulation of: 1. an assessment methodology on political influence (column b); 2. a comparative case-analysis to identify explanatory factors with regard to political influence (column c); and 3. judgement criteria to formulate recommendations for NGOs (column d). Each of these three tools are chronologically confronted with empirical material. These confrontations yield additional knowledge in any new round: descriptive knowledge on the NGOs' political influence (column c), explanatory knowledge on the backgrounds of this influence (column d), and prescriptive knowledge with regard to recommendations for NGOs (column e).

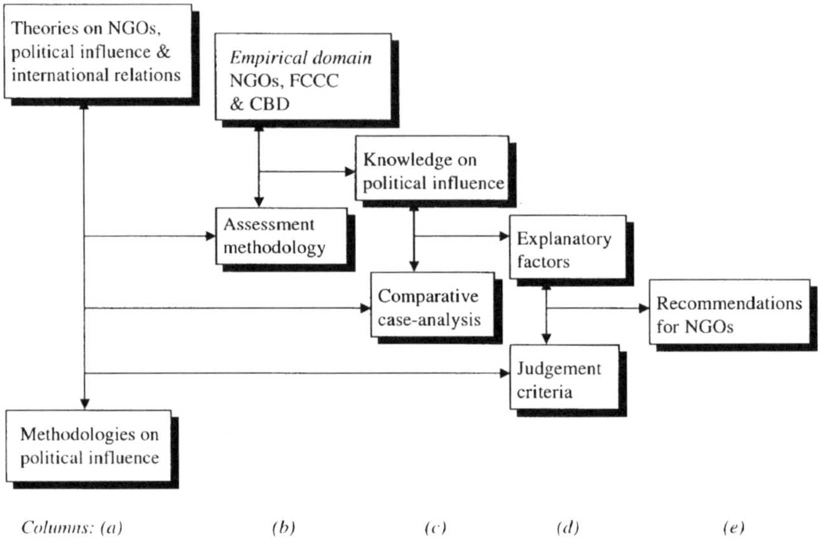

FIGURE 1.1 *Framework of the study*

C Research questions

The three kinds of knowledge – descriptive, explanatory and prescriptive – which cover the output of this study give reason to formulate three main research questions: one on the assessment of political influence, one on the explanation of it, and one on recommendations to NGOs. These main questions, though, should be unravelled into a number of subquestions. The latter need to be less complex than the former and guide the collection of information which makes the answering of the main questions possible. Given the search for an assessment methodology, the subquestions of the first main question are quite methodological in nature. The different elements of the method which shall be developed in chapter 3 – ego-perception, alter-perception, and causal analysis – are spelled out. Ego-perception refers to the extent to which key representatives of global NGOs themselves believe they were influential. Alter-perception refers to the extent to which key government representatives and officials in both arenas believe that NGOs were influential. Finally, causal analysis refers to the reconstruction of the FCCC and CBD negotiation and decision-making processes by the researcher. These three elements together lead to overall conclusions on the political influence of NGOs on the FCCC and CBD.

The subquestions of the second main question are based on the comparison of the different cases studied in the context of research question 1. Here the word 'cases' refers to topics of the FCCC and CBD which might have been influenced by NGOs. Such topics may be covered by the legal text of either treaty or by formal decisions of countries which are parties to the treaties. The selection of cases – or topics – is based on what was called ego-perception above. The several key NGO representatives who were interviewed gave examples of what they believe was their influence on the two treaties. On the basis of these perceptions, a list of topics possibly influenced by NGOs was made. It is referred to as 'possible influence' because there are reasons to believe that players who strive for political influence – and many NGOs generally do so – will overestimate their own impact on policies (Potter, 1996). The examples given by the NGO representatives nonetheless became the basic input of this study. Subsequently, the ego-perceptions were confronted with the alter-perceptions of government respondents and the researcher's causal analysis. After these confrontations were made, some views of NGOs with regard to their own political influence could be confirmed, but others had to be rejected. In other words, some cases became examples of the NGOs' successes, others of their failures. By comparing these cases on the basis of a number of relevant variables which will be deduced from the theoretical framework (see chapter 2), one is subsequently able to identify explanatory factors which co-determined the political influence of NGOs. This procedure, which shall be described in chapter 3, is briefly spelled out in the subquestions 2.a – 2.d. Apart from this comparative case-analysis, the respondents were asked to give their explanation of the NGOs' political influence (or lack of it). Their views on this are taken into consideration as well.

The subquestions of the third main question, finally, refer to a number of steps to formulate recommendations for NGOs. To be able to do so, it will be judged to what extent NGOs may anticipate on the explanatory factors, when they take part in similar policy processes in the near future. In addition, it will be judged whether findings from the climate and biodiversity arenas can be extrapolated to other arenas. Additionally, the respondents were asked to give their views on how NGOs might increase their political influence. Again, their views are taken into consideration here as well.

Having elaborated on the linkages between the main and subquestions, the questions are as follows:

1 To what extent did global NGOs influence the formation and implementation of the UN Framework Convention on Climate Change (FCCC) and the UNEP Convention on Biological Diversity (CBD) in the period 1990-1995?

a Which topics of the FCCC and CBD were – according to key NGO respondents – influenced by global NGOs (*ego-perception*)?

b Which topics of the FCCC and CBD were – according to key governmental respondents – influenced by global NGOs (*alter-perception*)?

c To what extent are the outcomes of ego- and alter-perception in subquestions a. and b. being confirmed by the researcher's reconstruction of the FCCC and CBD negotiation and decision-making processes (*causal analysis*)?

2 What factors explain the NGOs' political influence in the climate and biodiversity arenas?

a To what extent is there a resemblance or difference in the cases studied with regard to the extent of political influence of global NGOs?

b To what extent is there a resemblance or difference in these cases with regard to variables which might explain the political influence of global NGOs?

c Which explanatory factors are to be deduced from the comparative case-analysis covered by 2.a and b?

d How do the respondents explain the political influence of global NGOs?

3 How might global NGOs increase their political influence?

a To what extent can global NGOs be expected to anticipate on the explanatory factors identified in 2.c in the near future?

b Which recommendations are to be deduced from the above judgement (3.a)?

c To what extent can these recommendations be extrapolated to the climate and biodiversity arenas of the near future as well as to other issue arenas?

d What do the respondents recommend to global NGOs to increase their political influence?

The above questions are to be linked to the organization of the book. In short, research question one will be dealt with in chapters 4 and 5, on the FCCC and CBD respectively. Research question two will be elaborated on in chapter 6, and research question three will be dealt with in chapter 7. Chapters 1 to 3 are supportive to the answering of these questions and cover the research design, the theoretical model and the methodologies, to assess, explain, and 'predict' political influence respectively. Chapter 8, finally, will summarize the output of this study in terms of overall conclusions.

D *Central concepts*

The objectives and (sub)questions contain a number of theoretical and methodological notions which should be elaborated on in depth: NGO, political influence, ego-perception, alter-perception, causal analysis, convention, formation, implementation, negotiations, decision-making, comparative case-analysis, explanatory factors, anticipation, extrapolation, recommendations, policy processes, and policy arenas. To prevent unnecessary repetition, though, it was decided not to elaborate on these concepts here, but to do this in chapters 2 and 3 where the theoretical and methodology frameworks of this thesis are developed. Besides, some notions – such as policy, politics, NGO and political influence – were already briefly defined or described in the foregoing sections of this chapter.

1.5.3 Technical design

The technical design of a study consists of two elements: 1. the research strategy, and 2. the research materials. These elements will be dealt with in chronological order.

A Research strategy

A research strategy is a way of collecting and analyzing empirical evidence (Yin, 1991, p. 50). The following strategies are generally distinguished in scientific literature: survey, experiment, case study, bureau research, and historical/archival analysis. This research project is definitely a multiple-case study. Such a study is characterized by the following aspects (Swanborn, 1996; Verschuren and Doorewaard, 1995; Yin, 1991):

- 'how' and 'why' questions are addressed;
- more than one case is investigated, but the total number is still limited;
- A contemporary phenomenon over which the researcher does not have direct control is studied within its real-life context;
- it aims at collecting qualitative data and in-depth knowledge about a phenomenon, not at collecting quantitative data to generalize on one aspect of the phenomenon;
- multiple sources of evidence are used; and
- field work is included.

This study meets these criteria, except for the fact that one 'To what extent' question is addressed. In addition, one aspect of the cases – political influence – is prioritized. Nonetheless, the assessment of this aspect is based on a qualitative methodology by which process and context variables are addressed as well.

The choice for a case study was – given the initial research objective – no surprise. After all, the study aimed at an evaluation of UNCED, and this conference was indeed a specific case in the history of international environmental policies. However, it soon became obvious that it was too voluminous to be considered from a political influence perspective. Therefore it was decided to identify two or more smaller cases within the UNCED process. All its 'products' – Rio Declaration, Forest Statement, Agenda 21, Commission on Sustainable Development, Climate Convention, and Biodiversity Convention – seemed good candidates at first sight. Different options were investigated, including the addition of one 'non-UNCED' case to address the impact of different contexts on the NGOs' political influence. An example of such a case might have been a treaty which was adopted in another arena and at another point of time (e.g. the 1989 Basel Convention on Hazardous Wastes). Finally, two cases – the FCCC and CBD – were selected for the following reasons. First of all, it was believed that the study of two cases – instead of one – would yield additional knowledge through comparison (Swanborn, 1996). At the same time the study of more cases was considered inappropriate, given the limited extent of time and money. Secondly, both the FCCC and CBD are – contrary to the other UNCED outcomes – binding international law, or so-called *hard law*. Therefore any impact of NGOs on these treaties has to be taken seriously from a political perspective, at least more seriously than NGO influence on soft, non-binding law such as the Rio Declaration and Agenda 21. Thirdly, it was assumed to be best to select cases as similar as possible because an explorative study was undertaken (*literal replication*; Yin, 1991). Such similarity makes the drawing of more general conclusions easier. Therefore the FCCC and CBD were used in combination and not one of these with a 'non-UNCED' case such as the Basel Convention. Nonetheless, there is still one fundamental difference between the FCCC and CBD, namely the environmental issue dealt with (climate change versus biodiversity loss). The possible role of this variable can therefore be taken into account.

A next delineation refers to the period studied: 1990-1995. This limitation was based on substantive as well as pragmatic grounds. To start with the latter, a period of five years was considered manageable for one researcher. Moreover, a reconstruction of a policy process on the basis of, amongst others, respondents, should be limited in time due to the phenomenon of temporal distortion (Westerheijden, 1987). People generally have a short memory and, besides, their memories become more and more distorted as time goes on. So there are limits regarding the extent to which one can interview individuals or groups on what happened in the past, at least if reliable answers are to be expected. The substantive reason to choose the period 1990-1995 was the objective to particularly focus on policy formation and implementation, as it is generally believed in academic literature that NGOs are mainly influential in the policy phase of agenda-setting (Liefferink *et al.*, 1993; Mazey and Richardson, 1992). As Hooghe (1995, p. 5) has noted: "One might say that the environmental movement is indeed able to push an issue on the political agenda, but cannot affect subsequent decision-making." To 'test' this proposition, it was decided to highlight the subsequent policy phases in this study. As the phase of agenda-setting on both treaties ended the moment the formal intergovernmental negotiations started in 1990, this year was identified as the baseline. The formation phase was subsequently completed in May 1992, after which both Conventions could be signed at UNCED (June 1992), be ratified by the signatories (1992-1993), enter into force (1993-1994), and start to be implemented (since 1994). The end-target of 1995 was based on the fact that the first Conference of Parties (COP) of both Conventions were convened just before and after January 1995. These COPs were attended by the researcher and subsequently marked the demarcation line of the period studied.

Another delineation concerns the NGOs. Only non-profit and so-called promotional pressure groups were included (see chapter 2). This delineation stems from the original focus of the study: the impact of such groups – in particular environmental organizations – on UNCED. However, the FCCC and CBD do not only cover environmental issues, but conservation and development objectives as well. Therefore it was decided to also involve pressure groups fostering these goals. In addition, only so-called *global* NGOs were considered. These are organizations, as defined above, which are mainly active at the global level. This excludes NGOs which only try to influence their governments at home with regard to the FCCC and CBD. An example might be an Indian or a Canadian NGO which asks its government to defend position X or Y in international negotiations without being active in the international arena itself. True, it is definitely possible that local NGOs may have impacted the Climate and Biodiversity Conventions through such domestic channels, but at the same time it is simply impossible to include all NGOs around the globe which possibly have pressurized their governments on the climate or biodiversity issues. Therefore the set of NGOs was confined to only those which were active in the global climate and biodiversity arenas in the period 1990-1995.

The assessment of NGO influence on *all* aspects of the two treaties in the period 1990-1995 turned out to be still too ambitious. A further selection procedure had to be applied. In consequence, two sets of relevant *sub*cases were identified, one set concerning the FCCC and one concerning the CBD. As was outlined in section 5 above, this selection procedure was based on ego-perception. The NGO respondents in both arenas gave examples of how they perceived their own impact on the formation and implementation of the treaties. Consequently, two lists of possible NGO influence, on respectively the FCCC and CBD, were constructed. These lists gave an overview of subcases – or topics – which should be further analyzed. But even then, these lists were too long from a time management perspective (they contained approximately 55 items). Therefore examples of NGO influence which were referred to by only one non-government respondent were omitted, as it was assumed that the more respondents refer to a certain topic the more relevant and reliable the reference will be. Finally, 18 subcases remained, 8 on the FCCC and 10 on the CBD. This number was considered manageable by one researcher. In short, ever more detailed cases were identified: from the UNCED to the FCCC and CBD and from these two conventions to 18 subcases.

We have to acknowledge the fact that the selection of subcases is based on NGO respondents, not on a sample taken by the researcher on empirical or theoretical grounds. This implies that the sets of subcases may be biased and are probably incomplete. The NGO-bias, though, is not really a problem, as the ego-perceptions will be confronted with alter-perceptions and causal analyses. But examples of NGO influence may indeed have been lost by this selection procedure, for example where no NGO respondents mentioned a topic due to lack of knowledge or temporal distortion. Consequently, *this study does probably not give a complete overview of NGO influence on the FCCC and CBD*. It presents a number of illustrations at best, which nonetheless gives an impression of the whole. Therefore the fact that the number of subcases is unequal – 8 with regard to the FCCC and 10 concerning the CBD – is not considered a problem, as both groups simply cover a set of illustrations based on the similar selection procedure. Besides, *examples of failures as perceived by the NGOs themselves are not covered by this study*. Respondents were only asked to mention 'positive' examples of their impact and not 'negative' ones on their *lack* of influence. However, many ego-perceptions turned out to be wrong given the outcomes of alter-perception and the causal analysis, as was expected, so that the 18 cases included examples of both success and failure.

The analysis of (sub)cases in this thesis will be based on the so-called *hierarchical* method (Verschuren and Doorewaard, 1995). In this approach the two cases and their subcases will be enlaced as if they are separate studies, but on the basis of similar procedures and techniques. In a second phase the outcomes will be compared and general conclusions are to be drawn and explained (comparative case-analysis). These two phases are also recognizable in the format of this book. Chapter 4 is a case-analysis of the FCCC; chapter 5 of the CBD; and chapter 6 covers a hierarchical analysis of both cases. This approach is the opposite to the so-called *sequential* method, where the selection of a second case is dependent on the outcome of the first case-analysis, the third case on the outcome of the second one, etc. Consequently, procedures and techniques of the analyses may change while the study proceeds.

The case-study design has, of course, some drawbacks (Rosenthal and 't Hart, 1994; Swanborn, 1994, 1996; Wester, 1991). Firstly, the researcher is deluged by variables, because phenomena are studied in their real-life context. It is therefore a difficult task to manage these and produce standardized descriptions and explanations. Consequently, there is a risk that the outcome is very much dependent on the qualities of the researcher concerned (control-effect). And secondly, case studies are hard to generalize, because only a few cases of a phenomenon are investigated, not a representative sample of it (lack of external validity). Most methodologists therefore discourage researchers from extrapolating conclusions to cases other than the ones they studied (so-called statistical generalization). Nonetheless, several authors still claim that theoretical propositions may be tested or formulated on the basis of case-studies (theoretical generalization) (Hutjes and Van Buuren, 1992; Yin, 1991). As this study's foremost aim is describing and analyzing the cases themselves, the pretention to generalize to other domains is limited, although this (im)possibility will be discussed. In addition, the recommendations will be formulated very carefully. What is more, they will be considered possible inputs into an open debate on NGO strategies at best, not the output of such a debate in advance.

B Research material

To be able to answer the research questions on the basis of a multiple-case design, specific information and knowledge should be collected, for example on the participation of NGOs in the FCCC and CBD between 1990 and 1995, on the effects of these NGOs on the formation and implementation of both treaties, on those policy processes in general, as well as on the policy context of that period. Therefore relevant and multiple sources of information and knowledge had to be selected and retrieved. The main sources of this study are people and documents, in particular political players and policy documents. Less important sources for this study are: relevant literature, the media, and – to some extent – the reality of policy-making. These sources were retrieved by the following techniques: the semi-structured interview, document analysis, media analysis, the reading of relevant literature, direct observations, and participation. These sources and techniques shall be briefly dealt with below. In addition, triangulation of sources shall be argued for.

SEMI-STRUCTURED INTERVIEW — The most important sources of information were representatives of NGOs on the one hand and representatives of governments and officials of UN bodies – such as the secretariats of both treaties – on the other. The former were approached to gather information on how NGOs perceive their own influence (ego-perception), the latter on how policy-makers perceive the political influence of NGOs (alter-perception). In addition, both groups were – given their long involvement in the FCCC and CBD – considered experts on the formation and implementation of both treaties. The interviews were semi-structured in the sense that most questions were formulated in advance and posed to all interviewees. Variable questions were added depending on the experience and knowledge of the respondent concerned. Needless to say, the interview guides relating to NGO representatives on the one hand and (inter)governmental representatives on the other differed to some extent. Questions covered amongst others: the background and experience of the interviewee, perceived examples of NGO influence, causes of NGO influence, NGO co-operation and NGO-state interactions. There were, however, two problems regarding the interview guide. Firstly, it turned out that most interviews took more time than was stated by the researcher. Some even exceeded the time available in a rather extreme way, for example when a respondent had a lot to say on a particular item or when a respondent – in particular at international meetings – was under severe time-pressure. As information on the assessment of political influence was prioritized by the researcher – given the object of study – other questions sometimes had to be deleted. Hence, the range and depth of information varied with individual respondents. Secondly, the interview guide was renewed during field work, so earlier respondents had not been confronted with the newly added questions, and therefore the information on these has remained incomplete. The specific consequences of these two drawbacks will be highlighted in the empirical chapters.

The respondents were selected as follows. Appropriate NGO representatives were approached by the researcher at two major international meetings: the first Conference of Parties to the CBD (Nassau, Bahamas, November 1994) and the first Conference of Parties to the FCCC (Berlin, Germany, April 1995). The selection of individuals was based on two criteria. First, an appropriate NGO representative should have been involved in either the FCCC or the CBD at global level since 1990 (or even longer). Second, he or she had to be a key person in the NGO community concerned, in other words had to be an (in)formal leader. If these two criteria were met, it was assumed that the respondents' opinions, information and knowledge was indeed based on some experience. Furthermore, an attempt was made to interview NGO representatives from all parts of the world, so that differences in opinion, information and knowledge based on various regional backgrounds could be covered as well. Whether the criteria were met in the case of specific NGO representatives was checked by observation (who are the leaders in Nassau and Berlin), documents (participants lists of international meetings since 1990) and short inquiries with several people, both from the government and the non-government side (reputation of specific NGO representatives). Finally, 12 and 13 NGO respondents respectively were selected in Nassau and Berlin. Interviews of approximately 30-45 minutes were subsequently conducted and taped at both international meetings.

The selection procedure with regard to (inter)governmental representatives was slightly different. Although it was planned to interview these informants at the same international meetings, this objective was unattainable for two reasons. First, both meetings lasted for two weeks and there was not enough time to prepare and conduct interviews with both groups. Secondly, the (inter)governmental representatives were so busy with negotiations in Nassau and Berlin – at least the key people among them – that interviews were out of the question. Therefore this group of respondents was approached by telephone, although not before the study was introduced to them through a letter from the researcher. After the agreement of the possible informant to participate, the interview itself was conducted over the phone and taped with the help of a special recorder. The selection of respondents was nonetheless based on similar criteria: involvement since 1990, being a key person, and differing regional backgrounds. These criteria were checked – again – through participants lists of international meetings on the FCCC and CBD since 1990; by the formal position of individuals (for example ambassadors, heads of national delegations, high officials, key negotiators); by reputation; and by regional background. Finally, 11 and 12 (inter)governmental representatives respectively were selected as respondents for the FCCC and CBD. Interviews of approximately 30-45 minutes were subsequently conducted by telephone.

It should be mentioned, though, that not all of the above criteria on 'appropriate' respondents could be met. Firstly, not all respondents were involved in the FCCC or CBD since 1990, in particular some NGO respondents, although most were. Secondly, there was no equal division on regional backgrounds. Most respondents stem from one of the Northern countries, in particular from Western Europe, whereas about 40% live in developing countries. And thirdly, not all respondents were truly key people, in particular government respondents, but all of these were nonetheless representatives of key countries in the negotiations.

Alongside the semi-structured interviews with the above respondents, other people were questioned both at the beginning and the end of the research project. Some Dutch government and NGO representatives who were involved in the CBD were approached and interviewed before the actual field work started in the spring of 1994. These interviews were aimed at gathering some general information on treaty formation and implementation so far, and at testing the interview guide. In addition, several people were approached after the period of field work (since summer 1996). These short, specific though unstructured interviews were intended to gather information and knowledge which turned out to be still unclear, something which became obvious while the thesis was being written. In total, 63 interviews were conducted.

DOCUMENTS — The second important source for this study were formal and informal policy documents of global NGOs, governments and UN bodies. These documents give information on amongst other things the FCCC and CBD policy processes in general, the nature of the negotiations, the role countries and others played, the decisions adopted, the political contexts, as well as on the participation, views, goals and activities of NGOs. All this information supplied the researcher with building-blocks to reconstruct the political influence of NGOs on the basis of the so-called causal analysis (see chapter 3). Archives of the following organizations were consulted once or more:

1 Dutch Ministry of the Environment, Physical Planning and Housing (The Hague, Netherlands);
2 Dutch Society for Nature and Environment (Utrecht, Netherlands);
3 Greenpeace International (Amsterdam, Netherlands);
4 Climate Action Netwerk Europe (Brussels, Belgium);
5 Dutch Committee of the IUCN (Amsterdam, Netherlands); and
6 The IUCN Law Centre (Bonn, Germany).

No archive of any UN organization was consulted, although this was originally planned. This intention was however not realized due to pressure of time and a limited budget (as such consultations would involve a rather long stay abroad). Besides, many relevant UN documents were available in those archives already consulted, in particular at the Dutch Ministry of the Environment. Furthermore, the FCCC and CBD secretariats regularly sent documents to the researcher. Approximately 18 days were spent in these archives altogether, at least half of this in the Dutch Ministry. Relevant documents were selected and copied, so that analyses could be made later on.

One drawback of research in archives is generally the excess of documents, which makes selection and analysis of relevant ones a difficult task (Verschuren and Doorewaard, 1995). On the other hand, a definite advantage of archives is the 'fixation' of happenings, meetings, views etc. of the past, whereas the output of interviews is entirely dependent on the restricted and biased memories of interviewees. Therefore the documents were also used to verify the information given by the latter (triangulation).

MEDIA — Several players involved in the FCCC and CBD publish journals. Some of these appear regularly, others are only published during international meetings. These journals are important sources of, for example, background information, policy positions of groups, and opinions. The following were regularly consulted by the researcher:

1. *Environmental Negotiation Bulletin* (published by the International Institute for Sustainable Development at major international negotiation sessions on the environment, including on climate change and biodiversity);
2. *Eco* (published by NGOs attending major international negotiation sessions on the environment, including on climate change and biodiversity);
3. *Biodiversity Coalition* (published by WWF Australia);
4. *United Nations Climate Change Bulletin* (published by the FCCC secretariat); and
5. *Biological Diversity Information* (published by the CBD secretariat).

Other specialized journals such as the *Earth Times* were consulted occasionally. Furthermore, the daily press also reports on major conferences. Relevant articles on UNCED and COPs in Brazilian, Bahamian, German and Dutch daily journals were therefore collected by the researcher as well.

LITERATURE — It is more than obvious that academic literature on international environmental policies is an important source of knowledge for this research. But there is another branch of literature which is extremely relevant: publications on international environmental negotiations written by the key players themselves. Good examples are the books *Biodiplomacy* by Sanchez and Juma (1994) and *Nego-*

tiating Climate Change by Mintzer and Leonard (1994). Both publications contain first-hand observations and analyses of key negotiators and participants and therefore yield additional material.

DIRECT OBSERVATIONS — The researcher attended five international conferences of which three were intergovernmental in nature – two major ones and a small one – and two were non-governmental. These were respectively:

1 The First Conference of Parties to the UNEP Convention on Biological Diversity. Nassau, the Bahamas, 28 November – 9 December 1994;
2 The First Conference of Parties to the United Nations Framework Convention on Climate Change. Berlin, Germany, 27 March – 7 April 1995;
3 Workshop on Consultative Mechanisms for NGO Inputs into the United Nations Framework Convention on Climate Change. Palais des Nations, Geneva, 2 March 1996;
4 Global Forum on the Convention on Biological Diversity, organized by the World Conservation Union, the World Resource Institute and UNEP. Nassau, Bahamas, 26-27 November 1994;
5 Rio Is Not Enough. An NGO Conference organized by the Climate Action Network. Berlin, Germany, 26 March 1995.

The two COPs were the most important meetings from the perspective of this study. The main objectives to visit these were: the observation of NGO lobbying and pressure in order to get a tentative impression of possible NGO influence; the selection of NGO respondents; conducting interviews; the construction of a research network; the observation of negotiations and decision-making to get informed on UN rules; and the collection of useful documents, papers and publications. The NGO conferences in Nassau and Berlin which preceded both COPs were visited as well (numbers 4 and 5 above). These aimed at exchanging information between NGOs and other players – such as governments, science and business – on the CBD, and at disseminating NGO opinions on how to proceed with the FCCC. These meetings, though, were also utilized to observe which NGO representatives seemed to be the leaders in their community. They were approached by the researcher later on. Finally, the meeting in Geneva – number 3 above and organized by the FCCC secretariat – was visited to learn about the debates on formal UN mechanisms which make the participation and input of NGOs possible (or impossible, as some claim). Moreover, the behaviour of NGO representatives in the debates was observed.

It should, finally, be mentioned that the researcher visited the UNCED and the Global Forum in Rio de Janeiro in 1992. In addition, the ECE Bergen Conference in Norway in 1990 was also attended. This meeting was a preparatory one for UNCED organized by the UN Economic Commission for Europe. These conferences – UNCED and Bergen – took place before this research project was even planned, and therefore the visits did not directly contribute to this study. Nonetheless, the researcher was given a practical insight into international environmental affairs and the world of NGOs due to these attendances. This experience was useful afterwards.

PARTICIPATION — There were two (minor) ways in which the researcher actively engaged in the work of NGOs. Firstly, some contributions were made to the production and publication of the NGO journal *Eco* at the COP in Nassau. There were two reasons to do so. They could use a helping hand given the amount of work at that time and it was a good opportunity to learn about the NGO logistics. Moreover, contacts were made which yielded good interview opportunities. Secondly, the researcher participated in the Dutch Biodiversity Forum. This Forum brings together all kinds of NGOs from the Netherlands which are involved in the issue of biodiversity and provides contacts and interactions with Dutch policymakers. This participation yielded useful insights with regard to such contacts and interactions. Furthermore, an additional channel of communication and information was established.

— 1.6 Summary —

The United Nations Conference on Environment and Development (UNCED) which was held in Rio de Janeiro in June 1992 is considered a milestone in the course of international environmental policies. New principles and action plans were adopted by the UN members there. In addition, thousands of so-called Non-Governmental Organizations (NGOs) were involved in the UNCED and its preparation, amongst others environmental pressure groups. This NGO participation, though, provoked debates on its effectiveness and one aspect was the perceived influence of pressure groups on the outcome of Rio. But detailed knowledge on such political influence is lacking. Furthermore, the subject of NGO influence at global level has not received much attention by social scientists yet. Therefore a study on the assessment of political influence is relevant not only from an NGO point of view, but from the perspective of science as well. As the UNCED is a topic too voluminous to be managed by one researcher, two cases were identified: the Framework Convention on Climate Change (FCCC) and the Convention on Biological Diversity (CBD). Both treaties were designed parallel to the UNCED process and signed by more than 150 countries at the Earth Summit in Rio de Janeiro. The study focuses on treaty formation and implementation in the period 1990-1995.

The research may be labelled an *explorative, hierarchical multi-case study.* Explorative, because it concerns a rather new topic in the study of international relations. Consequently, a theoretical and methodological framework had still to be designed. It is a hierarchical study, because each case is individually dealt with on the basis of similar procedures and techniques, after which a comparative analysis will be employed. And it is a multi-case study, because both the FCCC and CBD are the objects of study. Furthermore, the research aims at collecting descriptive, explanatory and prescriptive knowledge on the NGOs' political influence. The sources of evidence are NGO representatives, government representatives, UN officials, policy documents, scientific and popular literature, the media, and the practice of policy-making.

CHAPTER 2

Theoretical Model

A political arena perspective on NGOs in treaty formation and implementation

One's assumptions about world politics profoundly affect what one sees and how one constructs theories to explain events.

Keohane and Nye, 1989, p. 23

— 2.1 Introduction —

The objective of this chapter is to develop a theoretical model on NGOs in treaty formation and implementation at global level. It functions as a lens through which the researcher observes his domain of study, identifies relevant variables and relationships among these, excludes irrelevant ones and explains what he observes. The framework can therefore be labelled a heuristic tool. The achievement of descriptive as well as explanatory insights into the domain of study is the primary objective of the model, not its testing (Frouws, 1993; Swanborn, 1992). Furthermore, the model enables readers to identify the assumptions and 'worldview' of the researcher.

The procedure to design the model is based on three steps: 1. an elaboration of the main concepts of the study; 2. a nesting of these concepts in relevant theories; 3. the linking of these concepts and theories in a comprehensive model. Identified as the main concepts of the research questions are 'NGO', 'political influence' and 'convention' (see figure 2.1). These notions will be elaborated on and limited below (sections 2.2, 2.3, and 2.4). In addition, the 'context' in which NGOs try to influence treaty formation and implementation will be taken into account (section 2.5). But each choice regarding clarification and definition of concepts implies a choice in theoretical perspective. For example, the identification of NGOs as relevant players in international politics positions one in the so-called transnational school (and vice versa). Consequently, all three concepts will be nested in specific theories. Finally, the concepts and theories are linked to each other and inserted into one comprehensive model (section 2.6). This model, though, is rather broad, as it has yet to be put into operation. Whereas chapter 3 will aim at making the model operative, section 2.6 briefly goes into its application in this study.

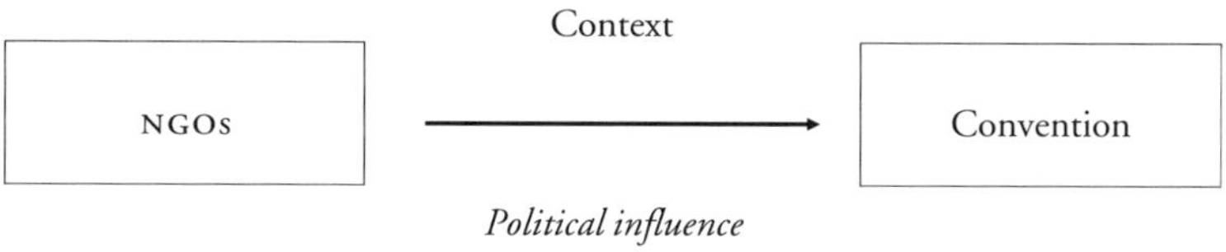

FIGURE 2.1 *The main concepts*

— 2.2 The concept of 'NGO' —

2.2.1 Definition

The UN defines an International Non-Governmental Organization (INGO) as follows:

> "Any international organization which is not established by intergovernmental agreements." (Feld and Jordan, 1983, p. 227).

This definition implies that a wide range of international organizations may be rightly considered to be INGOs: churches, scientific unions, sports clubs, trade unions, political parties, the Scouts, business organizations, transnational corporations (TNCs), hybrid organizations of governments and private organizations (like the International Labour Organization and World Conservation Union), environmental movements, peace organizations, liberation movements, terrorist groups, and even the Mafia. It is obvious that such a broad concept leads to confusion and disagreement. Therefore others conceptualize INGOs as *pressure groups*, which can be defined as organized groups of people who seek to influence political decisions (Willets, 1982, p. 1). According to Thompson-Feraru (1974, p. 32-33):

> "INGO's are international pressure groups (...) that have the capacity or desire to influence the course of international relations."

This definition is more specific than the UN's – since it also covers the nature and goal of an NGO – and excludes those organizations which do not generally wish to influence policies (such as churches, sport clubs and the Scouts). Also, this definition is more in line with the type of organization studied in this thesis. Therefore the basic idea of the latter definition is adopted here.

TABLE 2.1 *A classification of pressure groups (source: Willets, 1982)*

Pressure groups	
Sectional groups	*Promotional groups*
1 economic groups (companies, commerce, trade, agriculture)	1 welfare agencies (development associations)
2 professional associations (doctors, nurses, lawyers, scientists)	2 religious organizations (World Council of Churches)
3 recreational groups (Scouts, IOC)	3 communal groups (indigenous groups, women's groups)
	4 political parties
	5 issue-specific groups (environmental organizations, peace groups)

However, the concept 'pressure group' is still very broad. For example Willets (1982) distinguishes the following: three types of *sectional* groups which further the interests of a particular section of society and five types of *promotional* groups which defend more general interests (see table 2.1). This study focuses mainly on the second group, in particular on welfare, communal, and issue-specific organizations. Religious groups and political parties are excluded, as they do not operate as prominent pressure groups in the climate and biodiversity arenas. Besides, it is hard to accept the view that political parties are pressure groups, given the fact that their most important aim is to become part of a government. This is definitely not the case with other groups. To sum up:

> *In this study a global NGO is defined as a promotional pressure group which seeks to influence political decision-making on certain issues at global level. In the case of the FCCC and CBD we deal with welfare, communal and issue-specific organizations (respectively development, indigenous, environmental and conservation groups).*

Some remarks should be made with regard to this definition. Firstly, the 'I' of INGO is – compared to Thompson-Feraru's definition – omitted and replaced by the adjective 'global'. The reason for this is simple: the tendency for *national* NGOs also to participate in global arenas has been growing since the 1970s. This has even become a more or less normal phenomenon since Rio and was recently formally accepted by the UN system (Willets, 1996). Whereas the notion of 'INGO' refers to a non-governmental organization which has constituencies in two or more countries, 'global NGOs' refers to all non-governmental organizations which are active at the global level. Whether the latter have constituencies in one or in more countries is, however, not relevant. An example: the Environmental Defense Fund, which is based in the United States but which is nonetheless involved in the global climate and biodiversity arenas, is a *global* NGO, not an INGO. Secondly, the concept of 'pressure' should be elaborated on, as it stands for a specific approach to influence decision-making. According to Van Noort *et al.* (1987), pressure should be distinguished from protest. The former refers to efforts of NGOs inside political arenas to directly affect policy- and decision-making, whereas the latter concept covers indirect strategies of NGOs outside formal arenas to change policies. In other words, pressure groups try to influence decisions by participating or intervening in formal politics. They deliberately seek contacts with policy- and decision-makers, all the more so as they may have resources at their disposal to exchange with the latter (expertise, contacts, money, political support, etc.). In doing so, pressure groups can make use of their formal position in political arenas – for example by making oral statements at UN meetings – and act as public advocates of their cause (*advocacy*). At the same time, they may quietly lobby individual policy- and decision-makers in the corridors of meetings (*lobby*). Protest groups, on the other hand, try to influence decision-makers far less directly. They do not participate in policy- and decision-making, but make their demands known outside the formal political arena, for example through campaigns, the media, sit-ins, protest marches, or even violence. Although this study focuses on pressure, it is recognized that protest activities may be relevant from a political influence perspective as well. After all, various NGOs combine both types of activity and success in political arenas might be dependent on both.

Table 2.2 summarizes the several categories, objectives, targets, positions, strategies and resources of protest and pressure groups. This classification is based on a wide range of literature (Bader, 1991; Cox and Jacobson, 1973; Everts, 1985; Godfroy, 1981; Huberts, 1988: Kuypers, 1973; Van Noort *et al.* 1987; Van Schendelen, 1981). Therefore a general overview is presented here, while debates on details are avoided. The aim of this schema is to make clear that different groups apply different objectives, strategies and resources. In chapters 4 and 5 these elements will be translated into concrete activities of different NGOs in the climate and biodiversity arenas.

It should be stressed that many social movement and civil society theorists reject the notion of 'pressure group' (Friberg and Hettne, 1988; Johnson and Maiguashca, 1994; Melucci, 1988; Van der Loo *et al.*, 1984; Princen and Finger, 1994; Wapner, 1995; Zald and McCarthy, 1980). According to them, the importance of social movements – such as the environmental movement – is not so much linked to political influence on short-term politics, as well as to long-term political, social and cultural change. Examples of such change are processes of democratization and the greening of lifestyles. In addition, the concept of NGO is rejected by those theorists, as it only defines what an organization is *not* (non-governmental) and considers its relevance from a government perspective. It would therefore be better to utilize concepts such as 'social movement organization' (SMO). Nonetheless, the notion of NGO is maintained in this book. The reasons for this are both substantive and pragmatic in nature. To start with the latter, all involved in the UN system – even the social movement organizations themselves – are used to employing the concept. Therefore it would be foolish to choose a far less well-known alternative. Besides, conservation organizations such as the World Conservation Union (IUCN), Worldwide Fund for Nature (WWF), Sierra Club and Birdlife International, which were so relevant to the biodiversity treaty, can hardly be considered social movement organizations. After all, these do not aim at changing society, but at promoting the conservation of nature within the framework of current policies. But they can indeed be seen as NGOs – or pressure groups – because they try to influence political decisions.

TABLE 2.2 *Activities of pressure and protest groups*

Category	*1 Protest*	*2 Pressure*	
		a Advocacy	*b Lobby*
Objective	- achieve (radical) policy change	- achieve general policy goals	- achieve specific policy goals
Target	- politics in general	- political arena in general	- individual policy- and decision-makers
Position	- outside political arena	- inside political arena	- inside political arena
Strategy	- demonstration - contest - violence	- information - argumentation - promotion	- advice - mediation - persuasion - exchange of resources
Resources	- supporters - logistics - skills - money - public opinion - physical means	- expertise - skills - leadership - co-operation - money - logistics - legitimacy	- contacts - expertise - skills - reputation

On the other hand, groups which can be considered social movement organizations – such as Friends of the Earth International – also directly engage in politics to gain short-term successes. In other words, they sometimes operate as pressure groups. Therefore pressure on short-term political decision-making is one dimension of social movement activity, long-term social and cultural change another. The latter, though, is excluded from this study, although it is recognized that the participation of NGOs in the interstate system – e.g. through UNCED, FCCC and CBD – might lead to political transformations at that level in the long run. But this aspect is simply not studied.

2.2.2 NGOs and the study of international relations

One of the main themes in the history of the study of international relations is the question which players are relevant from the perspective of international politics (Hocking and Smith, 1990). Different theories in different era came up with various answers. Nonetheless, the mainstream in the study of international relations was – and is – *(neo)realism.* During this century realists successively focused on great statesmen, on the nation state as such and on the state system as a whole (Lieshout, 1993; Morgenthau, 1993; Waltz, 1979). They are pessimists, in the sense that they subsequently believed that either the imperialist nature of statesmen, the expansionist logic of the nation state or the anarchy in the international system led to insecurity, struggles for power among states, and ultimately violence and war. Nonetheless, some extent of security and co-operation among states is, according to them, still possible, but only under conditions of hegemonic rule (in case of one superpower) or a balance of power (in case of more than one superpower). Within such 'power games', there is little or no room for players other than (big) states. Non-state players such as International Organizations (IOs), Transnational Corporations (TNCs) and NGOs are therefore considered to be background participants in international politics at best. Willets (1982, p. 18), for example, quotes the realists Reynolds and McKinlay:

> "As far as INGOs are concerned it is evident that the consequences of the activity of many of them are trivial. (...) They may serve in some degree to alter the domestic environment of decision-makers, but with some exceptions their effect either on capabilities or on objectives is likely to be minimal, and in no way can they be seen themselves as significant actors."

Realism in all its expressions has been challenged by many students of international relations (Hocking and Smith, 1990; Koch *et al.* 1994; Kolk, 1996; Van der Pijl, 1992; Ray, 1987; Reinalda, 1997; Strange, 1988; Waters, 1995; De Wilde, 1991; Willets, 1982). So-called 'idealists' stressed the importance of international organizations – such as the League of Nations and later the United Nations – for peace and security. Marxists relaxed the importance of politics and states in international relations and emphasized the relevance of the world economy, transnational classes, social movements and transnational corporations, in explaining international trends and events. Functionalists and integration theorists emphasized the possibility of and need for technical co-operation among states through a multitude of players – for example in the fields of industry, communication and traffic – and they focused on the (relative) success of the European Community (which is the European Union today). The so-called 'transnationalists' of the 1970s analyzed the role of, amongst others, churches, economic interests, and revolutionary organizations in world politics. And more recently, globalization theorists put the role of nation states in perspective of the world economy, world politics and global culture.

With regard to the role of NGOs in international politics, the transnational debate of the 1970s and interdependence debate of the early 1980s are particularly relevant. Two authors were in the forefront of these debates: Robert Keohane and Joseph Nye. They published two influential and – to some – controversial books: *Transnational Relations and World Politics*, which they edited in 1971, and *Power and Interdependence – World Politics in Transition*, which they wrote in 1977. The former book criticizes the realist image of the international state system and introduces the transnational model. The ultimate authors of world politics are no longer the (big) states, a wide range of global players, including non-state participants, are. Although they relax their transnational arguments in their second book *Power and Interdependence*, they maintain their claim that the assumptions of realists are still inadequate. As traditional issues such as organized violence and security are no longer the main features of international relations, current developments in for example the economic and ecological realms can be better described in terms of the mutual dependence of states and non-state players in several issue-areas. Therefore they favour the concept of complex interdependence, which is – to mention just one feature – characterized by 'multiple channels' among players in world politics. Transboundary relationships are no longer solely intergovernmental in nature (among diplomats and heads of state), but transnational (among NGOs and states), transgovernmental (among sectoral departments of governments), and supranational (among international organizations and states) as well.

From the perspective of this study, one might conclude that the merit of Keohane and Nye's publications is that they 'took NGOs on board' as relevant global players in a theory of international politics. And although Waltz' neorealism dominated the academic debates in the 1980s, NGOs have remained on the agenda in the study of international relations (Willets, 1982). Today there even seems to be a revival of transnational thinking, so that non-state players – including NGOs – are back again in the discipline of political science (Klotz, 1995; Reinalda, 1997; Risse-Kappen, 1995; Rosenau, 1993).

2.2.3 A political arena perspective

Keohane and Nye conceptualize international politics not only in terms of a broad set of players, but in terms of a broad set of arenas as well. This arena metaphor is commonly used in the political sciences (Van de Graaf and Hoppe, 1992; Koppenjan, 1993). Literally, it refers to the delineated space in which for example gladiators or boxers collect to compete and try to win. In doing so, all players are bound by the rules of the game. Politics may be similarly imaged. Firstly, relevant political players – for example ministers, representatives of government agencies, civil servants, politicians, expert advisers, pressure groups – meet with regard to a specific policy issue, usually in distinctive rounds. Such meetings are generally organized within the context of a given political organization, which functions as the arena – e.g. the United Nations Environment Programme (UNEP) in international environmental matters. Secondly, the participants play their game and aim at specific outcomes, e.g. political agendas or policies. Thirdly, each of the players tries to win, which means: tries to further his interests and, hence, achieve his policy goals as best as he can. The players hope to influence the political outcome as much as possible and they make interventions to do so. These interventions may range, for example, from the use of authoritative powers in decision-making by governments to lobby-initiatives by NGOs. And fourthly, formal and informal rules of the game operate: who is in or out, how one can get in, which formal competencies players have, what the domain of the game is, how conflicts are settled and decisions are made, how players should behave, which strategies to influence the outcome are legitimate (or not), what the relationship with players outside the arena is, etc. In short:

> *In this study political arenas are defined as formal meeting places of political players who struggle, debate, negotiate and decide on policy issues and, in doing so, are bound by given rules (although these may be changed by the players as well).*

The composition of an arena in terms of the players varies, as different issues mobilize different agents. Nonetheless, some rather stable set of participants can be distinguished in any arena of international relations. For example Morss (1991) refers – just as Keohane and Nye – to four categories of global players: states, international organizations, NGOs and multinational companies. The fact that representatives of states – government officials, diplomats, ministers, heads of state etc. – are participants in any global arena is quite obvious. States are the sovereign entities on which the international system is built and the legitimate representatives of the world's nations, although some would prefer to abolish the state-as-player metaphor and distinguish between the state as a non-acting political structure on the one hand and government agents who indeed can act on the other. In this book both concepts – states and governments – are used interchangeably, as governments negotiate, decide upon, and sign treaties on behalf of their nation states (see and compare: Buzan *et al.*, 1993; Lieshout, 1993; Risse-Kappen, 1995; Waltz, 1979; Willets, 1989). International organizations, next, may be involved in global politics because they generally have competencies in specific issue-areas. These may be supranational to some extent and exceed some competencies of the participating states (in the case of the European Union), or they may refer to specific intergovernmental skills, knowledge and programmes only (in the case of UN bodies). Therefore such international organizations may participate as global players in global politics. However, these organizations can also function as *arenas* (Archer, 1983; Keohane and Nye, 1989). In that case, they form the organizational and institutional context in which policy-making takes place. For example the CBD was negotiated and adopted in the context of the UNEP and the FCCC in the context of the UN General Assembly (UNGA). Or in other words: the biodiversity arena is nested in the UNEP, whereas the climate arena is nested in the UNGA. The participation of NGOs and companies in international arenas, finally, is at first sight less obvious. Nonetheless, the practice of such participation has been developing since the beginning of the 20th century (see subsection 1.4.1 in chapter 1).

To conclude this section, it should be recognized that the arena metaphor for politics has not been welcomed by all. For example Melucci (1988) believes that the metaphor gives the incorrect impression of fixed players, roles, rules and outcomes. In reality politics is, according to him, much more fluid. Others would prefer to apply more recent concepts, such as the notion of *policy network* (Godfroy, 1981; Nelissen, 1992; Godfroy and Nelissen, 1993; Hufen and Ringeling, 1990; Liefferink, 1995; Van Tatenhove and Leroy, 1995; Van Waarden, 1992). A policy network is generally referred to as a more or less stable social system in which mutually dependent public and private players address policy issues and programmes. This concept seems appropriate for this study at first sight, but was rejected for substantive reasons in the second instance. Whereas policy network theorists emphasize the reciprocal relations between public and private players, this study focuses on a single relationship of political influence (NGOs influencing states). True, there are, of course, mutual connections and interactions, but it would go too far to see the NGO-state relationships at global level as one characterized by interdependence. States are definitely dominant in the international arena and, moreover, their governments are the formal policy- and decision-makers. Therefore their dependence on NGOs is generally quite limited. Whereas states at national level have recently handed over formal competencies to private players in accordance with the neo-liberal ideology, such is hardly the case at global level. All in all, the concept of political arena is more suitable for this study than the one of policy network.

— 2.3 The concept of 'political influence' —

2.3.1 Influence versus power

Most political scientists agree on the general definition of influence. For example Cox and Jacobson (1973, p. 3) write: "[I]nfluence means the modification of one actor's behaviour by that of another." But they add:

> "Influence is to be distinguished from power. Power means capability; it is the aggregate of political resources that are available to an actor. (...) Power may be converted into influence, but it is not necessarily so converted at all or to its full extent."

This distinction is not shared by all. Consider for example Dahl's definition of power (quoted in Clegg, 1989, p. 51): "A has power over B to the extent that he can get B to do something B would not otherwise do." This definition of power coincides with influence ('to get B to do something'). Obviously, political scientists disagree on the relationship between the two concepts. This disagreement in definition can be summarized as an emphasis on *exercising* power – or influence – on the one hand (Dahl) and an emphasis on *having* power – or capabilities – on the other (Cox and Jacobson): a distinction which might be characterized as respectively episodic and dispositional concepts of power (Clegg, 1989). Some favour the latter, as a player may be called powerful without actually exercising power. The USA is powerful because it possesses nuclear weapons, not because it uses them. Others, though, hold that it is impossible to assess the capability of a player if that capability is not applied. One cannot count someone's power as if it was money and, subsequently, forecasts outcomes. Good examples in this respect are the military adventures of the USA in Vietnam and the USSR in Afghanistan: both failed (Ray, 1987). Therefore, there is hardly a one-to-one relationship between resources and outcomes (Keohane and Nye, 1989). A concept of power without the notion of influence on outcomes is thus misleading.

This position is, of course, contestable as well. Resources may, according to Huberts and Kleinnijenhuis (1994), indeed give an indication of someone's power, at least if these resources proved to be effective in the past or elsewhere in the world under comparable circumstances. Following Kuypers (1973), they make the distinction between power as the ability to influence, and influence as the realization of an effect. This distinction is also followed in this book, though it is explicitly acknowledged that both concepts are intimately linked: *power is based on practices of influence whereas these practices sustain power.* Huberts' (1994, p. 39) definitions are good illustrations in this respect:

> "A player exercises political influence if his presence, thoughts or actions cause a political decision-maker to meet his interests or objectives more than

> would have been the case had this player been absent. A player has political power if he continuously succeeds in causing political players to meet his interests or objectives."

In other words, political power refers to a more or less permanent ability to influence policy outcomes, whereas political influence refers to an episodic effect on decision-making. Hence, in case my study confirms that environmental NGOs affected the FCCC and CBD in line with their policy goals, one might speak of political influence. However, to conclude that these NGOs are therefore powerful players in international environmental politics is premature. This distinction between power and influence also has the advantage of putting the position of NGOs *vis-à-vis* the one of, for example, the USA in the right perspective. The latter has a permanent ability to influence international politics and may be rightly considered a powerful player. But this cannot be said about the NGOs.

Although the basic idea of Huberts' definition on political influence is adopted in this study, some amendments should be made. Firstly, Huberts' definition refers to two players: an NGO A which influences a decision-maker B. Such an image of the process of political influence seems correct. However, the final aim of the NGOs is not the initial impact on decision-makers, but to influence the collective policy outcome, namely the treaties concerned (Stokman, 1994). In this study it is preferred to refer to this outcome, in the definition of political influence. Besides, influence is not necessarily exercised through bilateral contacts between A and B. For example a scientist C or an adviser D might have transmitted the NGOs' preferences to the decision-maker (Potter, 1996). Secondly, Huberts' definition includes unintended effects, as the possible impact of the *presence* of a player on a decision-maker is part of the concept. It therefore includes the possibility that B changes his behaviour because of the simple fact that A exists (anticipation), while the latter might have had no such intention. Although it is recognized that such effects may occur, these are excluded from this research project for pragmatic reasons. It is simply too difficult to assess such impacts (Van Schendelen, 1981; Bos and Geurts, 1994). Therefore a player should deliberately intervene to influence the outcome of policy-making. Finally, it was preferred to attune conceptual and methodological aspects of political influence as much as possible. As the causal analysis which shall be designed in chapter 3 is based on an assessment of the NGOs' goal-achievement caused by their own intervention, these notions are utilized in the definition below. Consequently, it runs as follows:

> *In this study political influence is defined as the achievement of (a part of) one's policy goal with regard to an outcome in treaty formation and implementation, which is (at least partly) caused by one's own and intentional intervention in the political arena and process concerned.*

It is also possible to rewrite this definition in terms of the so-called counterfactual (Lukes, 1974). Then political influence implies that the policy outcome concerned is more in line with the policy goal of the player involved than would have been the case had the latter not intervened. In other words, the player did matter and has indeed made a difference. Furthermore, it should be stressed that the achievement of one's goal might not only cover the realization of a desired outcome, but the prevention of an undesired one as well. In addition, it is obvious that political influence refers to effects on policy *outcomes* (which includes effects on intergovernmental decision-making). This however does not cover temporary effects of NGOs on global debates and negotiations. Another way to distinguish between these two sorts of effects is to speak of *product* and *process* influence. The former refers to effects on formal policy outcomes, the latter to those temporary effects on policy processes which are no longer recognizable in the outcomes. Finally, goal-achievement and political influence do definitely not coincide, as a player might attain his or her own goal due to the intervention of others (including the authorities themselves). Therefore political influence presupposes a causal linkage between specific outcomes on the one hand and specific player interventions on the other. This question of causality will be elaborated on in chapter 3. The nature of possible NGO interventions has already been referred to in subsection 2.2.1 (table 2.2 on protest, advocacy and lobby), but this will be empirically detailed in chapters 4 and 5.

Debates on dispositional power and episodic influence were, by the way, also held within the study of international relations (Guzzini, 1993). Whereas realists have focused on the former concept, the transnationalists have mainly stressed the importance of the latter.

2.3.2 Agent versus structure

The above definition of political influence definitely stands in the so-called *pluralist* tradition of the political sciences (Clegg, 1989; Self, 1985). After all, the concept is defined in terms of observable effects on decision-making and does potentially cover – given the object of study here (NGOs) – a wide range of state and non-state players. Such a voluntarist approach is criticized by many (Williamson, 1989). For example Bachrach and Baratz (1962) challenge the equation of political influence – or power – with the successful participation in decision-making. Instead, they conceptualize power as having *two faces*: 1. the face of participation in decision-making, and 2. the face of the so-called non-decisions. The latter refers to the exclusion of issues from the political agenda and therefore from decision-making by elites. Lukes (1974) goes even further and claims that power may also refer to subtle, often hidden, mechanisms through which the system favours the interests of the strong, apparently with the consent of all (the so-called third dimension of power along its two faces). Here the power of social structures instead of agent power is emphasized. Such an approach, however, might easily end up as the opposite of voluntarism, namely determinism. Therefore Giddens (1984) rejects the Lukesian idea that latent mechanisms may rule human beings as if they are playthings without a will. But neither does he wish to return to old pluralist concepts of power. In his structuration theory, Giddens tries to find a middle road between determinism and voluntarism. Human beings are assumed to be knowledgeable and capable agents, yet they are constrained as well as enabled by rules and resources in a given society. In terms of power and influence: agents are able to achieve outcomes in social and political practices, but the extent to which they can do so – or wish to do so – is co-determined by the rules in a given society on the one hand and by the distribution of allocative and authoritative resources in that society on the other (e.g. rules on 'good' and 'bad' behaviour; resources such as money and formal competencies etc.). Yet agents may in turn change the nature of these rules and resources while they act. Hence, there is no such thing as a pre-given and unchangeable structure in a society which unilaterally determines the social and political actions of agents, but neither are the latter entirely free to choose their course of action.

Criticism of pluralism is to some extent shared in this book and a few structural elements will be added to the model which will be designed in section 2.6 below. First of all, the arguments of Bachrach and Baratz are accepted in principle. It is assumed to be true that many issues are excluded from political decision-making. Consequently, political agendas are biased and selective, including those regarding the FCCC and CBD. However, the processes of agenda-setting are not the primary objects of study (see chapter 1). On the contrary, it focuses on treaty formation and implementation within the context of more or less given agendas, as most issues and elements which had to be included in the biodiversity and climate treaties had already been discussed and to some extent adopted at international meetings before the formal negotiations started in 1990 (see chapters 4 and 5). Therefore what is primarily studied is the influence of NGOs on policy- and decision-making *after* agenda-setting. Nonetheless, it is acknowledged that the struggle over the agenda might continue during subsequent policy phases. Hence, attempts of NGOs to change the agenda – or maintain elements of it – during the phases of policy formation and implementation are taken into account as well, at least in cases where it seems relevant. In short, the selective nature of agenda-setting is accepted in this thesis at the theoretical level, but barely covered by the empirical research.

Secondly, Giddens' arguments against voluntarist interpretations of influence and power are taken on board as well. It is accepted that NGOs are constrained by given rules and divisions of resources. After all, they cannot just do or achieve what they like. They are bound by the rules of the game, which produce some opportunities, but even more constraints. Likewise, the division of resources among the political players co-determines the outcomes they may achieve, and NGOs are definitely generally less well off than other players, such as states, international organizations and transnational companies. But to avoid any determinism, it is also assumed that individual agents or collective players are never unilaterally controlled by 'the system', whatever that may be. Agents are always able to reflect on the positions they take, interests they defend, and preferences they formulate in policy processes. Moreover, they are in principle able to intervene and achieve outcomes in political arenas. Hence, NGOs are never playthings of hidden power mechanisms.

Consequently, the *co-optation thesis* which holds that NGOs in fact unintendedly defended neoliberal and elitist interests by participating in UN initiatives such as UNCED, CBD and FCCC is rejected in this book (Chatterjee and Finger, 1994).

To conclude, the terminology on influence and power used in this study is as follows:

1 *political influence:* the achievement of a policy goal due to one's own and intended intervention;
2 *political power:* one's more or less permanent ability to influence policy outcomes;
3 *distribution of resources:* enables or constrains players vis-à-vis others in achieving policy goals;
4 *rules of the game:* co-determine whether and how agents may participate in politics and, consequently, have – or do not have – opportunities to exercise political influence.

— 2.4 The concept of 'convention' —

2.4.1 Definition

Treaties which are designed in the context of international organizations are generally referred to as *conventions* (Schermers, 1985a; 1985b). Such conventions are therefore – just like the 'classical' treaties designed by states – international legal contracts which set regulatory rules that are binding to parties (Kamminga, 1995). Parties are those states which have signed and ratified such an instrument. It is obvious that both the FCCC and CBD are conventions in this respect. The former was designed in the context of the General Assembly of the United Nations (UNGA), the latter in the context of the United Nations Environment Programme (UNEP).

Framework conventions are binding as well, but do, however, not set clear and stringent regulations (Hurrell and Kingsbury, 1992). They only formulate general principles and objectives. In a next phase parties to such a framework convention need to design and adopt protocols on specific topics and these do set detailed and binding regulations. An example is the 'Vienna Convention – Montreal Protocol' package on the protection of the ozone layer. The Vienna Convention defines general objectives such as the protection of human health, the control of human activities that have adverse effects on the ozone layer, and scientific co-operation among states, whereas the Montreal Protocol stipulates for example the phase-out of chlorofluorocarbons (CFCs) within a detailed time frame (Bergesen and Parmann, 1993). Both the FCCC and CBD are of the framework convention-protocol approach.

The legal literature is divided on the question whether 'convention' and 'treaty' are synonyms. Some indeed believe so (Kamminga, 1995). Others make the distinction that treaties are the traditional form of binding agreements among states and conventions are the legal instruments which are linked to international organizations (Schermers, 1985a; 1985b). A third group makes a distinction between comprehensive and detailed treaties on the one hand and convention-protocol constructions on the other (Hurrell and Kingsbury, 1992). Whatever legal experts think, the FCCC and CBD will be called both conventions and – sometimes – treaties in the text below, especially since both labels are already utilized in the literature with regard to these two instruments. Binding conventions and treaties should nonetheless be distinguished from non-binding legal instruments, such as resolutions, declarations, and recommendations. Whereas the former are referred to as *hard* law, the latter are also called *soft* law (Carroll, 1988). These labels do not, however, imply that the latter are not relevant. Although states are not bound by soft law, it may nonetheless guide their behaviour (Rittberger, 1993). Furthermore, soft law may be a source of binding law at a later stage.

This study focuses on both treaty formation and implementation. Formation is defined as the phase of formal negotiation and adoption of an international environmental convention by a number of states, a process in which other players – such as scientists and NGOs – also participate. Implementation refers to the phase of the initial putting into operation, the entering-into-force, and the putting into practice of the provisions of the convention concerned, both at national and international level. Again, agents other than states do play a role in this second phase. It should be recognized that the terms formation and implementation originate in policy science literature (Van de Graaf and Hoppe, 1992). The terms refer to two phases of the so-called policy cycle, which consists of the phases of issue-raising, agenda-setting, policy formation, policy implementation and policy evaluation. Although such policy cycle models are particularly designed for grasping policy-making at national level, they are employed in respect of international policies as well (Reinalda, 1997). However, the rationalist and phase-like nature of such models are criticized in some of the literature (Koppenjan, 1993; Van Tatenhove,

1993). For example, the distinct phases either overlap or are hard to distinguish in many policy processes. Furthermore, the so-called sequential process of issue-definition, agenda-setting, problem-definition and problem-solving is seldom followed in practice. But despite such criticism, the notions of formation and implementation are maintained in this thesis.

2.4.2 Conventions and regimes

There are at least two branches of the study of international relations that cover the study of conventions and treaties: international law and regime theory (or neo-institutionalism; Rittberger, 1993; Schermers, 1985a; 1985b). Whereas the former is quite descriptive in nature and focuses on legal, formal and procedural aspects of international law, the latter tries to assess the importance of rules and rule-making for state behaviour and state co-operation. Therefore regime theory is more political and analytical in nature than the study of international law and, consequently, more suited to be utilized in this thesis. Besides, regime theory addresses the role of *im*plicit, *in*formal and *non*-binding rules as well, at least more than international lawyers generally do. Regime theory also distances itself from positions that are too idealistic on the possibilities of firm legal rule in the international system. Additionally regime theory connects this study to a broad range of other environmental policy research which has already utilized this perspective (Humphreys, 1994; Rowlands, 1995; Young and Osherenko, 1993).

Regime theory was mainly developed in the USA in the late 1970s and early 1980s (Rood, 1987; 1984). The decline of US hegemony as well as the growth of global interdependence – two phenomena which were supposed relevant at that time – urged students of international relations to look for alternative concepts in theorizing about co-operation and security in world affairs. Regimes seemed good candidates. Keohane and Nye were among the first ones to use this concept in 1977. To repeat, world politics is, according to them, determined both by state power and by complex interdependence. States are mutually dependent, which means that each is sensitive or vulnerable to the actions of others. But these actions and their consequences are unknown and unpredictable most of the time. To avoid conflict in such circumstances, co-operation to exchange information, build trust and prescribe common action is very much needed. Hence, states design sets of governing arrangements to deal with the relationships of interdependence. Keohane and Nye define these sets of arrangements as international regimes (Keohane and Nye, 1989, p. 21).

The concept derives from the Latin words *regimen* and *regere*; the former means 'a rule', the latter 'to rule' (Aggarwal, 1985). It is generally applied with regard to national governments, mainly referring to forms of authoritarian rule (e.g. the

Pinochet regime in Chile during the 1980s). At international level, though, another meaning has become dominant (Krasner, 1982, p. 186):

> "Regimes are implicit or explicit principles, norms, rules, and decision-making procedures around which actors' expectations converge in a given area of international relations."

Principles are beliefs of fact, causation and rectitude; norms are standards of behaviour defined in terms of rights and obligations; rules are specific prescriptions or proscriptions for action; and decision-making procedures are prevailing practices for making and implementing collective choice. It is important to note the function of regimes: to shape and converge the expectations of the participants. As regimes prescribe roles and guide the behaviour of the participants, they increase predictability and security in international affairs. Consequently, regimes can be considered institutions (Keohane, 1993; Young, 1980). Due to such regimes, the anarchic nature of the international system – as proclaimed by realists – is softened. A final characteristic of the concept is its limited scope. Regimes refer to specific issue-areas in international relations. For example the environmental regime covers all those principles, norms, rules and procedures which are operative on this issue in international politics. It should be noted, however, that there are several regime theories. At least four variants are distinguished today, namely power-based, interest-based, knowledge-based and issue-based approaches (Haas, 1993; Haggard and Simmons, 1987; Humphreys, 1994; Junne, 1992; Rowlands, 1995; Young and Osherenko, 1993). In explaining the rise of international regimes, the various authors emphasize different variables. For example: the hegemonic power of a large state to force others to join a regime; the joint interests of states in an interdependent world; the role of knowledge in enhancing a regime; or the specific nature of an issue which makes a regime opportune.

Regime theory has received a lot of criticism, as most realists and Marxists reject it. They believe that regimes are epiphenomena and, hence, dependent on the power structure in international relations (Grieco, 1995). For example Strange (1982, p. 487) writes:

> "All those international arrangements defined by the label regime are too easily upset when either the balance of bargaining power or the perception of national interest (or both together) change among those states who negotiate them."

She believed – at least in 1982 – that the regime concept was a passing fad, because it is imprecise, value-biased, static and state-centred. It is imprecise, because it may refer to all interactions in international relations according to one theorist, or to one specific agreement in an issue-area according to another. It is value-biased, because it is linked to the liberal ideology of the need for world order (instead of, for example, world justice). It is static, because it focuses on stable arrangements in issue-areas and neglects short-term changes in for example monetary, trade and technological affairs. And it is state-centred, because it highlights arrangements between governments and neglects other important players in the international political economy, for example the transnational corporations. Although some of Strange's criticism is still valid, regime theory did, however, turn out not to be a passing fad (Kratochwil and Ruggie, 1986; Rittberger, 1993). One reason for this might be the fact that regime theorists did concede some of the objections.

Still one of the main controversies in neo-institutionalism is the delineation of the concept of regime. What is it that constitutes a regime? Humphreys (1994) refers to two possible applications: either in terms of one global agreement or in terms of all those institutions that refer to one issue. Take the example of forests: should a regime theorist be sensitive to one global agreement – a forest convention (which is, by the way, still lacking) – or include all those organizations and rules that already exist and cover the forest issue, e.g. Agenda 21, Commission for Sustainable Development (CSD), Convention on International Trade in Endangered Species (CITES), International Tropical Timber Organization (ITTO), Biodiversity Convention (CBD) and World Trade Organization (WTO)? Humphreys chooses the first option, whereas the latter is favoured in this thesis (see below). The strength of a regime is another controversial aspect of the delineation debate. For example Von Moltke (1995) believes that Agenda 21 of UNCED is not a regime, as it only nicely sums up weak principles and norms. Others, though, believe that UNCED did indeed come up with a regime (Spector *et al.*, 1994). Likewise, the effectiveness of regimes is discussed. According to some, regimes should: "(..) transcend the notion of established patterns of conference diplomacy and their outcomes, since these may turn out to be mere 'paper regimes'" (Rittberger, 1993, p. 10).

2.4.3 Regimes as frames of reference

The nature of Krasner's definition of a regime as given in the previous subsection is adopted here, as it is *the* consensus definition widely quoted in international relations literature. Nonetheless, some amendments are made:

> *In this study a regime is defined as a set of rules – principles, norms, regulations and procedures – which frame the (common) actions of states and non-state players in a specific issue-area in international relations.*

The amendments should briefly be elucidated. Firstly, the term 'rule' is utilized as an umbrella concept for different sorts of rules here: principles, norms, regulations (instead of 'rules' in the Krasner definition) and procedures. The reason for this stems from the adoption of a more broad concept of 'rule' such as employed in Giddens' structuration theory. Secondly, regimes 'frame' individual or common actions of states (instead of 'converge expectations' in Krasner's description). The above definition is therefore more action-oriented, although 'framing' is meant to be a weak term. It does definitely not imply that regimes determine when and how states and other players should behave. A third and final amendment refers to non-state players. These are considered relevant to conventions and regimes in this book as well (Haufler, 1993).

The scope of a regime is tied to an issue-area in this thesis, not to a single legal instrument. If the latter definition had been adopted there would be hardly any difference between a treaty and a regime and, hence, between regime theory and the study of international law. Therefore the surplus value of regime theory should be the attention paid to broad sets of implicit and explicit rules in international relations, not to formal contracts alone (Huizinga and Verbeek, 1995; Hurrell, 1993). Hence, the domain on the basis of which the concept of regime should be defined is an *issue-area*, not a specific legal agreement. The latter, though, is still part of the former. Therefore the CBD and FCCC are not considered regimes in themselves, but are assumed to be nested in the atmospheric and nature conservation regime respectively of international relations (Aggarwal, 1985). These regimes include a wide set of principles, norms, regulations and procedures which were adopted in different arenas (European Union, United Nations, Organization for Economic Cooperation and Development, World Trade Organization, Stockholm Conference, UNCED, etc.) and which are laid down in different binding and non-binding instruments (treaties, declarations, action programmes etc.).

Regimes frame the (common) actions of states and non-state participants. Such actions may include the formation of new legal instruments – such as a treaty – within the given issue-area of such a regime. Therefore regimes can be conceptualized as frames of reference for negotiators of international conventions. In the words of Jönsson *et al.* (1995, p 24) (although they focus mainly on agenda-setting):

> "The existence of an effective regime constrains agenda-setting, insofar as a certain framing of the problem is embedded in the regime. Alternative frames are then less legitimate and frame restructuring becomes difficult. Those international organizations that are created to administer a specific regime typically embody the legitimate frame and become guardians against alternative frames."

However, the existence of specific rules might not only constrain players, but enable them as well (Giddens, 1984). The given fact that specific rules exist in arena X legitimizes and strengthens the demands of those who wish to introduce and adopt similar ones in arena Y (Risse-Kappen, 1995). For example, rules already adopted within the context of ozone policies played a significant role in the negotiations of the FCCC and this worked in favour of those who propagated relative strong commitments (see chapter 4). In addition, new or alternative elements can be added to regimes by any negotiation round, as players might decide not simply to reproduce the given rules. Consequently, new instruments might, once adopted, enrich the given regime. So the FCCC is now part of and nested in the atmospheric regime and has therefore become a frame of reference not only for further negotiations on climate policies, but for related environmental arenas as well.

It should, however, be added that regimes of other issue-areas also guide the formation and implementation of environmental treaties. A good example is the international trade regime, mainly institutionalized in the World Trade Organization (WTO). Rules on trade played a significant role in the biodiversity negotiations, in particular with regard to access to and trade in genetic resources, and also with regard to technology transfer and intellectual property rights (see chapter 5). Consequently, environmental negotiations are guided by a set of relevant regimes, not by the environmental one alone.

— 2.5 The context —

2.5.1 The international system

Negotiations and decision-making on conventions take place in the political arena concerned. However, a wider context than this arena impacts such policy processes. Two domains are considered relevant here: 1. the nature of the international system and 2. the events and trends which happen to occur in the world. The inclusion of the first domain in the theoretical framework is based on the conviction that all actions of agents are shaped by the nature of the system in which these actions take place (Giddens, 1984; Keohane, 1984, 1986; Waltz, 1979). The second domain is included due to the fact that both policy and regime theorists emphasize the relevance of issue-related events and trends to policy-making (Rittberger, 1993; Sabatier, 1987).

According to the neorealist Waltz (1979), the international system consists of interacting units (states) and the system structure. The latter is constituted of three tiers: 1. the organizing principle of the international system (which is *anarchy* in the current system as it lacks a central authority); 2. the functional differentiation of units (which is *absent* in the current system, because all sovereign states are in fact 'like units'); 3. the distribution of the unit's capabilities (which is the *power* division among states in the current system). Because of these characteristics, the current state system is – from the perspective of all states – a system of 'self-help'.

It urges self-defence. Consequently, states are inclined to strengthen themselves or weaken others, either by peaceful or violent means. Put differently, all states face a security dilemma and engage in (military) power struggles – even against their own will. Consequently, it is the structural distribution of capabilities that primarily matters in international relations and determines outcomes in politics.

This view is challenged by others (Buzan *et al.*, 1993; Keohane, 1986; Powell, 1994; Wendt, 1987). Critics believe that Waltz's concept of the international system is too narrow, too static and too deterministic. Some wonder whether anarchy is the only organizing principle in international relations and whether a functional differentiation among states is indeed absent. Others, though, criticize the limited power concept of Waltz. Others again believe that Waltz excludes the possibility of change in the international system. A final group emphasizes the relevance of other structural variables in international affairs, such as regimes (Keohane, 1984; 1986). According to this last group, such variables definitely make a difference in world politics, as they impact the behaviour of states and therefore co-determine political outcomes. This particular view is, given the arguments in the previous section on regimes, of the utmost importance to this thesis. However, regimes are assumed to be additional variables to Waltz' system structure, not alternative ones, as the (uneven) distribution of resources is taken into account as well (see definitions on power and influence in subsection 2.3.2). Consequently, a broad concept of system structure is followed in this thesis, both in terms of rules (rules of the game and regimes) and in terms of capabilities (distribution of resources). Such a concept of structure is – again – legitimized by Giddens' structuration theory.

The three structural elements – rules of the game, regimes, resource division – co-determine both the opportunities and the constraints for players. As was outlined earlier, existing regimes function as frames of reference. They guide for example problem-definition, agenda-setting and rule-making with regard to new policies. Some players may feel constrained by such processes, others enabled. The rules of the game, next, give unequal access to political arenas and it is obvious that states are in a better position in this respect than NGOs in international politics. Finally, the distribution of resources also provides opportunities and constraints for players. It is obvious that it is easier for the USA to achieve results in international environmental politics it is than for a small Third World country such as Benin. However, it would be misleading to believe that these structural elements unilaterally determine outcomes in politics, as processes of interaction in political arenas intervene. The way negotiations proceed may for example close off opportunities to some and put constraints on others. Hence, between structure and outcome stands process.

It should be acknowledged that outcomes may in turn change structure (Giddens, 1984). For example, new conventions may redefine rules in the international system and even redistribute resources, such as among the developed and developing countries through agreements on financial aid. Negotiations may also include the reformulation of the rules of the game.

2.5.2 Events and trends

Policy processes may be impacted by major events such as the Cuban missile crisis, oil crises, the Chernobyl disaster, the discovery of the hole in the ozone layer, the fall of the Berlin Wall, and the Gulf War. Furthermore, major trends such as rising unemployment, good prospects for economic growth, growing environmental awareness or a continuing process of disarmament, do matter with regard to many policy processes. Therefore trends and events in the global economy, polity, and – of course – in the global environment itself may be important explanatory factors with regard to the success or failure of international environmental policies. In particular the relationship between the economy and the environment seems of the utmost importance (Arts and Mulleneers, 1995; Cramer, 1989). As long as the economy is in good shape and prospects are positive, environmental policies seem relative easy to design and implement. But as soon as negative economic events occur or negative trends come up, the environment as an issue seems to be rather easily dropped. The oil crises of the 1970s, for example, had that discouraging effect on environmental policies. On the other hand, the recovery of the world economy in the late 1980s seems to have stimulated the re-appearance of the issue on the global political agenda. But developments in global politics may be equally important. The end of the Cold War, for instance, seems to have opened new policy windows for other issues, such as the environment.

In addition, developments in science and technology matter. For example, the progress in computer technology enabled climate models to predict climate change in an ever more sophisticated and reliable way. This development contributed – together with hot summers and hurricanes in different regions of the world during the 1980s and 1990s – to the formulation of (initial) climate policies. The same goes for the discovery of the hole in the ozone layer by NASA or the biologists' conclusion that ever more species are endangered or becoming extinct today. Without such knowledge, it is less likely that environmental policies would have been designed the way they are now.

However, it is difficult to assess the precise impact of all these events and trends on policy-making and outcomes. It is nonetheless obvious that they raise opportunities for or put constraints on agents to undertake – or refrain from – some actions. Events and trends may for example encourage players to try to design and apply specific policies, such as to compensate for the negative effects of a certain shock event. In other circumstances, negative economic trends may encourage them to refrain from certain expenditure which benefits the environment. But a feed-back also operates, as an environmental problem may cause the adoption of a convention, which – in turn – may lead to actions which soften this problem. Furthermore, conventions may also encourage scientific research and technology development. In short, events and trends impact treaty formation and implementation, but events and trends may result from them as well.

— 2.6 The theoretical model —

Figure 2.2 presents the theoretical model of this study. It is based on the conceptual and theoretical elaborations of the above sections. Now the different elements are linked to each other in one framework. Given the fact that the political arena constitutes the central part of this model, it is called a political arena perspective on NGOs in treaty formation and implementation. To highlight the linkages of the different concepts, the several arrows in the framework are briefly elaborated on here.

Arrow (a) – As was outlined in the previous section, the context of treaty formation and implementation which consists of rules, resources, events and trends in the international system, raises both opportunities for and constraints on players who negotiate legal instruments. Arrow (a) should therefore be read as 'enabling/constraining'. However, it should be remembered that neither the context as a whole nor the opportunities and constraints are pre-given, objective entities. True, they are experienced as such by players in concrete situations, but they are nonetheless constructs of human action and human history (Giddens, 1984; Wendt, 1992). Moreover, they are not always fixed. What player A defines as a constraint may be defined as an opportunity by player B.

Arrow (b) – This arrow is a central one in this research project. It refers to the possible influence of NGOs on representatives of states who negotiate and decide on international legal instruments. Therefore arrow (b) should be read as 'political influence'.

Arrow (c) – The same applies to this arrow. Here the possible influence of other players – such as representatives of international organizations, business organizations and scientific associations – on country delegates is represented. Note that arrows (b) and (c) point one way only and, moreover, that an arrow connecting NGOs and these other players is absent. This was done on purpose, to prevent an image of a policy network in the model (see subsection 2.2.3). States are conceptualized as the dominant players in political arenas. Given their formal status and position, they possess the authoritative powers to decide on policies. Of course, the relationships are reciprocal and state representatives impact NGO and other representatives as well, but this relationship is of less relevance from the perspective of this study. Furthermore, linkages among NGOs and other non-state players are not represented in the model either, although these, of course, exist. Again, such relationships are of less relevance from the perspective of this thesis, unless NGOs form coalitions with such players to influence state representatives.

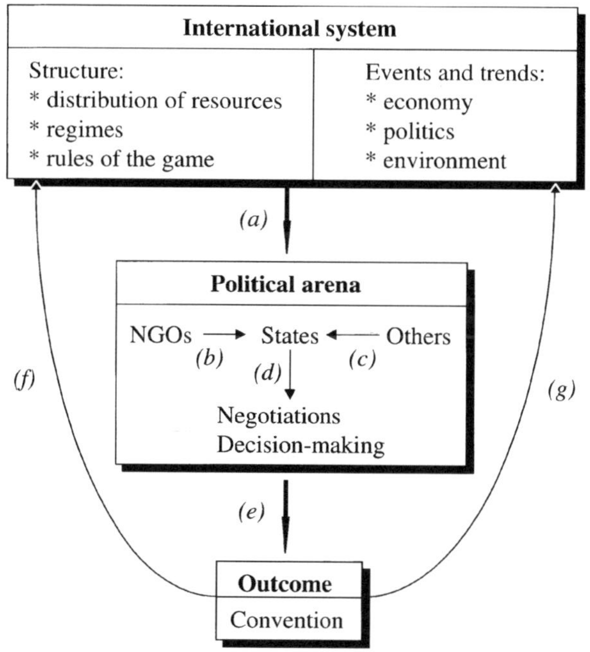

FIGURE 2.2 *A political arena model on NGOs in treaty formation and implementation*

Arrow (d) – This arrow simply reads 'engage in'. After all, state representatives engage in negotiations and decision-making procedures in political arenas in order to design and implement international conventions. Note that the nature of negotiations and decision-making may differ in political arenas (Young, 1975; 1980; 1993). States might for example clash on topics (confrontative negotiations). In other cases, consensus-building turns out to be much easier (integrative negotiations). Such different processes definitely have various effects on the outcome.

Arrow (e) – The policy processes lead to certain 'products': the adoption or implementation of a legal instrument. Arrow (e) should therefore be read as 'leads to'.

Arrow (f) – The outcome of a policy process is co-determined by the context in which it was produced. For example the content of a convention is generally a mirror image of the given regime in the issue-area concerned. For as far as the rules of a convention resemble those of the given regime, the former reproduces the latter. Nonetheless, between context and outcome stands the strategic interaction of agents. Besides reproducing existing rules, political players might decide to design new ones, or they may unintendedly change given rules. In that case the convention transforms the given regime. In short, arrow (f) should be read as 'reproduction/transformation'.

Arrow (g) – Trends or events might encourage players to produce conventions, but the latter might 'produce' trends and events as well, for example by mitigating environmental problems, stimulating scientific research or reframing ordinary happenings into policy-relevant events (for example hot summers in the context of the climate issue). Arrow (g) should therefore be read as 'produce'.

Chapter 3 aims at putting the theoretical model into operation with regard to the three research questions: the assessment, explanation and 'prescription' of the NGOs' political influence. Therefore the application of the model will be elaborated on in the next chapter. Nonetheless, a few words on this topic will follow here. Firstly, the model offers specific insights into NGO-state relationships, as well as into the process of treaty formation and implementation. Such insights enable the researcher to adopt a related methodology to assess the political influence of NGOs. Secondly, the model identifies relevant variables to explain aspects of treaty formation and implementation. The NGOs' political influence covers one aspect of these processes and may, therefore, be explained by factors identified in the model. A list of potentially explanatory variables based on the model will, consequently, be drafted in chapter 3. Thirdly, the explanation of influence enables the analyst to subsequently formulate recommendations to NGOs to impact similar policy processes in the near future. Therefore the model indirectly guides the formulation of recommendations as well.

To conclude this chapter, something should be said about one of the main limitations of the model: the fact that it does only cover the global level. This implies two important omissions: the domestic as well as the regional level. However, it is more than obvious that policy processes at these levels are also important if NGOs wish to exercise influence on global treaty formation and implementation. After all, government positions on global treaties are established in the countries themselves and at this domestic level national NGOs might influence the process of position formation. In addition, an intermediate level between the domestic and the global exists: the regional one. The best example in this respect is the European Union. Processes of regional position formation also raise opportunities for NGOs to influence global negotiations. In a case where an EU position is for example influenced by regional – or even national – NGOs and this position is maintained and adopted by the global community, the NGOs concerned were influential, irrespective whether they participated in global meetings or not. But although the domestic and regional levels are considered relevant to global treaty formation and implementation, this study still focuses on the global level only. The reason is simple: it was impossible to study the role of thousands of national and regional NGOs in the formation of policy positions on the FCCC and CBD in more than a hundred countries around the world and in several regional organizations. Therefore it was decided to focus on the global level: on *global* NGOs which try to influence the outcome of *global* meetings, however limited this focus may be.

— 2.7 Summary —

This chapter has yielded a political arena model on global NGOs in treaty formation and implementation. Its design was based on an elaboration of the main concepts of this study: 'NGO', 'political influence', 'convention' and 'context'. These concepts were nested in related theories and fused into one framework. The model positions global NGOs in treaty formation and implementation in a specific way. The basic assumption is that NGOs are able to impact policy outcomes in principle, but that they, in doing so, are either constrained or enabled by other players (e.g. states, firms, international bodies) as well as by contextual factors (e.g. rules of the game, regimes, distribution of resources). Although these theoretical premisses seem perfectly obvious at first sight, they are nonetheless the opposite of some of the mainstream views of the political sciences. For example, the model accepts the relevance of non-state players and regimes in international politics, something which is rejected by the realist tradition. The arena model will function as a 'lens' through which the researcher will observe and explain his empirical domain. As such, it will guide the selection of methodologies to assess the political influence of NGOs, as well as the selection of relevant variables to explain the phenomenon of political influence.

CHAPTER 3

Methodological Framework
The EAC Method

One of the most serious problems in the study of influence arises from the fact that, no matter how precisely one defines influence and no matter how elegant the measures and methods one proposes, the data within reach even of the most assiduous researcher require the use of operational measures that are at best somewhat unsatisfactory.

Robert A. Dahl (1961, p. 330)

— 3.1 Introduction —

The objective of this chapter is to develop a methodological framework to assess and explain the NGOs' political influence on the one hand and to formulate recommendations for NGOs on the other. In doing so, the theoretical concepts and model which were elaborated on in chapter 2 function as guides. First of all, the nature of these concepts and model urges one to choose related techniques. Therefore in section 3.2 the decision-making method will be identified as the principal point of departure for the assessment of political influence. A specific, qualitative variant will be presented: the so-called *intensive process analysis.* Although this method yields useful tools for an assessment of political influence, additional ones will be designed below, adapted to the characteristics of this study. Ultimately, the so-called *EAC* Method will be constructed, which is based on the respondents' perceptions on the one hand – *E*go- and *A*lter-perception – and a *C*ausal analysis by the researcher on the other. Furthermore, a formula will be designed to determine the extent of political influence. Note that the term 'assessment' is used instead of, for example, the term 'measurement', as any quantitative or qualitative determination of political influence remains after all an *informed guess.*

Secondly, the theoretical model distinguishes a number of variables which may explain the NGOs' political influence (e.g. interventions of specific players, nature of negotiations and decision-making, rules of the game etc.). These variables are therefore linked to the several examples of more and less NGO influence. This is done by way of a data-matrix in which – on a case-by-case basis – correlations between the specific 'scores' of these variables on the one hand and the extent of NGO influence on the other are investigated (comparative case-analysis). This technique will be highlighted in section 3.3. However, the term 'explanation' should not be interpreted as if universal laws on NGO influence are to be deduced from this study. On the contrary, NGO influence will be explained *within* the restricted context of this research project. Therefore the limitations of the explanatory analysis will be given due attention below.

Section 3.4 of this chapter will deal with the technique to formulate recommendations for NGOs. This procedure builds on the explanation of political influence. After all, if one knows which explanatory factors (co)determine NGO influence, one might reflect on the question of how those NGOs might anticipate on these factors in policy arenas in the near future. Specific judgement criteria to guide this reflection are distinguished below (criteria of anticipation and extrapolation). However, the formulation of such recommendations needs to be carefully done. The nature of this study does not allow for bold predictions. Finally, section 3.5 will summarize this chapter.

— 3.2 The assessment of political influence —

3.2.1 Classical methodologies

With regard to techniques to assess political influence – or episodic power – authors generally classify three classical approaches: the position, reputation, and decision-making methods (Clegg, 1989; Huberts, 1988; Westerheijden, 1987). Some students of power, though, distinguish more variants (Van Schendelen, 1981). For example Huberts and Kleinnijenhuis (1994) are able to list ten methods, but most of these are either quantitative or qualitative elaborations of the three classical approaches.

The *position method* assumes that political influence is based on the position of players in policy processes and decision-making. As participants in various (in)-formal positions have different competencies and unequal access to resources, information and authorities, different positions yield different opportunities to control outcomes. Protagonists of the position method therefore believe that political influence can be deduced from such dispositional power. A prime minister has generally more control over outcomes than a minister, a minister more than a member of parliament, such a member more than a lobbyist, etc. In consequence, the first-mentioned player can be assumed to be more influential than the second, the second more than the third, and so on. However, there are several varieties of the position methodology. One may for example focus on formal or informal positions; on individuals or networks; or on qualitative or quantitative research designs (Felling and Hüttner, 1981; Van Goor *et al.*, 1985).

The *reputation method* equates power with reputation. Its protagonists argue that what counts in social relations are the mutual perceptions of power (Westerheijden, 1987). If someone is defined by others as influential, then he will be treated as such. The same goes, of course, for the opposite: if someone has the reputation of being powerless, he will be powerless in the eyes of others. Or to put it in the famous formulation of the Thomas-theorem: "If men define their situation as real, it is real in its consequences" (Volkert, 1951). When using this method respondents selected from a political community are asked to assess the political influence of leaders and subleaders in that community. This is done through written questionnaires or interviews.

However, positions and reputations do not necessarily say much about who really influenced political decisions (Clegg, 1989; Dahl, 1961; Westerheijden, 1987). Although the position of an individual is certainly a relevant power resource, the critics argue that it does not tell us much about the control of outcomes. After all, not every member of parliament – to mention one formal position – is equally influential in specific cases of decision-making. Critics of the reputation method argue that it assesses perceptions instead of 'real' power, and perceptions are frequently incorrect. Besides, the methodology does not discriminate between different issues and political arenas, as respondents are asked to assess the power of leaders in general. Therefore the *decision-making method* aims at analyzing decisions in specific issue-areas, in order to reconstruct the contribution of players to the final outcome. It assesses to what extent the participants' interventions were successful in the decision-making process. On the basis of such an assessment, the relative influence of each participant can be determined. Because key issues and key decisions are studied, the players' influence is considered to be politically relevant. Techniques to collect the necessary data are generally interviews with decision-makers, analyses of policy documents, as well as direct observations of policy-making.

Although the decision-making method does discriminate among different issues and elites, it is criticized as well (Bachrach and Baratz, 1962). For example, the successful participation of one person in – let's say – the politics of urban development does not say much about the power of this person in the political community as a whole. Therefore the method has a bias in favour of *pluralist* theory, as many players in many arenas are easily considered influential (see subsection 2.3.2). Furthermore, power is not only exercised by participation in decision-making, but by the control of the political agenda as well. Therefore the making of *non*-decisions should be taken into account too. Moreover, the reconstruction of a complex decision-making process – including agenda-setting – is a rather difficult task. One might question the possibility of making past decision-making processes so transparent as to conclude on the political influence of individuals.

It should be recognized that the criticism is sometimes beside the point. If someone defines power in dispositional terms, then the position method may be valid. After all, such approaches define power in terms of capabilities, not in terms of outcomes. To equate position with power may be justified in these cases. However, if someone applies an episodic definition of power – as is done in this study – then the position method will not suffice. Then one needs to assess the exercise of power in specific cases, as the decision-making method does. Furthermore, it should be recognized that nearly all students of power combine elements of the three classical methods (Van Schendelen, 1981). Even Dahl (1961), who is considered the founding father of the decision-making method, promotes an eclectic approach in his famous book *Who Governs?*. He combines aspects of amongst others the position and decision-making methods.

3.2.2 Intensive process analysis

Given the fact that this study focuses on episodic power in political arenas – the actual influence of NGOs on specific policy outcomes – the decision-making method seems to be the most appropriate to employ (see the definition of political influence in subsection 2.3.1, page 57, and the theoretical model in figure 2.2, page 71). This also became obvious after consulting quite similar studies on pressure groups. A good example is Huberts' (1988) PhD-thesis *De politieke invloed van protest en pressie. Besluitvormingsprocessen over rijkswegen* (The political influence of protest and pressure. Decision-making processes concerning national highways). Huberts studied the influence of the Dutch environmental movement on policies concerning the construction of national highways. To do so, he designed a specific methodology, which he labelled the 'intensive process analysis'. This method has its roots in Dahl's decision-making method and Braam's process-effect technique, the former of which has been briefly dealt with above.

Braam's (1973) study regarding the influence of Dutch industry on local infrastructure policies paid due attention to the effects of private players on government decisions, and therefore to the issue of causality. According to Braam, a causal relationship between A and B exists if: A and B both change; A changes before B (asymmetry); the relationship is not apparent. The latter implies two things: that no other player caused B to change (no 'third player causation') and that B did not cause change in himself (no 'autocausation'). Although in empirical research it is difficult to know whether A indeed caused B's change and not another player or B himself, it is of the utmost importance to do so. Otherwise the changes of A and B may be misinterpreted as A's influence on B. To prevent such faulty reasoning, the possibility of third player causation and autocausation must be thoroughly assessed. Therefore Braam studied a large number of rather similar and simple decision-making procedures. The simplicity made causal inference possible, the great number of cases a quantitative analysis.

Huberts elaborated on Braam's approach and formulated twenty premises and rules for the qualitative assessment of political influence in a single, rather complex decision-making process. Third player causation is ruled out by assessing the influence of *all* relevant participants in decision-making. Autocausation is assumed to be excluded by covering the change in personnel at the side of decision-makers. In short, Huberts assumes that the chance that pressure group A indeed caused the change of decision-maker B increases if:

1 A really intended to change B;
2 A had access to B;
3 the time lag between A's attempt and B's change is short;
4 B's policy change is in A's interest (goal-achievement);
5 B remained the same individual during the decision-making process.

Therefore the indicators *intention, access, policy change, time-lag, goal-achievement* and *personnel* are central to the intensive process analysis. In addition, *resources* are – according to Huberts – important. If several players with similar objectives undertake similar (or collective) attempts to influence a political player and succeed, then one might discriminate among them on the basis of resources. The player who possesses or invested most resources is assumed to be most influential.

The reconstruction of political influence in the 'intensive process analysis' is guided by the above indicators. Furthermore, it is based on a thorough analysis of agenda-setting, decision-making, the decision itself and the political context. The data are mainly collected by document analysis and interviews.

3.2.3 Adaptations

Huberts' method is considered a relevant point of departure, as his object of study has a lot in common with the one dealt with here. Nonetheless, there are major differences which prevent a literal reiteration of the 'intensive process analysis'. Firstly, this study focuses on more complex forms of decision-making. Whereas decisions on highways are of an unequivocal and zero-sum nature, those on international conventions are not. In the former case, authorities have to choose between two or three different options for the stretches of a highway. Hence, the policy positions of the different participants are rather easy to identify and the outcome of the decision-making process is either X, Y or a compromise between these. The decisions which are studied in this thesis, though, are of a rather different nature. For example, the FCCC contains some 8,500 words, over which political players may have struggled one-by-one. Hence, the adoption of this convention might theoretically have led to thousands of 'sub-decisions'. Moreover, the end result is not a zero-sum, either X or Y, as the outcome of a treaty is not fixed at the moment a negotiation process starts. Consequently, the reconstruction of such a decision-making process is a complex task. *Therefore it was decided to select a limited number of relevant topics within the domains of the FCCC and CBD.* In other words, aspects of a policy-making process were studied, not the process as a whole. In consequence, this study does probably not give a complete overview of NGO influence on the FCCC and CBD, a conclusion already come to subsection 1.5.3. This is the first adaptation of the 'intensive process analysis'.

Secondly, the study focuses, contrary to the one by Huberts, on the global level, and international politics are of a different nature to national politics. As was outlined in chapter 1, the international political system lacks a central authority and, consequently, representatives of all states are the formal decision-makers in global arenas. Decision-making is therefore very complex and diffuse at global level. Moreover, international politics can be characterized as a 'two-level game': played domestically and globally. This implies that the number of players involved is immense (in case of the FCCC and CBD: more than 150 countries, hundreds of national NGOs, about a hundred international NGOs and at least 10 intergovernmental bodies). It is more than obvious that it is beyond the control of one researcher to reconstruct the intentions, goals and activities of *all* those players. *Therefore it was decided that the thesis' focus should be limited to the global level.* Consequently, the formation of national positions on the CBD and FCCC as well as the interventions of NGOs at domestic level were excluded from the study's domain. Therefore some additional channels of NGO influence will probably remain untapped, a consequence which has already been discussed in both chapters 1 and 2 (subsection 1.5.3 and section 2.6). This is the second adaptation of Huberts' methodology: a number of possibly relevant processes and players are excluded.

A third adaptation of the 'intensive process analysis' is the adoption of some additional tools. Huberts considers both the position and reputation method as poor instruments. Neither formal positions nor reputations do, according to him, say much about actual influence (see subsection 3.2.1). Nonetheless, Huberts agrees that in particular the latter might be utilized in addition to the process analysis. After all, the views of decision-makers on the political influence of the different players may yield additional and useful information. *Therefore it was decided to utilize the reputation method as an additional tool in this study*, all the more so since the 'intensive process analysis' could not be applied to its full extent given the complexity of global policy-making. Consequently, expert respondents from both the government and non-government communities of both treaties were asked to assess NGO influence on the FCCC and CBD. Subsequently, the examples they came up with – for example NGO influence on specific articles in the treaties or on decisions of the Conference of Parties to the FCCC and CBD – were subjected to a so-called *causal analysis*, an approach which builds on Huberts' intensive process analysis (see below).

As expected, the number of topics perceived by the respondents as being influenced by NGOs did *not* cover all the aspects of the FCCC and CBD decision-making in the formation and implementation phases. Therefore the reputation method also functioned as a means to select topics for additional causal analysis. This procedure, though, was already outlined in chapter 1 (subsection 1.5.3).

3.2.4 EAC METHOD

The following assessment technique was designed on the basis of an interplay between theoretical and methodological literature on the one hand and initial field experience of the researcher on the other (iterative approach, see subsection 1.5.1). First, a number of key respondents of NGOs were selected (for the procedure see subsection 1.5.3). These respondents were asked to assess their own influence on the text and the implementation of the CBD and FCCC in the period 1990-1995. Subsequently, they got the opportunity to elaborate on their examples of NGO impact. These interviews produced two lists of so-called *ego-perception*, one on the FCCC and one on the CBD, of which a number of relevant cases was selected (see subsection 1.5.3). Both the quality and quantity of these NGO claims were, by the way, taken into account in the assessment. This approach was reiterated with governments and intergovernmental bodies (bureaus and secretariats of the FCCC and CBD). Selected key respondents were also asked to assess the political influence of NGOs. These perceptions, in turn, resulted in two lists of so-called *alter-perception*. Again, the quantity as well as quality of those perceptions were given due consideration. These views could either confirm, reject or add to the claims of NGOs (first control). Finally, the relevant cases of NGO influence were subjected to a *causal analysis* (second control of the NGOs' claims). This was done to verify whether NGOs had indeed caused those effects on the policy outcomes which they had claimed. To be able to employ this causal analysis, a great number of respondents as well as government and non-government policy documents were consulted, while several indicators of Huberts' methodology were applied. In short, three methodological tools were utilized, referred to as the *EAC Method (E*go-perception, *A*lter-perception and *C*ausal analysis; see table 3.1).

The causal analysis is constructed as follows. First of all, it was checked whether the NGOs involved had achieved their goal – fully or partially – with regard to a policy topic under investigation. To repeat, such a goal may refer to the enhancement of a desired 'good' or to the prevention of an undesired 'bad' (see subsection

2.3.1). The check itself was done on the basis of a comparison of the NGOs' initial policy goal – deduced from documents and interviews – and the parallel policy outcome. In cases where NGOs had not achieved their goal, it was assumed that they had not been influential at all. In cases where they on the other hand had (partially) achieved their goal, the causal analysis continues. Then, secondly, it was checked that this goal-achievement was caused by the NGOs' own interventions and not by those of governments or other players (international organizations, science, business). In other words, it was verified whether the outcome was more in line with the NGOs' goals than if the latter had not intervened (different counterfactual). To check this, the policy- and decision-making process of the topic concerned was reconstructed as accurately as possible. Particularly analyzed were:

1. possible goal-achievement by governments and other relevant players;
2. access of NGOs to the political arena and to (key) countries;
3. interventions of NGOs and other players made with regard to the topic concerned;
4. number and intensity of those interventions;
5. adoption of NGO views/proposals by governments (policy changes by states).

TABLE 3.1 *The EAC Method*

Ego-perception	Views of NGO representatives with regard to their own political influence (claims)
Alter-perception	Views of government representatives with regard to the political influence of NGOs (first control of NGO claims)
Causal analysis	Researcher's assessment of NGO claims on the basis of policy documents and additional interviews (second control)

It should be recognized that not all indicators of Huberts' intensive process analysis are applied in the above (see subsection 3.2.2). Covered are: intention, access, policy-change and goal-achievement. But excluded are: time-lag, personnel and resources. Although the 'time-lag' between the NGOs' interventions and the adoption of their views by governments is implicitly taken into consideration, this variable was in many cases not useful. After all, many countries remain firm in their policy positions until the very last days of international negotiations. Then, under pressure, the final compromise on a treaty has to be made. Some options are excluded, while others are linked and adopted. Finally, the binding decision is made. NGO proposals, views or pressure can play a role in these final hours of decision-making, although the first NGO intervention on the topic concerned may have taken place a long time ago. Hence, political influence is still a possibility then, despite the fact that the time-lag between the first intervention of NGOs and the final change of government positions may be very long.

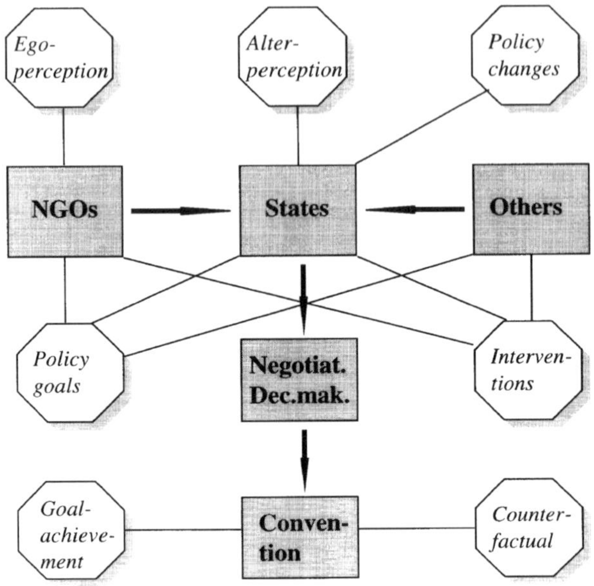

FIGURE 3.1 *Indicators of the EAC Method*

Secondly, the indicator of 'personnel' is excluded as well. To repeat, Huberts assumes that many changes in government personnel increases the chance of 'autocausation' (policy changes of governments caused by themselves). Although there were many changes of personnel in all the country delegations of the FCCC and CBD, the *key* negotiators remained more or less the same during the period studied (1990-1995). This became obvious from policy documents and participants' lists of international negotiation sessions. Therefore the personnel indicator is considered an irrelevant variable in this study. Finally, Huberts' indicator of 'resources' to assess the distribution of influence in successful coalitions is excluded, but in as far NGOs were part of a successful coalition it was (roughly) checked to what extent the outcome was due to the interventions of the NGOs themselves and to what extent to the interventions of others in the coalition concerned (see next subsection).

Figure 3.1 links the EAC Method to the theoretical model which was designed in the previous chapter. The several indicators of political influence as designed above are linked to the players in the political arena, the negotiation and decision-making processes as well as to the policy outcome (convention). Those indicators are: ego-perceptions of NGOs; alter-perceptions of states; policy goals of the players; interventions of the players; goal-achievement by the players; policy changes by states due to NGO interventions; and counterfactual (what would have been the outcome had the NGOs not intervened).

The application of three methodological tools, however, raises the question of how to construct a final judgement on the NGOs' political influence. It is obvious that a conclusion will be positive where ego-perception, alter-perception and causal analysis are all affirmative with regard to a specific policy topic. Then NGOs were definitely influential. The same goes for the opposite: where both alter-perception and causal analysis are negative, the NGOs were not influential, despite their claims (note that the indicator of ego-perception is always positive as the NGOs' perceptions of their own political influence form the basis of this study's assessments). The situation changes in cases where alter-perception and the causal analysis differ. Then one should decide which tool should prevail. In general, the causal analysis is considered dominant vis-à-vis the indicator of alter-perception, as the former is mainly based on 'fixed' policy documents whereas the latter builds on 'temporally distorted' views of respondents. But this relationship of dominance also depends on the quality of the data, as these differed case by case. As respondents have different experiences, access to different information and different interests to defend, the quality of their views on the NGOs' political influence varied. Some were quite detailed and turned out to be reliable, others were not and others again were in-between. At the same time the researcher had to work with a sample of policy documents with regard to each case to base his causal analysis and conclusions on. Given the complexity and global nature of the FCCC and CBD policy processes, it was unthinkable that *all* relevant documents could be available to the researcher. Consequently, alter-perceptions turned out to be more reliable than the outcomes of the causal analysis in a number of cases. Where this was the case, the dominance of the latter was questioned and conclusions were adapted. But generally, the causal analysis was the final judge, unless the quality of the data urged the researcher to follow another procedure.

To conclude this section, one should recognize the fact that any assessment of political influence is after all only an *informed guess* (Everts, 1985; Van Noort *et al.*, 1987; Van Schendelen, 1981). One can never be sure of having included all relevant visible and invisible (f)actors in the analysis, in particular not with respect to global studies. Even scientific judgements on political influence remain guesses. But they are 'informed' at least, as the political processes concerned are analyzed as thoroughly as possible.

3.2.5 Extent of political influence

The EAC Method also raises questions on the *extent* of political influence. Given its definition, the extent of political influence coincides with the extent of goal-achievement due to one's own intervention (see subsection 2.3.1). However, these two dimensions – 'goal-achievement' and 'own intervention' – may independently vary. For example, one achieves all policy goals in a specific case, but this appears to be due to only a small intervention by oneself and much more to interventions by others. The opposite may be valid as well: in that case one achieves less, but had to invest a lot more oneself in order to attain that outcome. In mathematical terms:

(1) $$PI = GA \times OI$$

PI: extent of political influence
GA: extent of goal-achievement
OI: extent to which goal-achievement is due to one's own intervention

This formula, which is only heuristic in nature, also makes obvious that 'goal-achievement' (GA) and 'own intervention' (OI) are not sufficient for political influence (PI) independently, but their combination is. As soon as either GA or OI ranges to zero, PI is nil as well. Therefore both need to produce at least some minimal score before one may positively conclude there is political influence in a specific case. Note that PI will be positive too, if both GA and OI are *negative.* However, this possibility is excluded here, since negative scores of these variables are out of the question in reality.

The application of the above formula urges one to choose specific scores and scales. Given the qualitative nature of this study, these need to be ordinal. For example Huberts (1988) distinguishes 'no', 'some', 'substantial' and 'much' political influence, a simple ordinal scale which is adopted here. Its application can be highlighted by a number of examples. If an player A achieves all his goals regarding a policy outcome, then A's goal-achievement will be 'much'. In case such an outcome is entirely due to A's own intervention, the variable OI will be credited 'much' as well. Then political influence (PI) will, of course, also range to the highest possible score. The situation will change if A only attains some of his goals, which is moreover mainly due to others' interventions. Then GA, OI and their combination PI will remain on the score 'some'. However, GA and OI may yield different scores in specific cases – for example 'some GA and much OI' or 'substantial GA and some OI' – and then the question is how one might combine these. One way of doing this is to translate the scores expressed in words into figures and multiply GA and OI in accordance with the above formula (1) (e.g. no = 0, some = 1, substantial = 2, and much = 3). The extent of political influence then ranges from 0 to 9 (the latter will only appear where GA = 3 and OI = 3). These combined figures may in turn be re-translated into non-mathematical expressions (for example: 0 = no political influence; 1 to 3 = some PI; 4 to 6 = substantial PI; 7 to 9 = much PI). *Please note that figures produced by this heuristic formula do not refer to objective quantifications nor to standard units of the NGOs' political influence.* On the contrary, these are relative data at best and function as a means to position the different cases with respect to each other. Consequently, the figures have no meaning whatsoever outside the restricted domain of this study.

However, the above formula (1) is considered too narrow, because it does not refer to the nature of the political issue which has been influenced by the player concerned. In other words, the extent of political influence in two cases may – according to formula (1) – be similar, while the relevance of the topics impacted may proportionally differ. As Van Schendelen (1981) has noted, such relevance should definitely be taken into account in assessing political influence. A bit of influence by player A on a very relevant topic might be as important as – or even more important than – much influence by A on a less relevant one. Therefore the factor of *political relevance* (PR) is introduced here. It refers to the extent to which a policy topic is a *key* topic in international treaty formation and implementation. The term 'key' refers to either core items, binding outcomes to parties and/or disputed decisions, which implies that real interests of players are at stake. If so, the political relevance may be considered 'high'. On the other hand, if a rather trivial topic, a non-binding outcome or a symbolic decision is impacted by NGOs, then the relevance of such influence will be 'low'. And, of course, some cases may be in-between ('moderate' relevance). Additionally the distinction between *direct* and *indirect* political influence is added to the above formula (1). It appeared that in some cases NGOs were able to directly influence the outcomes of negotiations and decision-making, for example by providing text elements which were adopted by governments and integrated into treaties or intergovernmental decisions. But in many more cases, NGOs were able to influence the outcomes only indirectly, through their pressure, for example by campaigning for specific objectives which reframed the debate at international sessions or by urging governments to compromise on topics. Although in these cases the outcome was still determined and written by governments, NGOs nonetheless made a difference. It was decided that direct influence of NGOs should be given more 'weight' than indirect influence. Therefore the notion of *directness* (D) has been added to the above formula as well. All in all, formula (1) has been transformed into formula (2).

(2) $$PI = C\,[\,(GA \times OI) + PR + D\,]$$

PI: extent of political influence (maximum yield = 14; no PI = 0; some PI = 3 to 6; substantial PI = 7 to 10; much PI = 11 to 14)

C: Counterfactual (similar = 0; different = 1)

GA: extent of goal-achievement (no = 0; some = 1; substantial = 2; much = 3)

OI: extent of one's own intervention (no = 0; some = 1; substantial = 2; much = 3)

PR: political relevance of the policy outcome concerned (no = 0; low = 1; moderate = 2; high = 3)

D: directness of NGO influence (no impact = 0; indirect impact = 1; direct impact = 2)

A few additional choices made with regard to formula (2) should be highlighted. Firstly, the C of counterfactual has been added as a necessary precondition for political influence. This was done because the variable PR (political relevance) is seldom zero, given the fact that any topic will generally have at least some relevance in treaty formation and implementation. But PR in itself does not coincide with political influence. The same may even apply to variable D (directness). It will appear in chapter 5 that there is one example of direct impact of NGOs on a small text element of the CBD, while the counterfactual is considered 'similar'. In other words, NGOs successfully delivered some legal language in that case, but they did not achieve more of their policy goal as a consequence (legal advice). Secondly, this formula might range from the minimum 0 (where C = 0) to the maximum 14 (where C = 1, GA = 3, OI = 3, PR = 3, and D = 2). Note that the minimal positive score is 3 (in case C = 1; GA = 1; OI = 1; PR = 1; and D = 1). Again, these figures are relative data and have no meaning outside the domain of this study. The following index scale for political influence has been designed: 0 = no PI; 3 to 6 = some PI; 7 to 10 = substantial PI; 11 to 14 = much PI. Thirdly, the (combined) variables GA/OI, PR and D are to be added and not multiplied in formula (2), as all three dimensions refer to 'more or less' influence, yet are based on different indicators.

To conclude this subsection, one should reflect on the crediting of nominal scores. After all, it is obvious that the demarcation line between the different units of scale are vague. In what case is for example the NGOs' goal-achievement 'some' and in what case 'substantial'? One way to discriminate is to *compare* the different cases on all the variables of formula (2). As these variables are by their very nature relative data to position each case vis-à-vis others, such an approach seems legitimate. Nonetheless, crediting remains very much dependent on the researcher's own impressions. This will be given due attention in the empirical chapters 4 and 5.

TABLE 3.2 *General format to assess political influence on a case-by-case basis*

Case X

A Introduction	- policy topic - negotiations & decision-making 1990-1995
B Ego-perception	- quantity - quality
C Alter-perception	- quantity - quality
D Causal analysis	- Goal-achievement by NGOs - Interventions of NGOs
E Extent of political influence	
F Final conclusion	

The foregoing elaborations as well as the hierarchical approach of this study (see subsection 1.5.3) give reason to design a general format to assess the NGOs' political influence on a case-by-case basis. Table 3.2 represents such a format. It will be employed with regard to all 18 cases in chapters 4 and 5.

3.2.6 Discussion

The EAC Method was designed during the initial phases of field research in 1994 and improved during the subsequent ones. Its objective is to produce plausible and transparent qualitative assessments of political influence, while addressing the complexity of the international policy processes concerned. The latter is done through interviewing expert respondents, which yields a quick introduction to the object of study, and a selection procedure of relevant topics for further analysis. Transparency is achieved by analyzing selected cases of possible NGO influence in accordance with the elements of table 3.2. Finally, plausibility is achieved by confronting the perceptions of the expert respondents with available data from (non)government policy documents and additional interviews. This last refers to what methodologists call triangulation of methodologies (Verschuren and Doorewaard, 1995; Wester, 1991). Different methods are employed with regard to a similar domain of study, either to verify each method's validity by mutual confrontation or to compensate for each method's shortcomings. The second form of triangulation is employed here, as 'ego', 'alter' and 'causal analysis' themselves are too limited as tools to assess political influence.

Given the foregoing characteristics of the EAC Method, it is believed that it is especially well equipped to study political influence of players in single, complex policy processes by one researcher. Policy processes at international level are good examples with respect to that. Of course, the method has some drawbacks as well. First of all, the EAC Method is biased towards *intended* influence. As NGO respondents are asked to assess their own impact on decision-making, they will probably assess the effects of their 'intended' interventions. But this one-sidedness is not really a problem, given the fact that unintended influence was already excluded from the theoretical domain of this study (see subsection 2.3.1). It may exist but it is simply not covered or assessed. In addition, the EAC Method is very much oriented towards the confirmation of 'success'. Respondents are asked to assess influence and, hence, to mention successes in this respect. Nonetheless, lack of political influence of NGOs ('failures') can be studied as well, given the fact that many NGO claims could not be confirmed, either by alter-perception or by causal analysis. These examples are subsequently considered 'failures' in the EAC Method. Furthermore, some respondents were explicitly asked to mention failures, at least when time was available during the interviews. Besides, some respondents spontaneously came up with NGO failures. Finally, the EAC Method does not assess structural aspects of power. But this is not a problem either, because the elements of structural power as distinguished in the model in chapter 2 – distribution of resources, regimes, rules of the game – are considered *independent* variables in this study. They should be treated as given and *explain* the NGOs' political influence (see next section).

As the EAC Method is rooted in both the reputation and decision-making method, the more general drawbacks of both approaches are valid as well (see section 2.1). With regard to the former method one should be sensitive to possible bias in ego- and alter-perceptions. Respondents may have reasons to respond strategically or reify general beliefs. NGO representatives may be assumed to overestimate their own influence, as such influence is generally valued positively by them. After all, it's the job of lobbyists to impact decisions. On the other hand, government representatives may have reason to overestimate the role of NGOs as well. It may make them look more democratic, or more sympathetic to the researcher, the NGOs and the public. However, they may also underestimate the impact of NGOs, for example to show that they as governments take their formal responsibilities seriously. Whatever may be the case, potential bias in the respondents' views should be taken into account.

The causal analysis, next, has – just as the decision-making method – a bias in favour of pluralist theory. After all, there is hardly any policy process in which players do not have a bit of influence. Therefore the possible conclusion that NGOs are influential to some extent is not really a surprise. More important then is the political relevance of such influence and that aspect should be given due attention (see above). Furthermore, influence does not coincide with the agents' power. To repeat, episodic effects (political influence) should be distinguished from permanent capability (power). If my study confirms that environmental NGOs affected the CBD and FCCC in line with their objectives, one might speak of political influence. However, to say that these NGOs are therefore powerful is premature. This might be the case, though, if these NGOs turn out to be influential on the whole range of international environmental politics over a longer time span. Then one could argue that NGOs exhibit a *permanent ability* to co-determine the outcome of such politics in general. It is obvious that the range of this two-case study is too small to draw such a conclusion.

Finally, the EAC Method does not particularly aim at an assessment of non-decisions, although it is not impossible to cover this item. As different players need to be interviewed, the struggle over issues can be reconstructed. Furthermore, perceptions of respondents on NGO influence may refer to agenda-setting. Hence, non-decision-making can be part of the EAC Method, but – again – is not the primary aim.

— 3.3 The explanation of political influence —

3.3.1 Explanatory factors

This section deals with the methods required to answer the second research question on the explanation of the NGOs' influence (see subsection 1.5.2). Such an explanation should be built on the analysis of several cases, as policy processes with and without NGO influence should be compared (Huberts, 1988). It appeared that it was possible to apply such a comparative case-analysis within the context of this study, as 18 subcases – 8 for the FCCC and 10 for the CBD – yielded different outcomes with regard to the NGOs' political influence (chapters 4 and 5). Note, by the way, that the preposition 'sub' in the word 'subcase', which refers to one of the 18 policy topics, is omitted below. Consequently, the comparative case-analysis covers the comparison of those 18 topics and is not restricted to a comparison of the FCCC and CBD alone.

One of the central questions with regard to any comparative case-analysis deals with the factors on the basis of which the cases involved should be compared. It is obvious that only those factors which are considered theoretically relevant from the perspective of the study should be included, in this case political influence. Such potentially relevant factors can be deduced from academic literature, from both empirical and theoretical studies, but may also stem from the informed assumptions of the researcher (Cox and Jacobson, 1973; Everts, 1985; Huberts, 1988; Huberts and Kleinnijenhuis, 1994). Therefore the list of factors which may explain the NGOs' political influence is built on the theoretical model which was designed in chapter 2. This model covers the political environment deemed relevant to NGOs, in which they have to operate and which enables or constrains them to achieve outcomes. Therefore it also covers those factors which might explain their influence.

Table 3.3 summarizes the possible explanatory factors of NGO influence. In accordance with the theoretical model, three levels are distinguished (see figure 2.2, page 71): 1. the level of the political arena (including the players, negotiations and decision-making); 2. the level of the policy outcome (specific elements of a treaty or a specific decision on the implementation of the conventions); 3. the level of the context (the international system). These levels will be briefly elaborated on below. First of all, policy processes in political arenas are shaped by agents. Therefore the extent of NGO influence may be explained by the nature and input of the players involved, including the NGOs themselves. For example, some NGOs might be more experienced in international environmental diplomacy than others and therefore have a better guarantee of success. Or some might be considered more legitimate or persuasive than others by the authorities concerned and their goals are therefore more easily achieved. The same applies to the nature of their interventions. After all, not all of their strategies will be similarly effective, irrespective of which NGOs are involved. It is for example possible that the strategy of directly lobbying individual delegates will turn out to be more effective than a general campaign to pressurize for certain outcomes. On the other hand, the NGOs' success will be dependent on the players to be influenced as well, in this case representatives of states. Some are probably more accessible and responsive to NGOs than others. Therefore the states which are dominant in each of the cases investigated are taken into account as well. Additionally, other players may also intervene. 'Countervailing powers', such as business organizations, can neutralize NGO efforts. Or other players may make the NGOs' attempts superfluous, such as scientific organizations which make similar claims to those of environmental groups.

TABLE 3.3 *Possible explanatory factors of the NGOs' political influence*

1	Arena level	a	Players involved
		b	Players' interventions
		c	Nature of the policy issue
		d	Nature of the negotiations
2	Level of the outcome	a	Nature of the policy outcome
		b	Likely policy effects
3	Context	a	Nature of relevant regimes
		b	Distribution of resources
		c	Nature of the rules of the game
		d	Relevant trends and events

But other factors at the level of the political arena are also relevant: the policy issue and the nature of the negotiations and decision-making process. Given the definition, political arenas are organized around policy issues (see subsection 2.2.3). Different issues such as climate change or biodiversity loss probably mobilize different players, constitute different arenas, and produce different policy processes. Therefore it may be assumed that different issues raise different opportunities for NGOs to intervene. If they are for example the only supplier of technical advice on the issue concerned in the arena, then they might rather easily become influential. Their impact, though, also depends on the nature of the negotiations and decision-making. These differ in different policy phases. The formation phase aims at the formulation of a binding legal text, the implementation phase at putting its provisions into operation at both global and national levels. It is obvious that these activities differ and one may assume that different policy phases raise different opportunities and constraints for NGOs (as is also regularly assumed in the literature; see subsection 1.5.3). But the nature of negotiations and decision-making may also vary within one policy phase. For example, it does not seem very likely that NGOs can successfully impact negotiations among states at the moment when these are tense and subject to rough power games. On the other hand, if negotiations are integrative and very much oriented on the issues themselves and, consequently, at technical expertise, then NGOs might make a difference.

Secondly, the expected outcome of any policy game impacts the interactions among the players. In that respect, the status of the intended policy instrument does matter. For example, the making of soft law (declarations, recommendations) probably leads to different processes of mutual influencing among the players than the making of hard law (binding instruments). After all, the latter generally has more consequences for nation states than the former. Likewise, the expected and more specific effects of any new rule play a role. If states expect a regime to change through proposed new rules, then they might think twice about adopting them, as things will change fundamentally. If on the other hand they believe that the status quo will not be affected by any new rule, then they might worry less. Consequently, it is possible to assume that soft law and customary rules are more easily influenced by NGOs than hard law and those rules which may invoke regime change. Similarly, it may be assumed that principles and norms in treaties may be more easily impacted by such groups than regulations and procedures, as the latter prescribe immediate action for nation states whereas the former only formulate the premisses of the legal instrument. Nonetheless, such relationships are not obvious in advance. But what this discussion does make obvious is that the intended or expected outcome may co-determine processes of influence, as the players may anticipate on the potential domestic and global effects of the treaties they negotiate.

Thirdly, the context – the final level – definitely matters. Generally, NGOs are constrained by the distribution of resources in the international system. States are ahead with regard to authoritative resources, whereas for example business organizations are ahead with regard to allocative resources. The chance that these players generally achieve more of their goals than NGOs is – as a consequence – substantial, although not necessarily pre-given. Failures of NGOs may therefore be explained, fully or partially, by the asymmetrical distribution of resources at global level. There may however be exceptions, situations in which NGOs dominate in terms of resources, for example expertise, support from public opinion, contacts, etc. Such unexpected and probably rare balances of resources could then eventually account for an NGO success. In addition, regimes are assumed to be important explanatory factors. As was outlined in chapter 2, these are frames of reference for policy players. Consequently, existing rules either constrain or enable players in defending their interests and achieving outcomes. If an NGOs' demand, for example, is close to the nature or content of a substantive rule already adopted in another arena, then this demand can be easily legitimized by referring to that rule. The opposite, of course, applies to radical demands without any precedent in international law or politics, which are hard to defend and legitimize. The third element of the international system refers to the rules of the game. These rules co-determine, at least formally, to what extent NGOs can participate in negotiations and decision-making. And such participation is a prerequisite for much, if not all, political influence. Hence, restrictive rules of the game may (partly) explain the lack of NGO influence just as supportive rules may (partly) account for successes.

Apart from the international system structure, the context of the theoretical model consists of trends and events in different subsystems, such as the environment, the economy and politics. It is obvious that an event such as an ecological disaster or a scientific breakthrough with regard to the solution of an environmental problem enables NGOs to make their point and legitimize their policy position. In addition, a political event such as UNCED may indirectly promote the NGOs' views. But other events, such as the publication of a scientific study which undermines the reliability of environmental data, definitely constrain them. The same applies to trends, for example in the economy. If there is a feeling of economic crisis, environmental arguments are not easily accepted. In short, events and trends may (partly) explain the successes or failures of NGOs too.

It is obvious that many of the above players overlap or are mutually linked. In addition, their number is quite extensive. An explanatory analysis is, as a result, quite difficult to make. Therefore a specific technique will be detailed next.

3.3.2 Comparative case-analysis

To be able to explain the NGOs' political influence, the 18 cases in chapters 4 and 5 will be compared with each other on the basis of the factors identified in table 3.3 in the previous section. This will be done in a data-matrix, the format of which is represented in table 3.4 below.

The rows refer to the 18 cases, the columns to the explanatory factors, and the cells to the 'scores' of the individual cases in those factors. Those scores were collected by the researcher with the help of policy documents, interviews, and – to a limited extent – observations during the whole period of study. The putting into order and analysis of the data in the matrix includes three procedures. First of all, the 18 cases will be positioned based on the extent of political influence. The aim is to have a hierarchy of cases in column 1 of the matrix, ranging from a high to a low level of NGO influence. This will be done on the basis of formula (2) and the nominal scale designed in subsection 3.2.5. Secondly, columns 2, 3, 4, 5, etc. are filled in, so that tentative conclusions on the possible correlation between the individual factor in each column and the extent of the NGOs' political influence may be drawn. The cases are then compared in each column and where a 'pattern' is discerned – for example a positive relationship between the extent of political influence and the factor concerned – the latter is assumed to (partly) explain the

NGOs' impact on treaties (*vertical* comparison of cases in the columns). However, the 'explanatory ability' of each factor may differ vis-à-vis others. Therefore, secondly, the factors are compared (*horizontal* comparison of factors in the rows). This is done to assess the potential overlap, the possible linkages between, as well as the relative strength of the several factors as identified in table 3.3. Finally, those factors which seem (un)able to explain the political influence of NGOs can be deduced. Some will turn out to be neutral factors, others enabling factors and others again constraining factors to NGOs.

TABLE 3.4 *Format of the data-matrix to deduce explanatory factors of political influence*

Factors / *Cases*	*Extent political influence*	*NGOs*	*NGO intervention*	*States*	*Others*	*Issue*	*Topic*	*Etc.*
case 3	much	NGO A, B, C	text proposal	State X, Y, Z	Players R,S,T	bio-diversity	objective of CBD	
case 2	substantial	etc.	etc.	etc.	etc.	etc.	etc.	
case 6	substantial							
case 8	some							
case 12	some							
case 18	some							
etc.	some							
case 1	no							
case 9	no							
case 10	no							
case 11	no							
etc.	no							
Column 1	2	3	4	5	6	7	8	9

To end this section, some methodological reflections should be made. Although the analysis is an attempt to draw conclusions on the origins of the NGOs' political influence in international environmental policies, one should be very careful with generalizations in this field. As the conclusions are based on a limited number of cases, so-called statistical generalizations cannot be made (Hutjes and Van Buuren, 1992; Yin, 1991), but *theoretical* generalizations can. One may for example consider the conclusions in chapter 6 of this book as theoretical hypotheses on the background of the NGOs' political influence. These hypotheses can then become the input of new research. A second reflection concerns the organization of the data-matrix, which is based on political influence. According to Swanborn (1994,

1996), one should be careful with such a comparison of cases on the basis of a *dependent* variable (in this case: political influence), as the *independent* variables (the explanatory factors) are hopelessly confused in most qualitative studies. This implies that it is difficult, if not impossible, to determine which individual factor has most explanatory ability. To face this dilemma, the horizontal comparison of factors was added to the analysis, although the problem of confusion remains hard to solve.

A final methodological reflection refers to the difference between *correlation* and *causation* (Rosenthal and 't Hart, 1994). If a relationship is discovered between a factor and the extent of political influence, it will be implicitly assumed that there is a causal link between the two. This assumption is based on the theoretical model, which is after all a causal model. However, the data-matrix produces correlations, which may be causal in nature, but can be based on luck, coincidence, an unknown third factor – or whatever – as well. This difference between correlation and causation, though, cannot be resolved on the basis of the analysis employed here. Nonetheless, this drawback can be kept in mind while formulating conclusions.

3.3.3 Respondents' views

The factors of success and failure are deduced from the researcher's assessment of political influence and the comparative case-analysis. However, respondents themselves do probably also have a view on the 'why' of the NGOs' political influence. Therefore non-government as well as government respondents were asked during the interviews to sum up those factors on which they believe the NGOs' political influence – or lack of it – is based. These perceptions yielded additional information for the comparative case-analysis. However, not all respondents dealt with this issue, because on the one hand a number of questions in the interview guide had to be deleted during some conversations, and on the other, the guide was renewed during the period of field research (see subsection 1.5.3).

The arguments of the respondents as formulated in their own language were, as far as possible, translated and categorized into the theoretical language of this study. This procedure was not that easy, because arguments of respondents are not always as obvious as they seem. Moreover, down-to-earth language and theoretical concepts do not always match well. Consequently, there is always a tension between the respondents' interpretations of events and the researcher's re-interpretations of this. Therefore a simple test was applied to consider this problem. The theoretical translation and categorization of the respondents' arguments was done twice at different times, and it was assessed whether differences between the two procedures occurred. They did, but with regard to some details only. The general conclusions therefore remained identical.

— 3.4 The formulation of recommendations —

3.4.1 The limits of prescriptive knowledge

Whereas the previous two sections covered research questions 1 and 2 on the assessment and explanation of political influence respectively, the technique to answer the third question, on recommendations, is dealt with here. Therefore in this section description and analysis of past events is replaced by prescriptions focusing on future events. This implies that a situation of relative certainty for the researcher (with regard to past events) is replaced by one of uncertainty (regarding future events). But even the knowledge of past events remains uncertain to some extent, given the nature of this study. Its qualitative focus on two policy issues (climate change and biodiversity loss) and the informed guesses on political influence, urge one to be careful with generalizations (see above). In consequence, the recommendations will be far from predictions – as in the formula '*if* A, *then* B' – and they will be rather limited in scope (to more or less similar negotiation processes in international environmental policies).

In addition, the level of prescription is limited. As the study primarily aims at an assessment of the NGOs' political influence, a middle level of analysis is implied, namely in terms of organizations, negotiations and decision-making. The individual level was therefore not thoroughly analyzed (although respondents were the units of observation). Therefore quite general recommendations at the middle level will be produced, not specific ones at the micro-level, such as suggestions on how an individual lobbyist can perform better. Recommendations instead will refer to, amongst others, NGO interventions and arena processes in general.

3.4.2 Judgement criteria

The technique to formulate recommendations is based on two so-called *judgement criteria*. These are: 1. anticipation and 2. extrapolation. First of all, it will be judged by the researcher to what extent NGOs might be able to anticipate on the constraining and enabling factors in the near future. If they can (partly) neutralize the former and enhance the latter, they should be able to strengthen their political influence in coming policy processes. To repeat, the enabling and constraining factors are deduced from the analysis elucidated in section 3.3 above. As these may refer to different aspects of policy-making – the players, the arena, the outcome, as well as the context – it is obvious that the extent to which those factors can be manipulated and therefore anticipated on by the NGOs definitely differs (see table 3.3). Some factors are more or less given and hard, if not impossible, to change – for example the given regime, distribution of resources and rules of the game in an arena – whereas others can be more or less managed by individual players, for example those involved, their interventions and the nature of the negotiations. Consequently, not all explanatory factors can be translated into useful recommendations, as some are simply outside the control of NGOs.

Secondly, it should be judged to what extent the set of explanatory factors which is potentially manageable by NGOs is relevant to other arenas and future policy processes. One cannot simply believe that for instance a successful NGO strategy in the biodiversity arena can be transferred to the international trade arena of GATT/WTO and guarantee success there as well. On the other hand, there is a chance – no certainty – that an example of success or failure may be reproduced under comparable circumstances. But the question is: what are 'comparable circumstances'? In the search for an answer, the second judgement criterium on extrapolation should be applied. It will be judged by the researcher to what extent events and circumstances in the climate and biodiversity arenas can be extrapolated to the near future as well as to other arenas. It is obvious that this may differ factor by factor, as some are probably more arena-specific than others. Ultimately, only those factors will remain which can be anticipated on by NGOs and which seem appropriate to extrapolate to other arenas and policy processes.

It should be recognized that this technique is not based on objective criteria, but on the researcher's (informed) judgements. Therefore the recommendations should be subjected to debate. Consequently, NGOs should not take them over literally nor implement them without reflection. The recommendations are meant to be the input for a strategic debate, not its output in advance.

3.4.3 Respondents' views

The technique mentioned above aims at producing recommendations for NGOs. But the views of the expert respondents may be taken into account as well. During the interview they were not only asked to explain the NGOs' political influence (see section 3.3), but also to reflect on the question of how NGOs might increase their impact on intergovernmental environmental negotiations. True, these questions are highly interdependent. If a respondent for example believes that NGOs are influential because of factor X, he or she will definitely believe that a lack of influence is caused by the lack of X, and that an improvement of the performance of NGOs can be achieved by increasing X. It follows that the answers to these different question will generally show a similar or complementary pattern. Again, it should be recognized that not all respondents were able to deal with all the questions. The number of respondents in each category of answer varies and the reasons for this variation were elucidated previously (see subsection 1.5.3). Still, the perceptions of the respondents may yield some additional recommendations for NGOs, in particular views of government representatives. After all, these players are the ones which need to be influenced by NGOs. Their perceptions of how the latter might perform better are therefore important sources of information.

— 3.5 Summary —

This chapter has yielded methods to assess and explain the NGOs' political influence, as well as a technique to formulate recommendation for NGOs to improve their performance in future policy processes. As far as assessment is concerned, the *EAC Method* was designed. This technique is based on respondents' perceptions on the one hand and an analysis of the policy process by the researcher on the other. The objective is to assess the effects of the NGOs on the outcome of policy-making – the conventions – and specific indicators were designed to measure these (ego-perception, alter-perception and causal analysis). The method to explain NGO influence, secondly, builds on the theoretical model in chapter 2 and is based on a comparative case-analysis. Examples of NGOs' successes and failures in terms of more and less political influence are compared on the basis of a number of factors which are considered theoretically relevant from a political influence perspective. This is done through the use of a data-matrix, in which correlations between these factors and the extent of NGO influence are looked for. Thirdly, the procedure to formulate recommendations for NGOs builds on the explanation of political influence. It is judged to what extent NGOs might anticipate on the so-called explanatory factors in the near future and, possibly, in other policy arenas. However, the nature of this study does not allow the formulation of bold predictions. The recommendations are therefore meant to fuel the strategic debate among NGOs. Finally, the views of all respondents on the backgrounds of the NGOs' political influence as well as on the possible improvement of the NGOs' performance are also taken into account.

CHAPTER 4

NGOs and the UN Framework Convention on Climate Change

— 4.1 Introduction —

The objective of this chapter is to assess the political influence of NGOs on the *UN Framework Convention on Climate Change* (FCCC). It is therefore an elaboration of research question 1 (see subsection 1.5.2). The assessment will follow a phase-by-phase as well as case-by-case approach. In doing so, the different elements of the EAC Method function as a format to describe and analyse all the cases in the same way. This similarity is needed to address the hierarchical nature of this case study, as was explained in chapter 1 (subsection 1.5.2). Such an equal format also provides transparency for readers. But before the cases are presented in sections 4.3 and 4.4, on the formation and implementation phases respectively, the nature of the issue of climate change, the content of the FCCC, as well as the entire policy cycle, will be briefly highlighted in section 4.2. The chapter will be concluded with section 4.5, which presents an overview of the eight cases, and section 4.6, in which the conclusions will be formulated.

The (inter)government and non-government policy documents which will be referred to in the text below are listed in Appendix 1 (page 341). Only the most relevant ones are included, so this appendix is not a complete listing of all the hundreds of policy documents which were consulted in the archives in the period 1993-1997. The documents are referred to in the text as follows: (Doc 1.5) or (Doc 1.32, p. 67). 'Doc' stands for document, '1' for Appendix 1, '32' for document number 32, and 'p. 67' for page 67.

— 4.2 The UN Framework Convention on Climate Change (FCCC) —

4.2.1 The issue of climate change

The so-called greenhouse effect was already understood more than a century ago (Mintzer and Leonard, 1994; Schneider, 1989). Some gases in the higher atmosphere – including water vapour (H_2O), carbon dioxide (CO_2), nitrous oxide (N_2O), methane (CH_4), ozone (O_3) and man-made chlorofluorocarbons (CFCs) – have specific characteristics regarding solar energy. They are transparent to the incoming short-wave solar radiation, but absorb infra-red radiation from the earth's surface. Subsequently, these gases release energy back into the atmosphere and warm it up. Without this feedback, the earth's climate would have been much cooler – by approximately 33 °C – and biological evolution would have been totally different, if it existed at all. This natural process is generally referred to as the greenhouse effect and the substances involved are so-called greenhouse gases (GHGs).

However, due to additional GHGs being emitted through human intervention – particularly CO_2 emitted by industry and the energy and transport sectors and CH_4 released by agriculture – the greenhouse effect is likely to be strengthened. This phenomenon is referred to as global warming. The Intergovernmental Panel on Climate Change (IPCC), which is the most authoritative body on climate sciences at present, predicts a doubling of CO_2 levels in the atmosphere before 2050 if current emission trends continue. This will result in an increase in global mean temperature in the range of 1.5 to 4.5 °C, with a best estimate of 2.5 °C (Doc 1.24). The possible impacts may be disastrous for many regions of the world, as both the overall and the regional climates will change. The IPCC refers to possible adverse effects such as more droughts in tropical regions and more rain in moderate ones, less output in agriculture and forestry, the disappearance of several species and habitats, the spread of malaria, and melting of snow, ice and permafrost and, consequently, a rise in sea levels (Doc 1.25). To prevent such adverse effects, the IPCC concludes that stabilization of today's levels of long-lived GHG concentrations is needed and this: "would require immediate reductions in emissions from human activities of over 60%."

This figure is a dramatic one and probably not feasible from an economic point of view, as the energy and transport sectors on which our economic prosperity is built need to be fundamentally restructured to achieve such targets on CO_2 (Hayes and Smith, 1993). For example: less use of coal and oil and more utilization of solar, water and wind power to fulfil our energy needs. Some also stress the option of revitalizing the nuclear sector, as it releases no GHGs (at least if the extraction and transport of uranium and the processing of nuclear waste are not taken into account). But this alternative is – according to most environmentalists – neither feasible nor desirable, given the risks with regard to radiation, the operation of plants and waste. Whatever position one takes, though, emission reductions of 60% or more will inevitably imply huge changes in production and consumption patterns. Therefore many industrialized countries are reluctant to accept such targets at this stage.

The same reluctance exists in developing countries, although for different reasons. These countries have emitted far less carbon dioxide than the rich ones, less than 25% of the total figure up to now. But future scenarios with regard to economic and population trends predict that they will soon catch up with the industrialized world. Therefore these countries fear that targets on CO_2 might hamper their development process in the near future. Besides, the emission of other greenhouse gases, for example methane from agriculture, is already substantial in the Third World. That is the reason why according to some reports countries such as China and India are already huge contributors to global warming. But many environmentalists and developing countries hold the view that one should not confuse the 'luxury emissions' of the North with the 'survival emissions' of the South (Agarwal and Narain, 1991). The North can make do with fewer polluting activities, the South cannot, as human survival and food security are at stake. Therefore the industrialized countries should take the lead in reducing GHG emissions and bear the financial burden, while developing countries should still have the right to increase their emissions.

The need to reduce GHG emissions is not only questioned on the basis of economic and development arguments, but on a scientific basis as well. The IPCC acknowledges that there are many uncertainties with regard to the assessment and prediction of climate change. For example the role of forests, clouds, oceans, polar ice, solar radiation, the Earth's orbit, the sunspot cycle, volcanic activity, atmospheric particles, and feedback mechanisms in the climate system are poorly understood. Therefore the so-called 'sceptics' in the scientific community challenge the theory of the enhanced greenhouse effect. Some do not doubt that the climate is getting warmer, but explain this phenomenon on the basis of natural variation (Karlen *et al.*, 1993). Recently, Friis-Christensen's theory on the relationship between the sunspot cycle and the earth's climate system has received attention (Rozendaal, 1997). This theory maintains that there are links between the magnetic activity of the sun which is cyclical in nature, the strength of the magnetic shield of the Earth, the influx of cosmic particles into the atmosphere, the build-up of clouds and Earth's mean temperature. The more active the sun is, the stronger the Earth's shield, the smaller the influx of particles, the fewer the clouds, and – consequently – the higher the temperature on the Earth's surface. This activity is in progress now, but will cease soon. Other sceptics, though, still maintain the claim, which was popular in the 1970s, that we may be in the first stages of a new ice age (Bray, 1991; Hecht, 1993). These theories notwithstanding, most climate scientists and environmentalists continue to support the conclusion of the IPCC that there are hardly any doubts left with regard to human-induced global warming (Doc 1.28, 1.63, 1.72).

4.2.2 Contents of the FCCC

The UN Framework Convention on Climate Change is the first global and legally-binding instrument to address the risks of climate change (Doc 1.38). As many details of the FCCC's articles and paragraphs will be dealt with in sections 4.3 and 4.4, it suffices here to summarize the contents. Article 2 contains the ultimate objective of the convention, which aims at a stabilization of greenhouse gas concentrations in the atmosphere at a level that prevents dangerous anthropogenic interference with the climate system. Although this objective does not set specific targets and timetables to reduce the emission of greenhouse gases by the parties to the FCCC, its 'spirit' seems to refer to the need for strong commitments in this field. After all, the IPCC had calculated – as was outlined in the previous section – that the stabilization of GHG concentrations in the atmosphere at current and, hence, relatively safe levels would imply substantial reductions in GHG emissions. However, the strong objective of the FCCC is not reflected in strong commitments in other articles, as will be shown below.

Article 3 contains five principles which should guide the actions of the parties. The first principle concerns equity and 'common but differentiated responsibilities'. It (implicitly) recognizes the specific responsibility of the developed industrialized countries in the creation of global warming and stresses that these parties should take the lead in combating climate change. The second principle asks for the full consideration of the specific needs and circumstances of developing countries and of areas vulnerable to climate change, such as small-island states. The third principle guides the parties in taking both precautionary and cost-effective measures. Where there are threats of irreversible damage, lack of scientific certainty should not be used as a reason to postpone measures, but such measures should ensure benefits at the lowest possible cost. The fourth and fifth principles refer to economic aspects. The former stresses the right to and need for sustainable development, the latter the need for a supportive and open international economic system based on free trade.

Article 4, paragraph 1 deals with the general commitments of all parties – both developed and developing countries. All parties shall make national inventories of sources and sinks of greenhouse gases (GHGs); design national programmes to mitigate climate change; develop technologies to reduce the emission of GHGs; conserve sinks and reservoirs for GHGs; prepare for adaptation to the impacts of climate change; minimize the impacts of climate change on the economy; promote scientific research in this field; exchange information; and educate and make the public aware. Paragraph 2 deals with the specific commitments of developed countries. This section contains the stabilization target, which states that industrialized countries should stabilize their CO_2 emissions at 1990 levels by the year 2000. However, this target is not formulated as bluntly in Article 4 as it is stated here. The language is ambiguous and therefore according to some lawyers not legally binding (Bodansky, 1994). But it is still included in the FCCC as a political message. Other paragraphs of Article 4 deal with financial assistance to the developing world by the rich countries to help the former implement the FCCC; with the transfer of environmentally sound technologies from North to South; and with the specific needs and circumstances of groups of countries.

Articles 5 and 6 cover intentions in the areas of research and of education on climate change respectively. The rest of the convention outlines the institutional structure and co-operative processes that must be developed to implement the FCCC. The institutional structure includes the Conference of the Parties (COP; Article 7), the Secretariat (Article 8), the Subsidiary Body for Scientific and Technological Advice (SBSTA; Article 9), the Subsidiary Body for Implementation (SBI; Article 10) and the Financial Mechanism (Article 11). Article 12, finally, deals with the review of national efforts to implement the convention. Parties should make available to the COP information on their national measures for implementation.

4.2.3 The policy cycle of the FCCC (1980-1997)

As was already outlined in chapter 2, a policy cycle consists of several phases, each of which produce different political struggles. The relevant phases with regard to the FCCC cycle to date are covered in this subsection: issue-raising, agenda-setting, policy formation and policy implementation (Van de Graaf and Hoppe, 1992). Again, it should be acknowledged that the phase-like reconstruction of a policy process has some shortcomings (Koppenjan, 1993; Van Tatenhove, 1993). The theory assumes a nice sequential order of distinct phases, but this cannot be observed in reality. For example, the struggle on the (re)definition of an issue, as well as on the (re)setting of the agenda, continues during the entire process. Nonetheless, the phases are still used here to facilitate a short description of the FCCC political process so far.

A Issue-raising and agenda-setting (1980-1990)

Although the phenomenon of the greenhouse effect had been known for a long time, there was little sense of urgency in the scientific and political communities originally. Human-induced global warming remained a theoretical possibility at best, irrespective of the fact that the World Meteorological Organization (WMO) had already organized a number of scientific conferences on the subject, such as the first World Climate Conference in 1979 (Paterson, 1996). But climate models were still rather primitive at that time and hard to rely on. This, however, changed dramatically in the 1980s and scientists were warned through their ever more sophisticated projections on climate change (Mintzer and Leonard, 1994). As a consequence, global warming entered into politics. But more factors than just improved computer technology played a role in this.

First of all, some scientists, diplomats and NGOs acted as entrepreneurs. They promoted the issue through conferences and reports and asked for the initiation of negotiations on a global, legally-binding instrument on climate change. And they were effective. For example Bodansky (1994) refers to the NASA scientist James Hansen, who gave testimony to the USA Senate in 1988 and whose message was frontpage news. Even presidential candidate Bush promised during his campaign to address the greenhouse effect. Boehmer-Christiansen (1994) notes the influence of the diplomat Crispin Tickell on Margaret Thatcher, who – for a Conservative – made a very 'green' speech on the atmosphere to the British Royal Society in 1988. This event was a breakthrough with regard to high-level support for the issue. Furthermore, the famous Brundtland Report *Our Common Future* (WCED, 1987) had taken up the issue and linked it to questions of scarcity of natural resources and energy efficiency. From there, it came to be included in the agenda of the Earth Summer (UNCED) in Rio in 1992. Besides, the business interest in the issue increased, at least from the nuclear industry, insurance companies and those interested in renewable energy and environmentally sound technology (while opposition became stronger simultaneously, mainly from oil and coal interests). Shock events – the hole in the ozone layer, droughts in the USA, floods in Europe and Asia, cyclones in the Caribbean and Pacific, higher temperatures in the 1980s – made the public and polity aware of the fact that something was happening. To sum up (Boehmer-Christiansen, 1994, p. 187):

> "A major convergence of environmental threat perceptions, science research agendas, and commercial interest had taken place, and provided the political energy for the climate negotiations."

The 1980s also witnessed early policy responses. A number of steps are important here (Bodansky, 1993; Mintzer and Leonard, 1994). Firstly, in 1988 governments requested the World Meteorological Organisation (WMO) and United Nations Environment Programme (UNEP) to establish the Intergovernmental Panel on Climate Change (IPCC). There had been scientific groups on climate change before, but these were non-governmental in nature and therefore viewed with suspicion by governments. Consequently, IPCC was given the mandate to re-assess scientific data on global warming from a policy perspective. Secondly, Canada organized a follow-up conference on the Brundtland Report in Toronto (1988). Its aim was to bridge the gap between science and policy-making on the global atmosphere. The conference – attended by some heads of state and dozens of government officials – recommended a 20% reduction in global CO_2 emissions at 1989 levels by the year 2005. This came to be known as the Toronto target (which was, by the way, set quite arbitrarily). Moreover, the conference recommended the start of negotiations on a global framework convention. Thirdly, Malta raised the climate issue in the United Nations General Assembly (UNGA) in 1988. A resolution was adopted in which climate change was perceived as a priority issue and as a 'common concern of mankind'.

Fourthly, the Netherlands organized two meetings in 1989: one in The Hague (jointly with France and Norway) and one in Noordwijk. These meetings endorsed the importance of the climate issue at a higher political level, as more heads of state and (prime) ministers were involved. The second meeting was the more important, as there the climate issue was interpreted in terms of economic development. It was now explicitly realized that emission reductions would imply drastic cuts in energy consumption, which in turn would limit the prospects for economic growth. Consequently, earlier ambitions were levelled down, mainly due to opposition of the USA, the Soviet Union and Japan against strong commitments. For example the Toronto target was omitted in the Noordwijk Declaration (Doc I.31). It only recognizes that a stabilization of GHG emissions at 1990 levels should be achieved 'as soon as possible'. Fifthly, global warming was on the agenda at the G7 meeting in Paris, discussed at the USA/Soviet summit in Malta and addressed by the Non-Aligned Movement in Belgrade, all in 1989, while in mid-1989 an intergovernmental decision to negotiate a climate treaty was also made. The UNEP was requested to prepare negotiations for such a treaty.

Finally, the Second World Climate Conference (SWCC) was held in Geneva in November 1990 (Doc I.35). This meeting was the largest ever held on global warming. Southern issues were now included in the full debate, with countries such as Brazil, India and China raising voices against an environmental approach to climate change without considering the needs and development priorities of the South. They also blamed the industrialized world for creating the problem of global warming, and asked for resources and technology to be transferred from the rich to the poor countries so that they could address the effects of climate change at home. Because of this Southern input, the Second World Climate Conference can be viewed as a dress rehearsal for the formal negotiations of the FCCC. In the meantime, the European Community (EC) had adopted the so-called stabilization target as Community policy ("to stabilize CO_2 emissions at 1990 levels by the year 2000"). This EC decision was welcomed by the SWCC, but not recommended to others. States were now asked "to establish targets and/or national programmes or strategies" (Doc I.29). Whereas the Toronto conference one year earlier had been firm on targets and timetables regarding carbon dioxide and other GHGs, the Second World Climate Conference made only weak references to such policy options. Obviously, high politics had entered the arena and caution on commitments emerged.

B Treaty formation (1990-1992)

Although the WMO and UNEP had been asked to prepare for the negotiation of a climate treaty in May 1989, these negotiations were finally held under the auspices of the UN General Assembly. This change was due to the interference of the developing countries, which tended – contrary to the industrialized countries – to perceive the climate issue as a development and political, not as a purely scientific and technical issue. Therefore the South believed that the UN was a more appropriate body to negotiate a climate treaty than specialized agencies such as the WMO, UNEP and IPCC. Consequently, on 21 December 1990 the UN General Assembly adopted resolution 45/212. This decision established the so-called Intergovernmental Negotiation Committee for a Framework Convention on Climate Change (INC) and all United Nations (UN) members were invited to participate. The INC held five sessions of two weeks each, in Chantilly (USA), Geneva, Nairobi, Geneva again and New York. An extra session was added to the last INC, because delegates could not cope with the workload. As the mission was to finalize the convention before the UNCED in June 1992, so that countries could sign the climate treaty in Rio de Janeiro, only eighteen month's negotiation time was available, which is not much for a complex treaty to be finalized. Nonetheless, governments succeeded in doing so after having convened an extra meeting in New York just before UNCED.

The negotiations proceeded slowly and with great difficulty (Bodansky, 1993; Mintzer and Leonard, 1994; Doc 1.12, 1.14, 1.17, 1.18). In 1991 most of the time was spent on procedural issues and substantive negotiations only started the final months before UNCED. Briefly, INC1 in Chantilly, February 1991, was totally dedicated to the organization of the negotiations. Two working groups were established, one on substance and one on legal and institutional matters. INC2 in Geneva, June 1991, faced a lengthy deadlock over which countries should chair those working groups. Nonetheless, debates on substance slowly started. INC3 in Nairobi, September 1991, continued to be difficult in terms of substantive progress. No single negotiation text for a treaty was available, only compilations of country positions, which became longer and longer as all delegations wanted to include their own preferences. Furthermore, most text elements became bracketed, as any new proposal introduced by one country faced opposition by another. The formulation of compromises was therefore extremely delayed. INC4 in Geneva, December 1991, hardly saw any progress as well. Most remarkable was the breakdown of the G77 and China, the group of developing countries. Three positions could now be distinguished there: 1. the Oil-Producing and Exporting Countries (OPEC) opposed a strong convention because of their oil interests; 2. the Alliance of Small Island States (AOSIS), on the contrary, opted for strong commitments, as they are potential victims of cyclones and rises in sea level due to climate change; 3. India and China were somewhere in the middle. Consequently, it was no longer possible to formulate a common G77 position. Finally, INC5a and 5b faced intense negotiations (New York, February and May 1992 respectively). Contact groups of key countries on specific and politically sensitive issues were established and these groups negotiated behind closed doors to formulate compromises. Now the group of the Organization for Economic Cooperation and Development (OECD), mainly Western industrialized countries, splintered. The USA was totally opposed to binding targets and timetables, while the EC hoped to include the stabilization target in the FCCC. The position of other countries – such as Japan, Canada, New Zealand, Australia (also referred to as JUSCANZ) – was somewhere in the middle. Consequently, an ambiguous compromise on the stabilization target was reached so that the FCCC could be finally adopted by the participating delegations at a special diplomatic conference, the last session of INC5b, but it was a narrow escape.

Number two on the list of controversial issues during the climate negotiations was finance (targets and timetables being number one). It was perfectly clear at the first session of the Intergovernmental Negotiation Committee for an FCCC (INC1) that the developing countries would link their support for the FCCC to the extent to which the industrialized countries were willing to provide them with 'adequate, new and additional funding' in order to implement the convention. Unfortunately, several rich countries – and in particular the USA – were totally opposed to new commitments in the financial area. The Americans would not countenance any increase in development aid, so assistance for climate projects in the Third World had to be budget-neutral. Nonetheless, at the end of the day the USA accepted the clause 'new and additional financial resources' for inclusion in the FCCC. The practical consequences of this wording are still unclear, especially as development aid in general tends to *de*crease. But climate projects have been funded in the South since the FCCC entered into force at the end of 1994.

Another controversy relating to funding was the financial mechanism to be established under the FCCC. The G77 favoured a new climate fund governed by the Conference of the Parties (hence, governed by both developed and developing countries). The OECD was in favour of making the UNCED-related Global Environmental Facility (GEF) the financial mechanism of the climate treaty. This fund, however, is governed by the World Bank, UN Development Programme (UNDP) and UN Environment Programme (UNEP) and, consequently, portrayed as a 'rich men's club' by the developing countries. The divergent positions on the financial mechanism almost resulted in deadlock at INC5, where a compromise was only reached with difficulty after long days and nights of tense negotiations. This compromise included the acceptance of the GEF by the South, but only on an ad interim basis (which is the concession by the North). Moreover, the industrialized countries promised to make the GEF more democratic and transparent. But the GEF has remained controversial until the present day, irrespective of some institutional reforms which were carried through in 1994.

C *Treaty implementation (1992-1997)*

The FCCC was open for signing at the UNCED in Rio. More than 150 countries did so. However, the convention would not enter into force until 50 countries had ratified it, something decided during the negotiations. The fiftieth country did so in March 1994. In the meantime, the Intergovernmental Negotiation Committee (INC) had continued to meet. To ensure a 'prompt start' of the convention machinery, delegates had decided to retain the INC after UNCED. It met six times before the first Conference of the Parties – the governing body of the convention in which all countries which have ratified the FCCC participate – convened in Berlin in April 1995, one year after the entry into force of the convention. Before this meeting, the INC had dealt with a lot of topics, such as: methodologies for national inventories on sources and sinks of greenhouse gases (GHGs); the Global Environmental Facility (GEF); climate projects in the South financed through the GEF; the adequacy of commitments in the treaty; the mechanism of joint implementation which enables parties to implement their commitments jointly; the preparation of the first Conference of the Parties to the FCCC (COP1); and a first review of national measures. Part of this last was the monitoring of the industrialized countries' commitment to implement the stabilization target, as was laid down in the FCCC text, at least as a political message. It soon became obvious that most OECD countries would not achieve this target in time. On the contrary, most future projections on CO_2 emissions predicted a substantial increase by the year 2000, compared to 1990 levels. This led to public condemnation of the industrialized countries' conduct by both developing countries and NGOs.

The first Conference of the Parties to the FCCC (COP1) in Berlin in April 1995 turned out to be a huge media event. Many journalists and NGOs – more than 160 – covered it. It was wrongly portrayed as 'the' follow-up of UNCED (Arts and Rüdig, 1995; Doc 1.5, 1.6). It was in fact the first meeting of the governing body – the COP – of the climate treaty. Now the machinery of the FCCC, for example the subsidiary bodies and the secretariat, could be formally installed and put into operation. In addition, substantive topics were dealt with as well. Most important was the negotiation and adoption of the so-called Berlin Mandate. This mandate legitimized the initiation of new negotiation rounds to adopt an additional legal instrument – probably a protocol – on further commitments to reduce greenhouse gas (GHG) emissions after the year 2000, a protocol which had to be adopted at the third Conference of the Parties in Kyoto, Japan (1-10 december 1997). The preparatory negotiations took place in the AGBM – Ad Hoc Group on the Berlin Mandate – which met regularly in the period 1995-1997 and included all the parties to the FCCC. Several policies, measures, instruments, targets and timetables on GHG reductions were considered. However, the positions of the various key countries diverged widely in advance of Kyoto and were well below the level of the original Toronto target. For example, the European Union (EU) proposed a 15%

cut of CO_2 emissions by 2010 compared to 1990 levels, Japan a 5% cut by the same date, and the USA did not want to go any further than a stabilization of emissions. Nonetheless, a compromise was recently reached in Kyoto on reductions in GHG emission compared to 1990 levels of respectively 8% (EU), 7% (USA) and 6% (Japan), to be reached in the period 2008-2012.

In the meantime, the second Conference of the Parties (COP2) had met in Geneva in July 1996 (Doc 1.80). It produced a Ministerial Declaration which confirmed the most recent assessment of climate change by the IPCC and witnessed the need for strengthening commitments in the AGBM process. Again, North and South clashed on the nature of policies and measures. Whereas the industrialized countries maintained their claim that all parties are responsible for reducing GHG emissions, the developing countries stressed the specific responsibility of the rich to do so. The change in position of the USA in Geneva was striking. It proclaimed itself to be no longer against a legally-binding agreement to be adopted in the AGBM.

4.2.4 NGOs in the climate arena

NGOs were able to participate in the FCCC negotiations as observers, both in the sessions of the Intergovernmental Negotiation Committee (INCs) and the meetings of the Conference of the Parties (COPs). To repeat, international NGOs can have consultative status with the UN Social and Economic Council (ECOSOC) (Feld and Jordan, 1983; Willets, 1982; Woods, 1993). This status is legitimized by article 71 of the UN Charter. NGOs which are able to contribute to the work of the UN are allowed to participate in ECOSOC meetings, make oral statements, circulate written statements, or may observe conferences which are organized under the banner of ECOSOC. However, not all NGOs have similar status, as there are different categories of NGO distinguished by the UN, and not all arenas raise equal opportunities to participate. But whatever the case in detail, nearly all UN organisations, bodies and conferences make arrangements for NGO participation nowadays. Furthermore, *national* NGOs which are competent and wish to contribute are generally allowed to do so. However, not all UN members are in favour of NGO interference, as was already outlined in chapter 1. Therefore the adoption of rules of procedure for any new UN initiative poses some difficulty with regard to NGO observers. A good example is the UNCED (Willets, 1996). NGO participation was fiercely opposed at the first meeting of the Preparatory Committee, but a compromise was agreed upon. Although NGOs were granted observer status, it was explicitly mentioned that they did not have a formal negotiating role.

This UNCED decision 1/1 was also adopted by the first session of the Intergovernmental Negotiation Committee (INC1) of the FCCC, paving the way for continuing NGO participation. Moreover, the regime at INC1 was quite liberal, which set a precedent for subsequent meetings. The Americans granted accreditation to all NGOs which wished to visit the meeting in Chantilly. Consequently, more than a hundred were present there. The subsequent INCs were visited by approximately 50-75 NGOs (Doc 1.12, 1.14, 1.17, 1.18). About one third of these NGOs were scientific in nature, one third were of business origin, and one third were environmental organizations. A large majority of these NGOs originated from the North, although the participation of Southern organizations increased during the INCs. This was due to extra funds from both governments and NGOs to increase the involvement of Southern counterparts. Important environmental organizations in these sessions were: World Wide Fund for Nature (WWF), Greenpeace International, Environmental Defense Fund (EDF), Bangladesh Centre for Advanced Studies (BCAS), Indian Centre for Science and Environment (CSE), Natural Resources Defense Council (NRDC), Environmental Protection Society, Friends of the Earth International (FOEI), Wahli, Sierra Club, World Conservation Union (IUCN), World Resources Institute (WRI) and others. Most of these NGOs collaborated under the umbrella of the Climate Action Network (CAN) and its regional offices (Europe, Central and Eastern Europe, North-America, Latin America, Africa, South-Asia, South-East Asia).

TABLE 4.1 *Activities of climate NGOs*

1 Protest	*2 Pressure*	
	a Advocacy	*b Lobby*
Organization of: - sit-in - march - ludicrous action - petition - disturbance	Publication of: - Eco journal - own report Dissemination of: - position paper - written statement - scientific data Presentation of: - oral statement Organization of: - parallel conference - workshop - lecture - round table - press conference - stand	Present to individual policy-maker: - oral statement - written statement - concrete proposal - draft text - legal or technical advice - compromise - deal (exchange of resources) - contact(s) with other delegate(s)

All these NGOs tried to intervene in the negotiation and decision-making processes in the climate arena in order to affect policy outcomes. Alongside these interventions of pressure organizations, some protest groups tried to indirectly influence the negotiations by organizing activities outside the formal political arena, for example during important international sessions (e.g. the international network 'a SEED' of youth organizations did so in Berlin). Table 4.1 summarizes the interventions of climate NGOs. The format of this table is built on table 2.2 (page 52). There several conceptual aspects of protest, advocacy and lobby were presented, whereas those concepts are translated into observable actions of NGOs here. This overview is probably not complete, as it is based on a sample of (NGO) documents on the one hand and on a few observations by the researcher on the other (at international UN meetings in Berlin and Geneva in 1995 and 1996 respectively). Nonetheless, the most important ones seem to be captured in table 4.1. Concrete examples of these activities will be dealt with in the case analyses below.

As far as the input of NGOs in the implementation phase after 1992 is concerned, the involvement has continued to increase. For example, ever more NGOs visited the Conferences of the Parties. But the key players remained those mentioned above. It seems as if a more or less stable set of environmental and development NGOs is now linked to the climate arena.

— 4.3 NGOs in the formation phase (1990-1992) —

4.3.1 The 'counterfactual'

In chapter 2 political influence was, briefly, defined as 'goal-achievement due to one's own and intended intervention' and it was added that such implies that the policy outcome concerned is more in line with one's goals than if one had not intervened. The second part of this definition refers to the so-called 'counterfactual'. In order to have a quick and general impression of the NGOs' impact on the convention text, the several key respondents from the climate arena were asked at the start of any interview to employ such counterfactual reasoning, but not before it was explained to them how political influence should be interpreted and which NGOs were covered by this study. In fact, they were asked to respond to the following question: "Assuming that all the NGOs had been absent during the negotiation process of the FCCC, would the convention text have been the same in your opinion?"

The answers of the respondents with regard to this question are summarized in table 4.2. Two trends can be distinguished here. First of all, those who confirm the NGOs' impact outnumber the ones who deny it. Most respondents from the climate arena – 18 out of 23 – do indeed believe that the counterfactual would have been different ('convention not the same' or 'no convention at all'). Even most government representatives believe that NGOs made a difference (7 out of 10). To quote one of these:

"The FCCC would have been different without the NGOs. The presence and pressure of environmental movements played an important role all the time."

TABLE 4.2 *The counterfactual of the FCCC as perceived by the respondents from the climate arena*

FCCC	*NGO respondents*	*government respondents*
convention the same	1	2
convention *not* the same	8	7
no convention *at all*	3	-
do not know	1	1
number of respondents	13	10

Secondly, NGO representatives are a bit more optimistic than government ones. What is more, three of them even believe that the counterfactual consists of an absence of the FCCC. As one of them put it: "Maybe I exaggerate a little bit, but I think that without the pressure of the environmental movement, there would have been no treaty at all." A minority, though, shares a different perception. For example one NGO respondent puts his and his colleagues' impact into perspective:

"The role of NGOs is quite marginal. Look at the convention which was negotiated. What influence did NGOs have on the final text? They could not change one word! So what is the big deal? OK, I think that certain NGOs on certain issues had some impact on government positions, but not necessarily on the entire negotiations."

One should, of course, reflect on the validity of the above answers. First of all, it is not always clear whether the respondents had correctly interpreted the notion of political influence. This could have been verified by asking additional questions, but an in-depth analysis was not the aim of this first introductionary question (although it was with regard to other items; see the cases below). Secondly, several respondents made it very clear that counterfactual reasoning is extremely difficult. Some even hesitated to respond or pondered the question for some time. But nearly all came up with a guess in the end. And thirdly, there are other mechanisms in operation which might undermine the validity of interviews in general (Vakgroep Methoden, 1994; Westerheijden, 1987). In particular the mechanisms of strategic conduct and reification seem relevant here. The fact that NGOs might (deliberately) overestimate their own impact was already referred to in this book, as the self-legitimization for lobbyists is, after all, their influence. But government representatives may over- or underestimate the role of NGOs for strategic reasons as well. In addition, reification is likely to occur. After all, NGOs are generally considered influential in international environmental policies since UNCED in 1992 (see section 1.4 in chapter 1). Respondents might tend to reify this general belief. Given these considerations, the validity of the outcome in table 4.2 cannot be considered 'high'. The chance that it is too optimistic in terms of NGO influence is substantial.

To give a general impression on the basis of a single question is one thing, to be more precise another. Therefore the respondents were asked to give examples of topics, articles, paragraphs or specific wording of the FCCC of which they believed NGOs had an influence. Furthermore, all the respondents were asked to elaborate on their examples in order to gather information on *what* had been influenced by *which* NGOs, to what *extent*, and *how* NGOs managed to intervene. On the basis of these examples, lists of ego- and alter-perceptions were constructed. These lists subsequently produced the eight cases which will be analysed in this chapter. In doing so, a format which was designed in chapter 3 will be applied (see table 3.2, page 86).

— 4.3.2 Case 1: The objective of the FCCC —

A Introduction

The objective of the FCCC was already referred to in section 4.2.2 of this chapter, but the full text reads as follows (Article 2 of the FCCC; Doc 1.38):

> "The ultimate objective of this Convention and any related legal instruments that the Conference of the Parties may adopt is to achieve, in accordance with the relevant provisions of the Convention, stabilization of greenhouse gas concentrations in the atmosphere at a level that would prevent dangerous anthropogenic interference with the climate system. Such a level should be achieved within a time frame sufficient to allow ecosystems to adapt naturally to climate change, to ensure that food production is not threatened and to enable economic development to proceed in a sustainable manner."

The first sentence is based on paragraph 10 of the 1990 Ministerial Declaration of the Second World Climate Conference (Doc 1.29), which urges parties to stabilize greenhouse gas (GHG) concentrations in the atmosphere at so-called 'safe' levels. This objective implies, at least according to the Intergovernmental Panel on Climate Change (IPCC), that countries need to reduce their GHG emissions sooner or later. The second sentence is based on paragraph 10 of the 1989 Noordwijk Ministerial Declaration (Doc 1.31). This sentence binds the objective of stabilization to a (non-specified) timetable which is based on both ecological and economic limits. Therefore Article 2 implicitly states that the time frame to achieve stabilization of

GHG concentrations should not be too loose, otherwise ecological damage due to climate change might already have been done. But it need not be too strict either, otherwise the costs might be too high and hamper economic development. Note that the article does not condemn climate change as such, just *dangerous human interference* with the climate system. 'Non-dangerous' change seems acceptable. Moreover, the legal status of Article 2 is, according to Bodansky (1993), unclear. The question remains whether it is belongs to the legally-binding part of the FCCC or not.

The wording of the objective was not smoothly accepted by all, at the Noordwijk Conference in 1989, the Second World Climate Conference in 1990, or at the negotiation sessions of the Climate Convention (INCs), in the period 1991-1992. For example some delegations suggested omitting the word 'concentrations' and replacing it with 'emissions', thus reformulating the ultimate objective in terms of the stabilization target. This reformulation, however, would imply a softening of the legal instrument, as a stabilization of GHG emissions leads to a *build-up* of concentrations in the atmosphere and, hence, to *more* global warming. Being aware of this, other delegations objected to this reformulation of the objective. They believed it would be foolish to have a Climate Convention which results in more global warming.

The specific language of the objective was proposed by the European Community (EC) during INC3, whereas the UK had already included it – at least the first sentence – in its own proposals for INC1 (Doc 1.13, p. 26). But it was immediately put in brackets, due to opposition from some nations, in particular the USA. These countries feared that the goal of stabilizing GHG concentrations at safe levels would burden their economies too much. Therefore several other forms of wording for an objective were considered as well. Some documents even contained 8 proposals for an objective, most put in brackets (Doc 1.16). Finally, the EC formula was included in the draft convention as proposed by chairman Jean Ripert of the Intergovernmental Negotiation Committee (INC) in May 1992 (Doc 1.19). The delegations had asked him to put forward such a proposal for the last negotiating session to overcome deadlock in the negotiations, and Ripert decided to add the EC's ultimate objective to the final draft of the treaty text. The objective nonetheless remained bracketed until the very last days of the negotiations. It seems as if the formula was finally accepted as part of a package deal: the EC – and others such as the AOSIS – got a 'strong' objective, whereas the USA – and for example OPEC – got 'weak' commitments in Article 4 of the Convention (see below). Additionally, the opponents may at the end of the day have agreed not only due to the pressure to compromise, but also because the objective itself does not set binding targets and timetables. Also, it does not stipulate what safe levels of GHG concentrations are, and it is open to different interpretations. The most important NGOs which focused on the nature and language of the convention's objective – among other things – were: Climate Action Network (CAN), WWF and Greenpeace.

B *Ego-perception*

Some respondents among the climate NGOs – 3 of the 13 interviewed – believe they were jointly responsible for the inclusion of the ultimate objective in the FCCC. One of their representatives said:

> "The ultimate objective, I think, was originally crafted with NGOs in 1990. That it survived was due to a number of key governments and to NGOs playing a role in defending that text."

Two other NGO respondents also refer to Article 2 as an example of NGO influence. Their references, though, differ. One believes, just as his colleague quoted above, that NGOs themselves participated in the drafting process of the ultimate objective during the Noordwijk Conference, the Second World Climate Conference, and the sessions of the Intergovernmental Negotiation Committee (INCs), whereas the other only refers to NGO pressure to defend the EC text. Hence, two NGO representatives claim *direct* political influence with regard to Article 2; one only *indirect* influence. Note, though, that the respondent quoted above *thinks* that NGOs were part of the drafting process. He himself was obviously not there. The same applies to the other NGO representative claiming direct influence. Consequently, these claims are second-hand and need further verification.

C *Alter-perception*

Three of the ten government respondents selected from the climate arena also believe that NGOs impacted the language of Article 2. One for example responded:

> "If I look at the key parts of the FCCC, and Article 2 in particular, there was an influence of NGOs. But I do not find it easy to be more precise."

It is obvious that this reference is rather general in nature. Another respondent also believes that NGOs played a role in the drafting process of the objective at the INCs. He credits some direct influence to NGOs, although he does not mention specific words or sentences of Article 2. The third respondent stresses the fact that NGOs were important in teaching several delegates the difference between the 'stabilization of concentrations' on the one hand and the 'stabilization of emissions' on the other. In doing so, NGOs indirectly influenced the outcome of the debate. In short, one government respondent refers to direct political influence, one to indirect political influence, and one to NGO influence in general terms.

D *Causal analysis*

GOAL-ACHIEVEMENT — NGO documents and publications make clear that NGOs more or less achieved their policy goal with regard to Article 2. An internal memo of one of the leading organizations says that the language of Article 2 'is basically OK' (Doc I.59, p. 3). Moreover, an assessment of the final outcome by one of the

NGO representatives refers to Article 2 as 'one of the key positive elements in the convention (...), which establishes a powerful ecological goal' (Doc 1.41, p. 2). NGOs also decided to literally adopt the FCCC's objective in an alternative treaty on climate change which they designed during the UNCED in Rio de Janeiro in June 1992 (Doc 1.67, p. 94). However, it must be acknowledged that earlier NGO proposals on an objective for a climate convention include maximum figures on GHG concentrations in the atmosphere (475 parts per million), on temperature increase (0,1°C per decade) and on rises in sea levels (2-5 cm. per decade) (Doc 1.71, p. 1). Such quantitative references to safe levels are lacking in the FCCC. Therefore NGOs did not achieve their goals entirely.

NGO INTERVENTIONS — The question remains, though, whether this outcome was due to the NGOs' own interventions. It seems not, as the text elements of Article 2 were drafted by government expert groups at the Climate Conferences in Noordwijk (1989) and Geneva (1990). NGO representatives were, according to both the key government and non-government players at that time, not involved in that drafting process. What is more, internal documents of NGOs also refer to the language of the objective as stemming from the Ministerial Declaration of the Second World Climate Conference. No reference is made to NGO sources as far as its origin is concerned (Doc 1.59, p. 3). Furthermore, it was the EC which introduced the ultimate objective in the negotiations at INC3. NGOs welcomed this move and defended the proposal *afterwards.* To conclude, direct political influence of NGOs with regard to Article 2 seems out of the question.

It may be still possible that NGOs influenced the wording and its adoption indirectly. For example, they may have pressurized the EC to come up with its proposal at INC3 or the key-players to compromise on Article 2 at INC5. But this seems unlikely as well. As more than one source confirmed, the language was already embraced by EC countries such as Germany, the United Kingdom (UK), and the Netherlands before the INCs even started. So the EC did not 'need' the pressure from NGOs to come up with its proposal. And although it is obvious that NGOs defended the text after its introduction, for example in their journal *Eco*, this pressure did not seem a necessary precondition for the compromise on the objective. Besides, the compromise was finally reached behind closed doors among the key countries. NGOs, as observers, were not part of that formal process. Of course, NGO pressure from outside can still be experienced by delegates in small and closed negotiation rooms, as several government representatives asserted during the interviews. But this was hardly the case here, as such pressure was focused more on targets and timetables during the final stages of the negotiations, and less on the objective (see case 3, page 121).

Still, it is believed that NGOs played a useful role in the debates on the objective, as the alter-perception instrument also indicates. As one government respondent noted:

> "We had long discussions on Article 2 and NGOs played an important role. Sometimes NGOs are useful in explaining the matter to delegates. For example, not all delegates had a clear distinction between emissions and concentrations of greenhouse gases. Through the work of NGOs some delegates became more aware of that. After all, NGOs are in a better position to go through all the ideas and books."

But given the initiatives from and very active role of the EC in this specific case, the valuable contribution of NGOs to the process seems not have impacted the product itself, namely the decision on Article 2 (for the distinction between process and product influence see subsection 2.3.1).

E Extent of political influence

In subsection 3.2.5, the following formula to assess the extent of political influence was designed: PI = C [(GA x OI) + PR + D]. In table 4.3 below, this formula, its variables and legends are applied to this case. This will be repeated with regard to each case in chapters 4 and 5.

TABLE 4.3 *Scores on political influence in case 1 (Objective FCCC)*

Counterfactual (C)	similar (0)
Goal-achievement by NGOs (GA)	much (3)
GA due to NGOs' own interventions (OI)	no (0)
Political relevance of topic (PR)	moderate (2)
Directness of impact (D)	no impact (0)
Extent of political influence: PI = C [(GA x OI) + PR + D]	none (0)

Legends
C : similar = 0; different = 1
GA: no = 0; some = 1; substantial = 2; much = 3
OI: no = 0; some = 1; substantial = 2; much = 3
PR: no = 0; low = 1; moderate = 2; high = 3
D : no impact = 0; indirect impact = 1; direct impact = 2
PI: no PI = 0; some PI = 3 to 6; substantial PI = 7 to 10; much PI = 11 to 14

As no political influence of NGOs is assessed here, an elaboration on its extent seems superfluous. Nonetheless, in order to have a complete overview of this case, the variables of the above formula and their scores are presented in table 4.3. The counterfactual of Article 2 would have been the same, as the outcome was not due to the NGOs' interventions, irrespective of the fact that they were very satisfied with it (which implies that Article 2 did 'much' respond to their goal). The topic of the objective is considered to be of moderate relevance, although it was indeed a key element in the FCCC negotiations. But it led to less controversy among the negotiators than other items. Besides, it does not directly prescribe specific actions for governments, especially as it has remained unclear whether the objective is binding upon them.

F Final conclusion

Although NGOs such as Climate Action Network, WWF and Greenpeace had some impact on the debates with regard to the objective of the convention, their influence on the outcome was negligible. Therefore one might speak of process influence at best in this specific case, but not of political influence. In short, Article 2 would have been adopted anyway, irrespective of the presence and interventions of NGOs at the INCs.

4.3.3 Case 2: principles

A Introduction

Article 3 of the FCCC contains five principles: on common but differentiated responsibilities of parties, on specific needs of developing countries and vulnerable areas, on precautionary and cost-effective measures, on the right to and promotion of sustainable development and on a supportive and open international economic system. The content of these provisions was already briefly highlighted in subsection 4.2.2. The principles, though, were subject to an intense North-South struggle during the negotiations. The developing countries were in favour of an article on principles right from the start and it was China which presented a first proposal at INC2 (Mintzer and Leonard, 1994; Doc 1.14). Most OECD countries however – and again mainly the USA – objected to the inclusion of such a chapter. According to them, principles should be part of the non-binding preamble of a treaty and certainly not go further than the binding commitments (which seems to be the case now if the principles in the FCCC are taken literally). They particularly objected to the inclusion of a 'right to development' principle in an environmental convention: the principle most fiercely defended by the South. A compromise was finally agreed upon. The OECD accepted an article on principles, whereas the G77 had to accept the omission of several paragraphs, among others the 'right to development' and the 'polluters pay principle'. To give an impression of the struggle, earlier drafts of Article 3 contained 16 different principles, whereas the final FCCC has only five (Doc 1.15). Moreover, an umbrella clause was added to the entire Article 3 which states that 'Parties shall be guided, inter alia' by these principles. This means that they might be guided by others as well and do not need to follow the principles exactly.

The confrontation on principles nearly ended up in a total failure of negotiations. Whereas key countries from both North and South were able to agree on the compromise, Iran threatened to reopen the whole package deal on the draft convention in the final diplomatic conference at the end of the second session of INC5, which was supposed to adopt the text by consensus before the UNCED. The reason for this threat was Iran's condemnation of the omission of the 'development principle' by those who had negotiated the compromise. Furthermore, Iran set a precedent at the diplomatic conference, as other countries followed in criticizing the package deal. It was difficult to prevent a crisis, but chairman Ripert, backed by a large majority of the delegations, ignored the objections in the end and the plenary finally adopted the climate treaty by a standing ovation (making the minority still objecting invisible).

B *Ego-perception*

Three NGO respondents believe they impacted the content and inclusion of Article 3. They particularly claim to have successfully pushed the precautionary principle on the one hand (paragraph 3.3) and the idea of equity between North and South and the principle of common but differentiated responsibilities on the other (paragraph 3.1). With regard to the latter, one respondent said:

> "The North was reluctant to admit the principle of common but differentiated responsibility, because they feared they had to pay for their historical role in the build-up of greenhouse gases. But we managed to push it and push it, because the rich cannot afford to live off the small and poor. A role of NGOs was pointing out that if the North is not going to accept that responsibility nor the historical nature of the threat, you will never have the developing countries on board."

It is important to note that none of the NGO respondents claim any direct impact on the language of Article 3. They only claim some effects of their pressure on the inclusion of certain ideas. In other words, they are convinced that without their pressure the precautionary principle as well as the equity issue would not – or not so well – have been addressed in this section of the Climate Convention.

C *Alter-perception*

Only one government representative gave the NGOs some credit with regard to this topic, although in rather broad terms:

> "I will not be able to pinpoint which sentence came from which NGO, but generally speaking, some of the principles were endorsed by the NGOs."

Another government respondent, though, spontaneously came up with an opposite view:

> "The entire area of principles was not due to NGOs. That was an essential part of the political discussion among governments themselves."

According to him, all the principles discussed during the INCs were already in several government draft proposals, even *before* the negotiations started. Moreover, many principles were incorporated from other environmental agreements which had already been adopted by governments at an earlier stage.

D *Causal analysis*

GOAL-ACHIEVEMENT — It is obvious that the section on principles coincides with some of the NGOs' views, in particular the clauses on equity, precautionary measures and sustainable development, although these are stated less clearly and bluntly than NGOs had wished. For example the precautionary principle was linked to the premise of cost-effectiveness, an addition which might soften the precautionary approach. After all, costly measures of which the long-term effects are either doubtful or limited are more easily rejected on the basis of that premise. In addition, NGOs 'lost' a number of principles they supported during the negotiations. An example is the Polluters Pay Principle, which covers the idea that all environmental costs should be reflected in the prices of raw materials and goods (Doc 1.46, p. 6). OECD and OPEC countries opposed this principle, as it would lead to higher prices for oil, coal, and energy-intensive products. Therefore the Polluters Pay Principle was deleted in the draft convention. Furthermore, NGOs opposed a principle concerning the consistency of the convention with GATT (General Agreement on Tariffs and Trade) rules, a provision which would make environmental standards subject to free trade premisses. Despite this NGO opposition, paragraph 3.5 holds that parties should in fact promote free trade (although a reference to GATT was omitted). Consequently, the picture of the NGOs' goal-achievement is ambiguous. They partly achieved their goals with regard to some aspects of Article 3, whereas they attained no result with regard to others.

NGO INTERVENTIONS — For as far as NGOs were able to achieve their goals, one should reflect on the causes. Again, it seems unlikely that NGOs themselves were responsible for this outcome. The following arguments underpin this proposition. Firstly, NGOs initially did not campaign for a distinctive section on principles in a Climate Convention. Their proposals particularly focused on objectives and greenhouse gas (GHG) reduction requirements in a treaty. Not until government element-papers for a convention included a section on principles did NGOs start to react and to formulate their own positions. This will become obvious if one for example reads the 1990 NGO Draft Convention from the Climate Institute or compares Climate Action Network (CAN) position papers addressing INC1 with those addressing INC4 and 5 (Doc 1.44, 1.46, 1.71). Secondly, several NGOs were reluctant to emphasize the need to include principles in the FCCC, as they preferred – to quote one NGO paper – "countries to negotiate specific obligations rather than generalized principles" (Doc 1.47, p. 2). Thirdly, several governments were already in favour of those principles which were endorsed by NGOs. The G77 and China – the group of developing countries which introduced the section on principles – immediately emphasized their right to develop their economies unconditionally, even under a Climate Convention, and the need for equity among the poor and the rich in addressing climate change. And some industrialized countries introduced the principles of sustainablity and a precautionary approach. Fourthly, the final compromise on Article 3 was to a large extent based on the wording of the proposals for the Rio Declaration which was negotiated in the UNCED process (Johnson, 1993; Mintzer and Leonard, 1994). True, NGOs may have influenced this wording in the UNCED arena, but this was not assessed by this study. As far as the climate NGOs were concerned, they did not take first initiatives in this field. They just followed the government track and delivered some comments.

However, there were two exceptions. First of all, some Southern NGOs such as the Indian Centre for Science and Environment (CSE) and the Bangladesh Centre for Advanced Studies (BCAS) already stressed the equity principle in the climate arena even before the INC had formally started. Their publications and letters to Southern leaders emphasized the need to secure justice among the poor and the rich on the climate issue (Agarwal and Narain, 1991; Doc 1.43), otherwise climate change policies would become another case of 'environmental colonialism'. Secondly, Greenpeace promoted the precautionary approach right from the start (Doc 1.57). This organization also 'instructed' the small island states (AOSIS) to support this principle, in order to further their specific interests as potential victims of climate change (Doc 1.66). Nonetheless, it would go too far to conclude that without those NGOs the equity and precautionary principles would not have been addressed – or would have been differently addressed – in the FCCC. Developing countries themselves fiercely stressed the need for equity – and have done so in international environmental policies since the 1970s –, just as some European countries strongly emphasized the precautionary approach, in their case since the 1980s. Therefore the contributions of the above NGOs were not a necessary prerequisite for the outcome. Even NGO representatives themselves admitted, while referring to the input on equity: "This is not to say that these issues would not have come into the negotiations at a later time and in some other way without the NGO debate" (Rahman and Roncerel, 1994, p. 250-251). Nonetheless, it is believed that the Centre for Science and Environment (CSE), the Bangladesh Centre for Advanced Studies (BCAS) and Greenpeace still played a role in the negotiation process on principles (process influence). They contributed to reframing quite general principles such as equity and precautionary measures, which were already adopted by several governments years before, in terms of the climate issue. For example CSE 'helped' India and Greenpeace 'helped' AOSIS with respect to that. But such a reframing process would have taken place sooner or later anyway.

E Extent of political influence

Table 4.4 gives an overview of the assessment of the NGOs' political influence in case 2. The scores parallel those of the first case, except for the extent to which the NGOs achieved their goals (due to others however). Obviously, the principles reflect their preferences far less than does the objective. The political relevance of the topic is – again – considered moderate. Although the negotiations were very tense in this case, as much of the North-South clash was reflected in this topic, the principles cannot be considered binding outcomes. States do not need to implement them nor can they force others to do so. To repeat, principles are just non-binding guides for action. While the North-South controversy urges one to consider principles highly relevant, the non-binding nature of Article 3 points at the opposite, namely low relevance. Consequently, a middle position was chosen.

TABLE 4.4 *Scores on political influence in case 2 (Principles FCCC)*

Counterfactual (C)	similar (0)
Goal-achievement by NGOs (GA)	some (1)
GA due to NGOs' own interventions (OI)	no (0)
Political relevance of topic (PR)	moderate (2)
Directness of impact (D)	no impact (0)
Extent of political influence: PI = C [(GA x OI) + PR + D]	none (0)

F Final conclusion

Although some NGOs such as the Indian Centre for Science and Environment and Greenpeace were able to contribute to the debates on principles, their political influence on Article 3 was very limited, as governments themselves were responsible for the outcome. In addition, the majority of NGOs barely focused on principles. They were mainly involved in other topics.

4.3.4 Case 3: target

A Introduction

Subparagraphs 4.2(a) and (b) of the FCCC contain the target and timetable with regard to the reduction of GHG emissions. In international environmental law the phrase 'targets and timetables' refers to legally-binding and quantitative limitations of the emissions of substances which damage the environment. It has become a popular phrase in the international regulation of atmospheric pollution (Bodansky, 1993). If one looks at the Montreal Protocol on the ozone layer or at the Geneva Convention on Long-Range Transboundary Air Pollution (LRTAP) on acid rain for example, both include reduction targets and timetables on chloroflorocarbons (CFCs) and SO_2 respectively. Although there are, therefore, precedents in international environmental policies, targets and timetables on GHGs were nonetheless not very popular in the climate arena, and most controversies which arose during the negotiations in the INCs had to do with subparagraphs 4.2(a) and (b).

Delegates disagreed on several aspects (Mintzer and Leonard, 1994; Doc 1.13, 1.30, 1.33). First of all, countries could not agree on when to adopt binding commitments on GHG reductions. Some preferred to negotiate a *framework* convention without any substantive targets and timetables first, and then – after the convention had come into force – adopt protocols with substantial commitments in the second half of the 1990s. Others, though, were in favour of immediately adopting clear and strong targets in a Climate Convention, given the urgency of the problem. Secondly, countries disagreed on the desirability of the instrument itself. Japan and the USA, for example, rejected one general target to which all parties would be bound. The latter country for example commented:

> "Specific commitments for emissions reductions should not be included in the framework convention because of the need for flexibility in nations' choices of their own measures." (Doc 1.13, p. 37)

Japan, on the other hand, introduced the 'Pledge and Review' instrument at INC3. This approach is as follows: all parties unilaterally and on a voluntary basis make a pledge for a specific target and timetable to be achieved at home, and implementation is subsequently reviewed at multilateral level. This Japanese proposal, though, was not at the end of the day adopted by the INC.

Delegates also clashed on the substance of targets and timetables. Countries for instance disagreed on which greenhouse gases (GHGs) should be covered by possible commitments: all GHGs or mainly CO_2. Developing countries urged a focus on the latter, as this GHG is mainly emitted in the North (approximately 75% of global emissions). Some industrialized countries also agreed to adopt this one-gas approach, as carbon dioxide is one of the main and best known causes of global warming. But the USA – responsible for 25% of global CO_2 emissions – rejected this focus and, instead, favoured a so-called comprehensive approach including *all* GHGs, so that for example a bad performance with regard to CO_2 could be compensated for by good performances in other fields (such as methane and CFCs). Furthermore, the USA and some other Western nations favoured a comprehensive approach based on *net* emissions. According to this approach, absorption of GHGs by sinks such as forests should be subtracted from gross emissions. This implies that for example a reduction or stabilization of net emissions could be achieved by reforestation. Others, though, rejected this, because it is a disincentive to really cut emissions from energy plants, transport and industry and to improve energy efficiency in those sectors.

Yet another debate dealt with the question whether only industrialized countries should commit themselves to (binding) targets, policies and measures under a Climate Convention, or whether developing countries should as well. It is no surprise that the G77 and China favoured the former option, whereas a number of rich countries wanted the latter. In the end developing countries indeed committed themselves to some general measures, but under Article 4.1 of the FCCC. It is Article 4.2, however, which covers the specific commitments for developed countries. Another hotly debated substantial issue was the strength of targets and timetables. Some more progressive countries and victim-states such as those in the Alliance of Small Island States (AOSIS) tabled the Toronto target at the first session of the Intergovernmental Negotiation Committee (INC1). This target, formulated at the Toronto conference in 1989 and already unilaterally adopted by some countries (e.g. Australia, Austria, Canada, Denmark, Luxembourg, New Zealand), demands that industrialized nations reduce, as a first step, their emission of carbon dioxide by 20% of 1989 levels by the year 2005. The EC, on the other hand, tabled the stabilization target, which had become Community policy while the climate negotiations went on. To repeat, this target states that industrialized countries should stabilize their CO_2 emissions at 1990 levels by the year 2000. This European position was not only due to environmental considerations, it was an expression of self-interest as well. In order to maintain the competitiveness of European business in the world market, environmental standards had to be the same worldwide. Therefore the EC wanted all countries to adopt the stabilization target in Rio de Janeiro.

Given the opposition from the USA (and 'its' business lobby) to reduction requirements, it soon became obvious that the option to include the Toronto target into the FCCC was not feasible, especially as several developing countries were reluctant to accept binding reduction targets for the industrial world as well (in particular China and the OPEC countries). The negotiations therefore focused entirely on the pros and cons of the inclusion of the stabilization target in Article 4, and were extremely complex and tense. The USA remained totally opposed to any targets and timetables until the very last day of INC5. At the same time the EC remained firm in their position, although there was strong pressure to give in in order to keep the USA on board. In addition, the EC was, under the surface, divided on the stabilization target. The United Kingdom (UK) preferred a different target year – 2005 instead of 2000 – and the Southern European countries were less than enthusiastic, although under Community policy they were still allowed to increase their emissions to help their economic development (see also section 3.7 on joint implementation). Consequently, the stabilization target was slowly eroded in the course of negotiations. Finally, a compromise was agreed upon under the leadership of INC chairman Jean Ripert, the USA and UK, and through silent and high-level crisis diplomacy.

This compromise on specific commitments for industrialized countries in the FCCC contains only a quasi-target and a quasi-timetable and covers all GHGs (except for CFCs). It no longer refers to stabilization, but to the commitment to 'return by the end of the present decade to earlier levels of anthropogenic emissions of carbon dioxide and other greenhouse gases not controlled by the Montreal Protocol' (Doc 1.38, p. 8). This implies that those emissions are allowed to increase again after the target year. Moreover, the baseline and target years of the timetable – 1990 and 2000 respectively – were included in separate subparagraphs. Article 4.2(a) of the final convention text deals with the commitment itself and refers to the obligation for developed countries to return GHG emissions to earlier levels 'at the end of the present decade' (i.e. by the year 2000). But the baseline year 1990 is not mentioned in subparagraph (a) and therefore a party which is for example emitting GHGs at 1998 levels in the year 2000 is still implementing this subparagraph. Moreover, he is not bound to anything after 2000. Article 4.2(b), on the other hand, deals with the review of this commitment and does, indeed, refer to the aim of returning GHG emissions to 1990 levels. However, in this subparagraph the target year of 2000 is not mentioned, so the aim might be achieved in *any* future year (for example in 2005 or 2010). Moreover, paragraph (b) deals with the review of the commitment, not with the commitment itself. Therefore the reference to the baseline year 1990 in Article 4.2(b) has no legal relevance for paragraph (a). In short, the stabilization target disappeared as a binding rule in the FCCC, but remained, as some assert, as a political signal.

The NGOs were utterly shocked by this result. 'GARBAGE' reads one of *Eco*'s headlines on the compromise (Doc 1.84, p. 1). NGOs had hoped for a strong convention, but left New York more or less with empty hands. But not only the NGOs were disappointed. For example EC Environment Commissioner Carlo Ripa di Meana, who was in office at the time, considered the outcome unacceptable.

B *Ego-perception*

Despite this result, some NGO representatives – at least two of those interviewed – still believe they had some influence on Articles 4.2(a) and (b). But others believe the opposite. Two ways of reasoning can be seen here. Those who believe they had some influence are convinced that the presence of NGOs pushed governments to compromise on (quasi) targets and timetables and adopt a political signal. As one respondent noted:

> "The important thing was that despite the incredible opposition of countries like the USA we did get language that set the framework for returning to 1990

levels. That is not even a first step, that is a political declaration that the developed world has to take the lead and will begin to look at measures to start to shift its energy path. And NGO traces are still there in terms of a reduction trend."

Those who believe that NGOs had no influence on article 4.2(a) and (b) refer to the *wording* of it. It is indeed almost certain that no NGO traces can be distinguished there, given the 'closed-door' diplomacy which was followed by France, the UK, and the USA. In the words of one representative:

"The text finally adopted was written by the USA. So it made no real difference. The USA got what it wanted."

Obviously, no NGO respondent claims any direct influence on the compromise.

C *Alter-perception*

Two government respondents also perceive NGOs as influential in the debate on targets and timetables. One said:

"I am sure that delegations from Western Europe were under strong influence of green NGOs, for example in setting a target for emissions, or even a reduction target. I am sure that NGOs were behind that."

Apart from in Europe, these respondents – and some other government representatives as well – believe that NGOs exerted some influence through the Alliance of Small Island States (AOSIS). The fact that victim-states were pushed by NGOs to ask for strong commitments counterbalanced the position of the USA to some extent.

D *Causal analysis*

GOAL-ACHIEVEMENT — It seems more than obvious that NGOs did not achieve their goals in this case, despite the fact that they had spent most of their energy and resources on this topic during the INCs. They had for example campaigned for a 20% cut in CO_2 emissions at 1988 levels by the year 2000 as a first step for OECD countries, but got nowhere (Doc 1.45). Therefore it seems that NGOs cannot be credited with any political influence here, either direct or indirect. However, some respondents claim that without NGO pressure the outcome would have been even worse. After all, the stabilization target is retained in the FCCC as a non-binding political message, and that is more than for example the USA had wanted. What is more, the coal and oil lobby was furious that the text had gone as far as it did (Doc 1.84). Therefore one might argue that NGOs – together with other players – had prevented worse, as an entirely 'empty' convention without any target or timetable could have resulted as well (a 'lowest common denominator' outcome based on the USA position). Given the elaboration of the definition on political influence in chapter 2, such a prevention of an undesired 'bad' is also covered by the notion of goal-achievement. Therefore NGOs achieved the goal of having a treaty with at least some sort of target.

NGO INTERVENTIONS — Was the compromise on the stabilization target due to NGOs? It seems more appropriate to point at the pressure of the EC, AOSIS, G77 and the UNCED process in this respect. The EC really wanted their proposed CO_2 policy adopted worldwide, the AOSIS kept pushing for strong commitments, the G77 finally urged the rich world to take the lead in climate policies through substantial commitments, and governments in general hoped for a successful Earth Summit in Rio de Janeiro. A Climate Convention 'with teeth' had to be part of that success. All these factors definitely put pressure on the USA. But it is believed that the role of NGOs cannot be entirely denied. First of all, NGOs reminded the EC of their earlier promises on green policies and of their green electorates at home (with elections in the UK, France and Germany ahead). These arguments were probably among those that kept the EC relatively firm in its position, despite US pressure and despite the internal division among EC members. Secondly, NGOs mobilized 'victim-interest' even before the INC negotiations started. They closely co-operated with the AOSIS, raised their awareness of their ecological vulnerability and helped them to formulate their position (Doc 1.66). This position gave another dimension to the negotiations, as several government respondents confirmed, because the G77 was no longer united in its initial resistance to substantial commitments. Moreover, the possible victims of global warming – the small island states – got a voice at the negotiating table. Thirdly, NGOs tried to convince developing countries that targets and timetables were definitely needed, but should only have consequences for the North. At the same time they tried to convince industrialized countries that they should take the lead, otherwise the South would not join a Climate Convention.

Fourthly, the NGOs brought public opinion into the INCs, which was still rather in favour of green issues at that time. The public expected governments to agree on something *substantial* in Rio and the Climate Convention was an important symbol in this respect. Hence, leaders could not afford to have a total failure and NGOs reminded them of this. Fifthly, NGOs made article 4.2 the basic criterion on which to judge the success or failure of the climate negotiations. 'We need strong and binding commitments now, not a vague framework treaty' was their message right from the start. Although other players tried to focus on other aspects, for example on principles (G77 and China) or science and technological co-operation (USA), the ultimate criterion of success became the targets and timetables, in the press as well. Finally, NGOs were able to soften an early attempt by Japan, France and the UK to get rid of general and binding targets and timetables. As was outlined above, at INC3 Japan proposed the procedure of 'Pledge and Review' for the FCCC. NGOs perceived this proposal as 'Hedge and Retreat' from binding commitments (Doc I.82, p. I). This reformulation of 'P&R' into 'H&R' made a big impression among delegates (Mintzer & Leonard, 1994, p. 136). The EC tended to support the Japanese proposal in the first instance, but decided to bury it at the end of INC3.

We have to be careful not to exaggerate, but NGOs definitely played a role in creating an environment in which the key political players 'had' to accept (quasi) targets and timetables. In other words, the NGOs' clear message on the need for strong commitments, their pressure on the EC, their co-operation with the AOSIS, their mediation between North and South and their use of the media just before UNCED, in the end impacted the outcome of negotiations.

E *Extent of political influence*

Given these elaborations, the counterfactual is considered different: an 'empty' FCCC without a target (see table 4.5). The NGOs nonetheless barely achieved any of their initial goals, but they – jointly with others – prevented worse. These others were: EC, AOSIS and G77. Therefore the outcome was only to some extent due to NGOs. The political relevance of this topic should be considered very high. After all, it touches upon high state interests – energy consumption, economic growth, development – and it ultimately led to high-level crisis diplomacy. Furthermore, unambiguous targets and timetables adopted in this part of the FCCC would be binding. Finally, it is obvious that the (small) impact of the NGOs was indirect. None of the NGO proposals and formulas were adopted by governments. Their impact was therefore indirect in nature and was realized by public and private pressure (both advocacy and lobbying).

TABLE 4.5 *Scores on political influence in case 3 (Target FCCC)*

Counterfactual (C)	different (1)
Goal-achievement by NGOs (GA)	some (1)
GA due to NGOs' own interventions (OI)	to some extent (1)
Political relevance of topic (PR)	high (3)
Directness of impact (D)	indirect (1)
Extent of political influence: PI = C [(GA x OI) + PR + D]	some (5)

F *Final conclusion*

All NGOs, and in particular the umbrella Climate Action Network, were very active in the field of targets and timetables, as they perceived this topic as the cornerstone of the Climate Convention. But despite their intensive campaigns and lobbies for substantial commitments, they failed to achieve their initial goals. NGOs wanted a strong reduction target in the FCCC, but they got a very weak and ambiguous version of the stabilization target. However, they were part of a movement that prevented worse: an entirely *empty* treaty. Therefore NGOs are credited with some indirect political influence on Article 4.2(a) and (b).

It is interesting to note at this stage of the analysis that there is no direct proportional relationship between goal-achievement and political influence. Whereas NGOs achieved some of their goals with regard to cases 1 and 2, they were not credited with any influence on the outcomes. The opposite is true for case 3. NGOs did not achieve their initial goals, but they nonetheless indirectly impacted the outcome.

4.3.5 Case 4: provisions on implementation

A Introduction

The FCCC includes several provisions on implementation, among others:

1 Paragraph 7.2 which defines the role of the Conference of the Parties (COP) in keeping the implementation of the convention under regular review;

2 Article 10 which covers the establishment and the functions of the Subsidiary Body for Implementation (SBI);

3 Article 12 which deals with the obligation of each party to the convention to regularly communicate to the Conference of the Parties (COP) information relating to national implementation of the FCCC;

4 Article 13 on the possible establishment of a multilateral consultative process for the resolution of questions regarding implementation.

This set of provisions was the outcome of tense negotiations between, mainly, developed countries on the one hand and developing countries on the other (Doc 1.30). The former preferred relatively strong implementation mechanisms, whereas the latter opposed these. The developing countries emphasized their sovereignty with respect to national implementation of the requirements under Article 4.1 and rejected any interference of the COP in internal affairs, in particular with regard to energy, industry and transport policies. Consequently, the provisions on implementation were diluted during the negotiations. Even the idea of being obliged to regularly submit national reports to the COP – a relatively common instrument in international environmental law to review implementation at the multilateral level – was objected to by India, China and other developing countries. What was left was article 12 on 'Communication of information related to implementation'. The word 'report' obviously had the wrong connotation for the G77 and China. Nonetheless, developing countries accepted a limited reporting requirement, in exchange for a differentiation in obligations – the North's reporting requirements are stricter than those of the South – as well as financial compensation under Article 4.3. The latter provision states that industrialized countries shall entirely finance the South's communications to the COP through the Global Environment Facility (GEF).

An open-ended Subsidiary Body for Implementation (SBI) was also established with hardly any authority, however, although the Intergovernmental Negotiation Committee (INC) had discussed several far-reaching proposals on a strong Implementation Committee. But these proved to be unrealistic, again mainly due to G77 pressure. The SBI as established therefore only assists the COP in reviewing the communication of information by the Parties, but it cannot for example verify data within individual countries or assess the steps taken to implement the Climate Convention. Besides, the body is open-ended to 'government representatives who are experts' instead of open to experts *à titre personel.* This means that debates can be easily politicized. Furthermore, the SBI is formally closed to NGOs. Finally, debates on how to deal with a party's non-compliance by the COP remained unresolved. Proposals to have, for example, the Implementation Committee consult with and assist countries which do not comply with their obligations were still a bridge too far. What is more, the entire item was postponed for future deliberation on the basis of Article 13 of the FCCC, which is in fact an empty provision. Nonetheless, during COP1 in Berlin (1995) an ad hoc working group was set up to deal with the possible establishment in the near future of a so-called multilateral consultative process on implementation, compliance, and non-compliance.

B *Ego-perception*

To quote one NGO respondent on this specific case:

> "I think we had some influence on institutional issues such as the mechanism for reporting, the implementation body, dispute settlements, voting, participation of NGOs and others."

Two of his colleagues agree, but with regard to reporting and the Subsidiary Body for Implementation (SBI) only. But their references are – again – unspecified. They simply claim they pushed for the strongest provisions on national reports and an Implementation Committee possible, and that they had some limited and indirect influence on the outcome.

C *Alter-perception*

No single government respondent referred to the implementation mechanisms as an illustration of NGO influence. Therefore the above NGO claims are not confirmed by alter-perception.

D Causal analysis

GOAL-ACHIEVEMENT — Most NGOs in the climate change arena focused on the substance of the FCCC, but some dealt with the procedural and institutional affairs in depth as well, for example the Dutch Society for Nature and Environment (SNM) and the US-based Natural Resources Defense Council (NRDC). In addition, several CAN papers which commented on government draft proposals for a Climate Convention and which were circulated among delegates at INCs included non-government views on implementation mechanisms (e.g. Doc 1.45). Not surprisingly, NGOs opted for strong mechanisms under the convention. Their proposals included among others things: 1. the obligation for each party to annually report to the COP their progress in implementing their commitments under the FCCC; 2. the establishment of an independent Advisory Committee on Implementation which grants access to NGOs and has both an advisory and a compliance-oriented role (Doc 1.46). The latter function covers the authority to verify information from parties through fact-finding on the ground. It is, however, obvious that NGOs did not achieve these goals, given the far weaker provisions on reporting, the implementation body and NGO participation in the FCCC.

NGO INTERVENTIONS — Still, NGO pressure might have, just as in the previous case, prevented worse: a 'lowest common denominator' outcome based on the G77 position. But this is quite unlikely for a number of reasons. First of all, NGOs were barely players in the way they were in the case on targets and timetables. As they in general did not spend too much energy or resources on legal and institutional matters, the pressure NGOs mobilized for these procedural negotiations was limited. Secondly, NGOs had to influence the G77 and China to ensure stronger provisions in this specific case, as compared to the EC and the USA in the earlier one. To move the South's position, though, is an even tougher game, as the NGOs' constituencies in developing countries are far smaller than in the industrialized world. In addition, many NGOs supported or declared their solidarity with the G77 and China in the climate arena and were therefore reluctant to criticize these countries. Thirdly, the OECD was relatively united on institutional affairs and stood for some sort of multilateral review mechanism as a minimum (Doc 1.30). Some countries even came up with far-reaching, 'NGO-like' proposals during the first INCs (Doc 1.13). Besides, the OECD was obviously prepared to 'buy off' the South's commitment in this field through both differentiation and compensation of reporting requirements. Consequently, NGOs were not needed to ensure the minimum on implementation mechanisms. This was different in the third case, where both the OECD and G77 were internally divided: a minimum provision on targets and timetables was therefore not guaranteed and NGOs could, consequently, play a (minor) role.

The fact that NGOs did not impact the outcome at the end of the day was also implicitly admitted by one of the NGO respondents, who nevertheless in the first instance claimed influence. He said:

> "We regularly approached delegations with requests to introduce our views on implementation mechanisms in the negotiations and we succeeded in a number of cases. But if I look at the text which was finally adopted, then I will have to conclude that the language became extremely diluted during the final days and nights of tense negotiations. That is a pity, but not a surprise really."

In other words, NGOs were, according to this respondent, able to influence individual delegates, but their views were in the end not adopted by governments or taken over into the text of the Climate Convention. Therefore at best NGOs influenced the debate on implementation provisions (process influence), not the outcome, either directly or indirectly (political influence). But one may even question the possibility of process influence by NGOs in this specific case. Of course, the above respondent may be right that individual NGO representatives and delegates exchanged views on Articles 10 and 12, and NGO proposals may even have been considered by delegations here and there. But given the fact that no government respondent experienced an NGO impact on this topic (see alter-perception), NGOs *in general* are obviously not considered relevant to the negotiations and adoption of implementation proviions. Hence, process influence seems out of the question as well.

E *Extent of political influence*

Table 4.6 summarizes the scores in this case. Obviously, no impact of NGOs could be traced. Only the variable of PR (political relevance) yields a positive score, as the implementation mechanism of any treaty is a highly relevant topic. After all, this mechanism determines how far states are obliged to fulfil the provisions of the convention and to what extent they might be forced to do so by other parties and international courts. This topic therefore touches upon the sovereignty of states and, hence, their vital interests.

TABLE 4.6 *Scores on political influence in case 4 (Implementation FCCC)*

Counterfactual (C)	similar (0)
Goal-achievement by NGOs (GA)	no (0)
GA due to NGOs' own interventions (OI)	no (0)
Political relevance of topic (PR)	high (3)
Directness of impact (D)	no impact (0)
Extent of political influence: PI = C [(GA x OI) + PR + D]	none (0)

F Final conclusion

Only a few NGOs did focus substantially on institutional matters, including implementation mechanisms. Although it seems as if these organizations were sometimes able to contribute to the negotiations, the effects of these interventions on the institutional debate in general as well as on the treaty text are negligible. In short, none of the NGO claims could be confirmed here, neither by alter-perception nor by causal analysis. What is more: whereas NGOs had at least some (temporary) impact on the negotiation process of all previous cases, even process influence is excluded here. Obviously, both the debate on and the content of the implementation provisions were determined by the governments.

— 4.4 NGOs in the implementation phase (1992-1995) —

Since treaty formation and implementation differ, the nature and outcome of NGO lobbying and pressure differ. Whereas NGOs aim to influence the treaty text in the former policy phase, they try to influence its elaboration and interpretation during the latter. They will also try to monitor the ratification and implementation of the convention at domestic level. Furthermore, they hope to have an impact on the agenda, negotiations and decisions of the Conference of the Parties (COP). It should, however, be taken into account that many of the decisions taken by this governing body are not *real* decisions. Such decisions either postpone actions (for example 'the decision to take a decision at the next meeting of the COP') or refer to *interim* measures (for example a mandate for a working group). Therefore the influence of NGOs cannot be assessed with regard to many *end* products, as the scope of this study does not extend beyond the first COP. One should reflect on this, as NGO influence on an interim product (e.g. a mandate for a protocol) does not necessarily imply some impact on the end product (the protocol itself).

In addition, it should be taken into account that the FCCC is a framework convention. This implies that several provisions are still rather broad and vague and that the Conference of the Parties (COP) needs to elaborate on them. Although the Climate Convention defines the way in many instances, there is still a lot of open space with regard to its interpretation and putting it into operation. In consequence, parties will – again – wrestle over topics or words and the negotiations will often take the form of renewed *agenda-setting* (although we are in the implementation phase now). This phenomenon of policy phases that overlap was already highlighted in subsection 2.4.1, although it should be kept in mind that agenda-setting in the implementation phase is less open, given the fact that the agreed treaty text limits the possibilities of interpretation and elaboration.

4.4.1 Case 5: AOSIS protocol

A Introduction

The Intergovernmental Negotiation Committee (INC) decided to continue to meet after Rio to realize a prompt start of the convention, to bridge the gap between UNCED (1992) and the first Conference of the Parties (COP1 in 1995), and to prepare for the latter. It met six times (INC6 to INC11). However, the committee was divided into two camps on the question of how to proceed. Many just wanted to make the *current* commitments and mechanisms of the FCCC operational, while others favoured a process aimed at immediately *strengthening* the convention (in casu article 4.2(a) and (b)). The former camp was composed of the 'conservative' group within the OECD, the Russian Federation and most developing countries, in particular the OPEC group – backed by the oil lobby – and China. The arguments used to underpin their positions are well known. The second camp was composed of those who were disappointed by the final result of the climate negotiations: the 'progressive' industrialized countries, the Southern victim-states, the environmental NGOs observing the INC, as well as several members of the Intergovernmental Panel on Climate Change (IPCC). This second camp was convinced that the FCCC commitments were inadequate to handle the threat of global warming and they made a plea for the negotiation of new instruments on additional GHG emission reductions, for example: a reduction protocol, an amendment to the FCCC, or a decision to re-interpret the convention. Some even opted for the immediate of a protocol at the first Conference of the Parties (COP1).

Given these opposing views, the readiness among the signatories and other country participants in the INC to discuss the need for additional measures was extremely small just after UNCED. Many realized that this topic would revive all the political controversies of the last INCs before Rio and few were willing to deal with it again. Therefore many were in favour of focusing on apolitical issues, for example the technical side of putting the treaty into operation. Nonetheless, the topic of the (in)adequacy of the commitments did appear on the agenda of the INC. A few factors were decisive in this. First, the FCCC states in article 4.2(d) that the COP 'shall, at its first session, review the adequacy of subparagraphs (a) and (b) above'. This provision in the treaty could easily be mobilized by those who were in favour of further commitments. Besides, the INC had to reflect on this FCCC provision sooner or later, as it had to prepare for COP1. Secondly, Bush left the White House in favour of the Clinton/Gore administration a few months after UNCED. The new President had a more positive view on the environment and Clinton had for example criticized Bush for not accepting the stabilization target during the days of the Rio Summit (Doc 1.36). His victory created some negotiation space in the INC as well as more unity in the OECD group. The latter could now agree on a common position, which was that the commitments were inadequate. Thirdly,

Germany had offered to organize the first COP in Berlin. This country's view on the environment was also a positive one and, moreover, it was in favour of strengthening the convention. Therefore Germany was determined to have strong outcomes at COP1. At INC11 it tabled an element paper on a draft protocol for further commitments (Doc 1.10). This proposal included the stabilization target, the principle of negotiating further reduction targets for the period after 2000, and ideas on new policies and measures, for example on energy efficiency, renewable energy, traffic and transport, forests and GHGs other than CO_2. Fourthly, some developing countries shifted position, although OPEC and China continued to block the adequacy debate. Whereas nearly all G77 countries started the follow-up of Rio being fairly hostile towards new commitments under the FCCC on the basis of 'old' arguments (the fear for a future drag on their economic development), a group of them – including Brazil and India – ended up quite the reverse at COP1. They now slowly started to support negotiations on additional measures for industrialized countries, convinced that this would not immediately create precedents for them in the near future. Finally, the AOSIS countries took an important initiative. They tabled a draft protocol to strengthen the convention at INC11, the so-called AOSIS Protocol (Doc 1.37). This proposal included the Toronto target on carbon dioxide for industrialized countries, the obligation to adopt targets and timetables on other GHGs at a later stage, and a provision on technology transfer in order to mobilize the support of developing countries. In putting forward this proposal, AOSIS provided the parties with concrete materials to negotiate new commitments. INC11 finally decided to take this proposal – together with the German element paper – into consideration. However, these texts were *not* formally adopted before the Berlin conference. There was no basis upon which to do so given the objections of, mainly, China and oil-exporting countries. In addition, the OECD group opted for negotiations on a legal instrument – and not necessarily a protocol – to start *after* COP1.

In short, the debate after UNCED slowly shifted from purely technical and institutional topics to substantive ones on the (in)adequacy of current commitments and the need for additional protocols – or other legal instruments – to strengthen the convention. The AOSIS Protocol was one factor which contributed to this shift. What is more, the tabling of this proposal at INC11 forced the parties – including those which were reluctant to do so – to consider the need for new commitments at COP1. Had no such proposal been tabled, this debate would probably have ceased, if it ever had started.

B *Ego-perception*

More than half the NGO respondents – seven out of thirteen – claim that they impacted the AOSIS Protocol. Some refer to substantive contributions of NGOs to the content of the instrument itself, others to the mobilization of government support for this protocol particularly by Southern NGOs. Yet others refer to both: to direct influence through the drafting process itself and indirect influence through pressure on delegations. One respondent stated bluntly: "The AOSIS Protocol was written by environmental institutes and NGOs." Another respondent elaborated on the way NGOs mobilized support:

> "When we went to New York to attend INC11 we were told that the AOSIS Protocol was dead. That there was no G77 support for it and that AOSIS might not even introduce it. But some effective work was done, especially by the Southern NGOs. They explained that if developing countries did support AOSIS, it would not mean commitments for them. That was done a lot by Southern colleagues and subsequently you had positive intervention from India. India started to have a positive key role in it. Then ever more developing countries started to support the AOSIS Protocol. A lot of that, I think, was due to NGO work."

It should be mentioned that most of the seven respondents refer to indirect influence through the mobilization of support. In addition, they disagree on the question of whether NGOs contributed directly to the substance of the AOSIS proposal. Some believe so, others not.

C *Alter-perception*

Almost half the government representatives and officials – four out of ten – also believe that NGOs impacted the AOSIS Protocol. Two of these confirm that NGOs were involved as drafters. The other two agree that NGOs successfully mobilized support for it in the INCs. One respondent commented:

> "The AOSIS countries very much worked together with a couple of legal-type NGOs. And one milestone was closely related to NGOs, namely the AOSIS Protocol. It was even drafted by an NGO. Of course, I cannot say that officially, because it was submitted by a government on behalf of others, but we think we know who drafted it."

Another government respondent, who himself was involved in this AOSIS initiative, agrees with former view (although he does not want to be quoted). NGOs were definitely involved as drafters.

D *Causal analysis*

GOAL-ACHIEVEMENT — One may wonder whether NGOs achieved their policy goals with regard to the AOSIS Protocol. The verdict is not entirely. First of all, the protocol includes the Toronto target, whereas NGOs had preferred more stringent commitments to be adopted in the treaty or in a protocol (see case 3). True, NGOs dropped these demands at a later stage – just before INC9 in 1994 – and accepted the Toronto target as the policy goal to campaign for in the INCs (compare for example Doc 1.45 and Doc 1.53, Doc 1.56). The fact that this target was subsequently formally tabled by AOSIS can therefore be considered a form of goal-achievement. Secondly, NGOs had campaigned for an immediate adoption of a so-called 'Berlin Protocol' at COP1 in April 1995 (Doc 1.50, 1.70, 1.74). They entirely failed to achieve this goal. Governments refused to decide on a protocol at that stage and only adopted a *mandate* to start negotiations on a new instrument after Berlin (see case 8 below). What is more, the inclusion of a reference to the AOSIS Protocol in this Berlin Mandate was a very narrow escape from total failure. Finally, the AOSIS Protocol remained a *proposal*. It has not been formally adopted by the parties as a protocol to date, only as a document which 'should be included for consideration in the process' (Doc 1.8, p. 3). To conclude, NGOs did not achieve most of their goals with regard to the AOSIS Protocol, although the fact that it was formally tabled and considered may be called a minor success. As one NGO respondent put it: "We do not have any sure feeling on whether we will get the Toronto target or not, but the fact that we have moved so far means that the NGOs had some good impacts."

NGO INTERVENTIONS — But did NGOs themselves cause this 'success'? It seems they did, at least in part. There is no doubt that the *Foundation for International Environmental Law and Development* (FIELD) was involved – together with AOSIS representatives – in the drafting process of the protocol. As the organization itself notes (Doc 1.55, p. 2): "The AOSIS Protocol was drafted with assistance from FIELD to reflect the most politically feasible and environmentally progressive approach to strengthening the convention." This assistance was confirmed by several sources, including a representative of the small island states. However, some NGO respondents are obviously not aware of this, given the fact that they believe that the draft protocol was entirely an AOSIS initiative. Or they may consider legal organizations such as FIELD not to be NGOs. The latter is a possibility, but remained unclear to the researcher. Nonetheless FIELD is considered an NGO here, given its mission and staff. This organization tries to achieve progress in international environmental law, offers legal assistance in particular to the economically disadvantaged on a *pro bono* basis, and its staff consists of 'NGO-like' members. However, it is more than obvious that this organization fundamentally differs from, for example, Greenpeace International.

Besides being co-drafters of legal text, NGOs were also important pressurizers. They stimulated the small island states into raising their voices both in the INCs and COP1, resisting the pressure of China and the OPEC countries in the G77 to refrain from subgroup action, and undertaking specific initiatives, of which the AOSIS Protocol was one (Doc 1.66). Subsequently, nearly all NGOs under the umbrella of the Climate Action Network (CAN) supported this proposal and, moreover, they tried to mobilize additional government support. For example, Northern NGOs organized a European tour for AOSIS representatives to mobilize support for the protocol in the EU just before COP1 (Doc 1.75), while Southern NGOs such as the Bangladesh Centre for Advanced Studies (BCAS) and the Indian Centre for Science and Environment (CSE) mobilized support among key developing countries, such as India, to compensate for the 'negative' pressure on AOSIS from China, oil-exporting countries and the business lobby. That these efforts had some success is confirmed by several government respondents, including one connected to AOSIS. Finally, there was enough support in the INC to convince the small island states that they could safely table and stand by their proposal. Of course, it was not through the NGOs alone that support was mobilized. Other players – AOSIS itself, progressive OECD states, UN officials, IPCC – had roles as well. But NGOs definitely played their part.

There is a third way in which NGOs may have impacted the protocol debate, although this possibility is less thoroughly assessed here. At INC9 the German *Öko-Institut* had disseminated a working paper on several possible protocols among delegations (Doc 1.73). It included proposals on CO_2 targets, energy taxes, energy efficiency, renewable energies, and education and training. It seems as if Germany may have taken over some of the elements of this working paper in its own proposal, as one NGO respondent asserted. This claim, though, has not been verified.

E *Extent of political influence*

It seems plausible to conclude that NGOs had substantial influence on the content and tabling of the AOSIS Protocol (see table 4.7). Firstly, the counterfactual is considered different, although it is difficult to point at its substance. Maybe there would have been no proposal at all, maybe a different one, or maybe it would not have been tabled. Anyway, the input of NGOs is valued to show that the outcome would definitely not have been the same without them. Secondly, NGOs substantially achieved their (second-order) goals (tabling of AOSIS Protocol, inclusion of the Toronto target, consideration of it in the Berlin Mandate), although initial goals were not attained (stronger commitments, Berlin Protocol). Thirdly, the outcome was due to NGOs to a relatively large extent, in particular compared to the foregoing cases. FIELD co-drafted the proposal and functioned as a secretariat for the AOSIS, whereas Greenpeace and CSE gave advice. Besides, CAN as a whole supported the initiative. Fourthly, one has of course to take the status of the AOSIS Protocol into consideration. Given the fact that it is still a non-binding proposal, its political relevance is considered moderate rather than high. The topic in itself would justify the latter, as – again – vast state interests are at stake (compare case 3), but the legal status urges one to consider its relevance to be less. Fifthly, the impact of NGOs is definitely direct, as FIELD co-drafted the proposal. All in all, the score on political influence is 'substantial'.

TABLE 4.7 *Scores on political influence in case 5 (AOSIS Protocol)*

Counterfactual (C)	different (1)
Goal-achievement by NGOs (GA)	substantial (2)
GA due to NGOs' own interventions (OI)	substantial (2)
Political relevance of topic (PR)	moderate (2)
Directness of impact (D)	direct impact (2)
Extent of political influence: PI = C [(GA x OI) + PR + D]	substantial (8)

F *Final conclusion*

A wide range of NGOs – from Greenpeace to FIELD – was engaged in the adequacy debate. Part of their strategy was to assist the AOSIS in coming up with a good draft protocol to strengthen the convention. What is more, they were able, both directly and indirectly, to influence the content and tabling of this AOSIS Protocol. Therefore the extent of NGO influence was quite substantial. Although the AOSIS Protocol has remained a proposal so far, it fuelled the debate on further commitments at INC11 and COP1. Without this draft protocol, this debate would probably not have been started at all.

4.4.2 Case 6: joint implementation

A *Introduction*

Article 4.2 of the FCCC states that the so-called Annex 1 parties – the industrialized countries – may implement policies and measures 'jointly' (subparagraph a). This includes the aim of returning anthropogenic emissions to 1990 levels (subparagraph b). Furthermore, subparagraph (d) notes that the first Conference of the Parties (COP1) shall take decisions regarding criteria for joint implementation (JI). It is the preparation for these decisions at INC8, 9, 10 and 11 that provoked fierce debate among the parties. The background to the concept of JI has several aspects:

– Given the global nature of the problem, it makes little difference to the greenhouse effect *where* emissions are reduced or removed, either at home or abroad. Therefore countries should be able to jointly reduce or remove GHG emissions were it is easiest or most cost-effective.

– It is more effective to invest in environmental technologies and measures in less developed and, hence, less efficient economies than in highly developed countries. As the marginal costs of environmental gains rise proportionally with efficiency standards, it is less expensive to make such gains in the East or in the Third World rather than in the West, and these two regions are therefore the best places to reduce GHG emissions.

– Experiments with tradeable emission permits in the USA prove that dirty and clean sectors of industry can jointly decrease emissions. Some emit more substances and, hence, need to pay more money to get hold of the necessary permits. Others emit less and, consequently, may even earn money, as they can sell some of their permits to their dirty colleagues. Overall, emissions decrease, because there is a permanent incentive to invest in clean(er) technologies to save or earn money on permits. A similar global programme, which divides permits among big and small emitters of GHGs (respectively rich and poor countries), might have the same effect. The poor can sell their surplus of permits to the rich, earn money and invest in clean and efficient technologies. The rich, on the other hand, need to reduce their emissions in order to save money. The subsequent global net result is positive: fewer GHG emissions in general.
– The European Community (EC) had decided to adopt the stabilization target at the Community level – and not at the individual country level – in advance of INC1. This implied that EC members would jointly stabilize CO_2 emissions. Southern member states such as Spain, Portugal and Greece are still allowed to increase their emissions, while Northern EC members such as Germany and the Netherlands have to decrease theirs.

Aspects of all these ideas and experiences were also fed into the climate debate. Therefore joint and cost-effective measures were already an issue at the Noordwijk Conference (1989) and they have been advanced by the Dutch since then, mainly on technical grounds. However, the issue was brought into the negotiations by Norway at INC3, especially as the Norwegians were convinced that they would not themselves be able to individually achieve the stabilization target at home. Therefore they introduced the concept of a clearing house mechanism, under which parties should be able to co-operate and implement aspects of the climate treaty together (Doc 1.32). For example, a rich country might start a project on energy efficiency or sink improvement (e.g. reforestation) in a developing country and the resulting gains in terms of less GHG emission or more GHG removal should be shared and booked as *credits* under the FCCC. This idea seemed very attractive to all parties. Developing countries would be able to implement the treaty on the basis of additional foreign aid, including both money and technology, while rich countries would be able to implement their commitments abroad at lower costs. Furthermore, their own GHG emitters – transport, energy, industry – would be spared.

The Norwegian proposal was in the first instance warmly welcomed, also by many developing countries and even by some (American) NGOs, but it was considered incomplete and premature. A clearing house was therefore not established, but the idea that parties should be allowed to implement their commitments jointly remained, given the language in Article 4.2. Most NGOs, however, immediately criticized the mechanism of a clearing house and the concept of JI (Doc 1.48, 1.83). They feared that the industrialized countries would buy off their domestic commitments, and so prevent painful interventions at home. Therefore many NGOs perceived JI as a rich men's attempt to legitimize the emitting of ever more GHGs at home and unload the burden on the poor. Moreover, they feared that JI might eventually lead to *more* GHG emission, as *new* projects would be implemented under a clearing house, projects which might otherwise have been non-existent. Alternatively, such projects might have been planned *anyway* ('double counting'). NGOs also drew attention to the danger of fraud, as the contribution of JI projects to combating global warming is difficult to measure against what could have been done with the same amount of money at home or in other countries. Finally, they criticized the mainly American focus on sinks rather than emissions. The idea of compensation for (increasing) GHG emissions in the North through reforestation projects in the South was fiercely opposed by NGOs (Doc 1.86).

Although developing countries were initially in favour of the concept, most of them slowly turned against joint implementation as a general mechanism to put the convention into operation (Doc 1.9, 1.20). Whereas for example India welcomed the original Norwegian proposal at INC3, it criticized JI as a means to implement the FCCC at INC9 (Doc 1.30). The reason for this shift is threefold. First, Third World countries feared that JI would be mixed up with other commitments of industrialized countries under the FCCC, such as the obligation to provide financial resources and technology to assist developing countries to implement the convention. Moreover, they realized that JI might also lead to new forms of interference in their internal affairs. After all, a large number of JI projects would be executed in the Third World, whereas the nature of these projects and the credits would be determined and claimed by the donors (i.e. the industrialized countries). Secondly, since the South had become in favour of substantial domestic commitments for industrialized countries (see above), the NGO concern that JI might imply the 'softening' of these commitments became theirs as well. And thirdly, the G77 and China used the JI issue as a bargaining chip vis-à-vis the North. They knew that the industrialized countries were very much in favour of it, but were at the same time dependent on them for its realization. This situation made it possible for the South to make additional demands, either on JI or other topics.

Joint implementation was the second important issue at COP1 (after the Berlin Mandate). It was finally decided to establish a pilot phase. Projects should be among so-called Annex I countries – industrialized states – only, but developing countries might participate as well, on a voluntary basis (Doc 1.7). Moreover, crediting was excluded, which means that the reduction of GHG emissions realized abroad cannot be added to that realized at home under the obligations of the FCCC. Finally, the financing of JI projects should be additional to the outflow of money from the Global Environment Facility (GEF) and Official Development

Assistance (ODA). This outcome is obviously a compromise between North and South. After all, the mechanism *is* established under the FCCC, but so far subject to many restrictions.

Apart from this pilot project, several regional JI projects on GHGs are already in place. For example the USA and some Latin-American countries, as well as Eastern and Western European states, do have common projects, amongst others on energy efficiency and reforestation in countries such as the Czech Republic, the Russian Federation, Belize and Costa Rica (Doc 1.39, 1.87). An example of JI in the private sector is an initiative of the Dutch electricity generating companies. To compensate for a share of their carbon dioxide emissions, these companies finance reforestation projects around the globe through their FACE programme (Forest Absorbing Carbon dioxide Emissions). Projects were started in for instance the Netherlands, Malaysia, Ecuador and Uganda.

B *Ego-perception*

Again a relatively large group of NGO respondents claims some influence on this topic (6 out of 13). They refer to their critical assessment of the concept, which was to some extent taken over by the G77 and China. Some NGO representatives also claim success with regard to pressure on Northern countries to soften their position on JI. According to one respondent:

> "NGOs pushed very hard that joint implementation must not be used to address emission reductions abroad. Now we are leaving Berlin we can say that we are leaving in a way satisfied, because there will be no credits allocated in the pilot phase. And only Southern countries that are willing to join on a voluntary basis will do so. At least NGOs pushed the governments from the South not to accept JI entirely, because it is not a good concept. Well, it is good in a way, but it is not good the way developed countries want to use it: to reduce their emissions abroad and particularly target soft or cheaper options in the South. For example, sink projects must not be used in JI activities. Today, I think that developing countries leaving Berlin really appreciate the role NGOs have played."

All NGO respondents refer to indirect influence on the JI topic. No one claims to have been part of the drafting process of the final COP decision on the pilot phase.

C *Alter-perception*

Three out of ten government respondents are convinced that NGOs played an important role in the whole debate on JI, in particular with regard to clarification of the concept and its potential drawbacks for developing countries. Some believe that this was a positive contribution to the debate. Others, though, have a more negative view, as NGOs according to them were among those players who blocked an immediate and unrestricted adoption of the concept by the COP. A respondent from a key developing country said:

> "Environmental organizations very clearly helped us to understand that JI was looked at by the industrialized countries as a means to avoid reductions in their own countries. Of course, JI has significant benefits, but you have to be very careful, otherwise it becomes the wrong mechanism."

This respondent nonetheless continued as follows:

> "I do not think there was a great role for NGOs in Berlin. The level of delegations was much higher and the really tough negotiations and decisions were actually held and made by governments. This process took place among high-level government delegations in small contact groups, up to 6 o'clock in the morning. My country was involved, so I know. Germany locked up two groups in different offices throughout the night and Germany was going back and forth from one group to another to produce a compromise. In that process the NGOs were actually excluded."

In short, the NGOs' views on JI were, according to government respondents, background factors at best. They pushed the G77 and China to redefine their position, but governments subsequently proceeded at the COP as they wished.

D *Causal analysis*

GOAL-ACHIEVEMENT — Whereas NGOs were relatively united on the need for strong targets and timetables, they disagreed on joint implementation (JI). Therefore it is difficult to distinguish 'the' NGO goals. Roughly two groups can be classified. On the one hand some organizations – including the Environmental Defense Fund (EDF), Natural Resources Defense Council (NRDC), WWF and some Latin-American groups – considered JI as a potential mechanism to contribute to the implementation of the convention, but combined with substantial reduction requirements for industrialized countries (Doc 1.54, 1.76). Hence, most of these NGOs believed that JI might become a relevant model, but was not feasible yet given the then weak commitments. Some therefore even rejected a pilot phase, but others defended such a programme to develop and test the mechanism. A second group – including the Bangladesh Centre for Advanced Studies, Friends of the Earth, Greenpeace – was more fundamentally against JI (Doc 1.60, 1.65, 1.85). They stuck to their argument that industrialized countries should in the first instance restructure their economies and change their lifestyles to reduce GHG emissions, rather than reforesting a few areas in poor countries. The combined Climate Action Network (CAN), despite these different opinions, still managed to operate with reasonable unity on this topic. Its position was that JI could only be an option after 2005 – the target year of the Toronto target – and for only those industrialized countries which would achieve the 20%-reduction (Doc I.49). Furthermore, only a fraction of the commitments should be realized through JI and this should not include the export of nuclear technology. But *if* a pilot phase was established, according to CAN, it should be among Annex I countries and without crediting. This 'if word' makes clear that most NGOs were not very enthusiastic about a pilot phase. The fact that it nevertheless was established implies that they did not achieve their main policy goal. Moreover, the chance that JI will be entirely legitimized and institutionalized through the pilot phase is considerable, especially as it is pushed by all OECD members and many Latin-American countries. So NGOs may have delayed the unrestricted introduction of JI, but the fact that it *has* been introduced cannot be overlooked. On the other hand, they achieved some of their second-order objectives, given the substance of the pilot phase which was decided on in Berlin (no crediting; JI only for Annex I countries).

NGO INTERVENTIONS — It is obvious that the policy outcome – the COP decision on the pilot phase – is a compromise between advocates and opponents of JI. As their positions were more or less known in advance of Berlin, it does not seem that NGOs could have played a significant role there, as the compromise and decisions were made behind closed doors. Nonetheless, the role of NGOs *before* COP1 is difficult to deny. They were among the first ones to formulate criticism on joint implementation, views which were adopted by others at a later stage. Examples are several key developing countries which were in favour of the concept at an earlier stage. However, it is difficult to *prove* that NGOs were indeed 'the' initiators of JI criticism. Not all policy documents are available to reconstruct the debate in such detail. Nonetheless, the three government respondents referred to above, several government policy documents, as well as a second check among a few other key negotiators confirm the claims of the NGOs that they inspired and pressurized others – mainly the G77 – to criticize JI (Doc I.30). Therefore it is concluded that NGOs indirectly impacted the negotiation and decision-making process on joint implementation. Without their interventions before the Berlin summit, there would have been less opposition at COP1 and the policy outcome – the pilot phase – would probably have been more liberal on joint implementation criteria.

E *Extent of political influence*

Again, the counterfactual is considered different in this case. The restrictions on JI would probably have been fewer without NGO interventions in the period 1993-1995. But as JI is still legitimized and institutionalized by the COP decision and as most NGOs had not wanted to go that far, their policy goals were only to some extent attained. This was, however, substantially due to their own interventions, as NGOs were among the original critics of the whole concept and persuaded the G77 to (partially) adopt that criticism. The political relevance of this topic is definitely high. JI was a key topic at the INCs and COP1. OECD states perceived it as one of their main interests, because this mechanism enables them to fulfil some of their commitments abroad at relatively low cost. Some developing countries also saw it as their key interest, namely as a mechanism to get hold of additional technology and resources. And those developing countries which turned against JI considered it as an important bargaining chip. Last but not least, the political status of the JI pilot phase is definitely higher than that of for example the AOSIS Protocol, as the former concerns a binding decision whereas the latter is only a proposal to be considered. Therefore, the relevance of the JI topic is considered high. Finally, it is obvious that the impact of NGOs was indirect. They did not co-draft the COP decision. In short, NGOs had some influence on the topic of joint implementation in the period 1993-1995 (Note that PI = 6 almost amounts to 'substantial political influence').

TABLE 4.8 *Scores on political influence in case 6 (Joint Implementation)*

Counterfactual (C)	different (1)
Goal-achievement by NGOs (GA)	some (1)
GA due to NGOs' own interventions (OI)	substantial (2)
Political relevance of topic (PR)	high (3)
Directness of impact (D)	indirect (1)
Extent of political influence: PI = C [(GA x OI) + PR + D]	some (6)

F *Final conclusion*

Nearly all NGOs in the climate arena were engaged in debates on joint implementation (JI) after Rio, but they disagreed on the substance of it. Some were in favour (Environmental Defense Fund, Natural Resources Defense Council), others rejected the concept (Greenpeace, Centre for Science and Environment). Despite this difference, NGOs were able to impact the debate mainly through their arguments against it. In particular the G77 and China shifted their position on the basis of some of the NGOs' arguments. They therefore indirectly influenced the decision on the pilot phase made at COP1. NGOs were important *initiators* regarding the formulation of criticism on JI, as well as important *pressurizers* to make the pilot phase subject to many restrictions. But NGOs did not achieve all their goals. Therefore the extent of political influence is still limited.

4.4.3 Case 7: first review of national communications

A *Introduction*

Articles 4.2(b) and 12.2 imply that Annex I parties – the industrialized countries – shall periodically communicate to the Conference of the Parties (COP) information on policies, measures, and their effects. The first 'national communication' had to be made available within six months after the convention had entered into force and – according to Article 7.2 – be reviewed by the first COP. As the date of enforcement was 21 March 1994, the Annex I parties had to transmit their first communications to the secretariat of the Climate Convention in September 1994. Only 17 of the 36 did so. The secretariat made a compilation and synthesis of these national communications – except for two which arrived too late to be included – and distributed the review among all parties at the end of 1994 (Doc I.21). This review indicates that Annex I parties indeed designed policies and took measures in line with paragraph 4.1 of the FCCC, but that specific targets on GHGs would not be met in many cases. For example CO_2 emissions continued to increase in most countries. Therefore the review implicitly suggests that achieving the stabilization target would require additional measures.

NGOs did their own reviews of policies and measures in Annex I countries at the end of 1994 and included – beside the FCCC commitments – domestic targets. Evaluations of national plans in OECD countries and Central and Eastern Europe were published and reviewed several times by the Climate Action Network (Doc I.51, I.52). Their conclusions are more or less similar to those of the FCCC secretariat, but stated more bluntly. The NGOs believed that parties were not on track to fulfil their domestic and international commitments and therefore concluded (Doc I.51, p. 4):

> "(...) the First Communications fail to demonstrate that industrialized countries are taking the lead in addressing the climate change threat."

To add some figures, Greenpeace estimated on the basis of official country reports to the UN that the aggregated projected fossil fuel CO_2 emissions of OECD countries would increase to 4% above 1990 levels by the year 2000. Hence, agreed climate policies were – even in the light of the stabilization target – extremely deficient. Additional measures were therefore needed. Moreover, the NGOs criticized some of the national communications themselves, something the secretariat could not do. For example, the Netherlands and Denmark were criticized for adjusting their levels of CO_2 emissions. The former had experienced an extremely hot summer in 1990, the latter had imported more electricity than usual in the same year. Both facts *decreased* the domestic energy consumption and, hence, CO_2 emissions in both countries in 1990. As this implied that the stabilization target would be more difficult to achieve, both countries adjusted their figures. "Unacceptable", according to NGOs.

B *Ego-perception*

Two NGO representatives claim that the above CAN evaluations played an important role in the review of the national communications by the INC. One reference, though, is quite vague ("...NGOs contributed..."), the other is more precise and claims that NGOs delivered arguments to deepen the review debate:

> "The second most important input NGOs made apart from the AOSIS Protocol was the evaluation of the national reports. This helped a lot to make these reports transparent to people from other countries. I think that the thorough evaluation INCII made of the national reports was partly due to NGOs. Our independent evaluation made the committee ask the right questions and look to the right numbers. It was so much stuff to digest and in a way it was predigested by the NGOs."

C *Alter-perception*

No government representative nor any UN official mentioned the CAN evaluations as an example of NGO influence. The NGO claim is therefore not confirmed.

D *Causal analysis*

GOAL-ACHIEVEMENT — The NGO evaluations of national plans aimed at providing a detailed and complete assessment of national communications in order to contribute to – among other things – the debate on the implementation of agreed commitments (Doc 1.51). This goal was more or less attained given: 1. some references to these NGO evaluations in government policy documents (Doc 1.30); 2. the views of some government respondents who were – at a later stage – explicitly asked to assess the role of these NGO evaluations in the INC. Two respondents confirm that the NGO reports yielded additional information on countries' performances. Assuming that these documents and respondents are correct, the NGOs' above mentioned goal was achieved.

But NGOs mainly published their evaluations to make clear that additional measures were immediately needed and, moreover, to publicly embarrass (most) OECD countries for not achieving the FCCC's stabilization target. However, COP1 in 1995 did not make any authoritative statement on this lack of implementation. In fact, it only installed the procedures for an in-depth review process of first national communications to be concluded at COP2 in 1996 (Doc 1.5). So NGOs did not achieve their goals in the strict sense, as OECD countries were not formally urged to perform better, nor formally condemned for not achieving the stabilization target. Nonetheless, the OECD did become a bit embarrassed. They were accused of undermining the FCCC by some G77 countries as well as by the press (Doc 1.78, 1.79). But this had not so far led to a change in climate policies in the OECD.

NGO INTERVENTIONS — The NGOs seem, according to policy documents and additional interviews, to have attained their aim of contributing to the adequacy debate through their evaluations of national communications. This impact on the debate was, of course, caused by the NGOs themselves. However, it is difficult to speak of political influence in this specific case, as no obvious policy outcomes have so far been impacted by NGOs, except for the depth and transparency of the debate. Therefore it seems better to use the label 'process influence' here. What is more, the secretariat's review of national communication also implicitly drew the conclusion that OECD countries fell short on implementation and that additional measures were needed. Therefore the question remains whether the NGOs caused the general feeling in the INC that more should be done and the answer is probably not. As one of the former government respondents said:

> "Actually, the secretariat's evaluation of national communications had by far the most impact on the INC. The difference is that the secretariat has to be polite and plays according to the rules. The NGOs can do what they like. But substantially, there are no differences between the two statements, although the NGO document is more transparent and clear."

The latter aspect, though, refers to process influence again. In addition, no formal decision was made on the performance of OECD countries by COP1. Therefore it is hard to conclude that NGOs exerted some political influence on this topic. They co-framed a debate on something which was already obvious in advance, namely that OECD countries would generally not succeed in stabilizing or decreasing GHG emissions at home.

E *Extent of political influence*

It has become obvious from the above that there is no political influence to detect on the part of the NGOs in this specific case, although – again – they attained some of their goals. The counterfactual is therefore considered similar, especially as concrete policy outcomes are lacking. The topic of the first review of national communications is however considered very relevant, as it touches upon the core of the FCCC: whether OECD countries will take the lead in combating climate change and whether they are able and willing to implement the target of the convention.

TABLE 4.9 *Scores on political influence in case 7 (First review)*

Counterfactual (C)	similar (0)
Goal-achievement by NGOs (GA)	some (1)
GA due to NGOs' own interventions (OI)	no (0)
Political relevance of topic (PR)	high (3)
Directness of impact (D)	no impact (0)
Extent of political influence: PI = C [(GA x OI) + PR + D]	none (0)

F Final conclusion

The NGO evaluations of national plans in OECD countries contributed to the debate on the inadequacy of existing policies and measures on global warming (process influence). But these initiatives yielded some additional information for delegates and a bit more embarrassment for the OECD at best, especially as the FCCC secretariat's review of national communications implicitly came up with the same message. Moreover, no substantive decisions or bold conclusions on this item were formulated by COP1, and no OECD country has so far strengthened its policy. To conclude, NGOs cannot be credited with political influence in this case.

4.4.4 CASE 8: BERLIN MANDATE

A Introduction

The first Conference of the Parties to the FCCC (COP1) convened in Berlin from 28 March to 7 April 1995. On the agenda were issues such as the rules of procedure for the COP, matters relating to commitments, the financial mechanism (GEF), appointment of a permanent secretariat, and implementation (Doc 1.3). The fifth item on the agenda – matters relating to commitments – was the most controversial, particularly the "Review of the adequacy of article 4.2(a) and (b), including proposals related to a protocol and decisions on follow-up". The parties – including the USA – had agreed that the obligations under Article 4.2 were inadequate to meet the ultimate objective of the convention. However, opinions differed on when and how the commitments should be strengthened. The AOSIS – and NGOs – had pressed for the adoption of its draft protocol in Berlin, but that option was not available given the opposition of many G77 countries, most OECD countries and the business lobby (Doc 1.30, 1.42). A mandate to start negotiations on further commitments *after* Berlin was the most realistic possibility. Furthermore, parties disagreed on the legal instrument to be negotiated – an amendment to Article 4.2 or a protocol – and on the way forward for the process. For example, should COP1 already decide on possible targets and timetables?

The positions when the delegates arrived in Berlin were as follows (Arts and Rüdig, 1995; Doc 1.2). The OPEC countries intended to block any mandate. The JUSCANZ group – Japan, the USA, Canada, Australia, New-Zealand – seemed to be willing to negotiate a mandate, but only one which would be very open-ended and not refer to possible reduction targets and timetables. Moreover, developing countries should be subject to binding commitments after the year 2000 as well. Not surprisingly, the G77 and China opposed this JUSCANZ position and, hence, they were reluctant to accept any mandate for further negotiations. The AOSIS countries asked for an immediate adoption of their protocol. The EU, next, aimed at the adoption of a mandate in Berlin, which would contain some guidelines on policies, measures, reduction targets and timetables at best (Doc 1.1). Furthermore, Europe also did not exclude the possibility of new commitments for non-Annex I countries after the year 2000. Finally, the CEE block – Central and Eastern Europe – while silent, was divided on the issue, with Slovakia on the 'progressive' and the Russian Federation on the 'conservative' side.

These conflicting positions made a successful outcome at Berlin very unlikely, and indeed, there appeared to be deadlock in the negotiations for days. Nevertheless, the parties agreed on a Berlin Mandate on the last day of COP1. The decision reads as follows (Doc 1.8, p. 1):

> "The COP (..) agrees to begin a process to enable it to take appropriate action for the period beyond 2000, including the strengthening of the commitments (..) through the adoption of a protocol or another legal instrument."

This mandate does not refer to specific targets and timetables, although the process will aim "to set quantified limitation and reduction objectives within specific time frames, such as 2005, 2010 and 2020". The negotiations will not introduce new commitments for developing countries, although the Berlin Mandate did reaffirm the existing and general commitments under Article 4.1 to which all parties – including developing countries – are subjected. Moreover, the AOSIS Protocol will be included for consideration in the process.

The breakthrough in Berlin became possible after India, China, Brazil and the USA shifted their position. India, Brazil and China had, firstly, to accept a split in the G77 between OPEC and non-OPEC countries. The latter group became referred to as the G72. Secondly, this group had to be convinced that a new protocol, or any other instrument, would not impose new commitments on them, although they also had to accept that the Berlin Mandate reconfirmed Article 4.1. Then a group of key developing countries – including China, India, and Brazil – delivered what came to be known as the 'Green Paper' at the end of the first week (Doc 1.11). It urged the COP to begin consultations on the strengthening of the commitments for Annex I countries, based on among other things the AOSIS Protocol. Consequently, the G77 block split and OPEC remained rather isolated. Although the OECD initially objected to the Green Paper, it nonetheless created room for manoeuvre, especially as many elements in the G72 proposal and the EU draft for a mandate were similar. Then the USA changed its position at the end of the second week. Although Clinton and Gore were already in favour of mentioning possible reduction targets and timetables in the mandate, the US delegates seemed mainly to listen to the Republican Congress and the coal lobbyists in Berlin. This, however, changed when Undersecretary of State for Foreign Affairs Timothy Wirth arrived at COP1 in the second week. Finally, a compromise among the G72, the EU and the USA could be reached, probably also due to external pressure.

B Ego-perception

Five out of thirteen NGO respondents perceive themselves as important causal factors in the breakthrough in Berlin. They believe they were able to isolate the so-called *dirty triangle* – JUSCANZ, OPEC and the coal industry lobby – and persuade the G72 to support a possible mandate. Furthermore, they pressurized the EU to hold its position on the need for a strong mandate in Berlin and pressurized the USA to move away from its more conservative position. Finally, they claim they had a substantive input in the Green Paper as well as the Berlin Mandate. One respondent refers to several of these aspects:

> "In all my years of involvement, Berlin was a high point of NGO influence. I see the creation of the Green Group being intimately linked to the role of NGOs in the preceding two years or so. INCII was a bit of a watershed. For the first time NGOs were able to convince a number of developing countries that they should withdraw their opposition to further commitments for developed countries. This led to the situation in Berlin where the Green Group was able to form and separate from OPEC. There are also several pieces of text in the Berlin Mandate which I think stem from NGOs. The reference to enhancing the Article 4.1 commitment of developing countries came from NGOs. It was a way of softening the 'no new commitments' position of the G77. Its acceptance by the G72 is politically significant, but had to be explained by NGOs to many OECD delegations. The US government for example failed to appreciate the significance of that until it was explained to the highest level of the delegation. So overall I think the NGOs had a very big impact in Berlin. I doubt whether the outcome would have been so strong without the NGOs there."

Obviously, the NGO respondents claim both direct and indirect political influence on the Berlin Mandate which was adopted at COP 1.

C Alter-perception

Surprisingly, nearly all government respondents and officials – 8 out of 10 – confirm that NGO pressure had an impact on the Berlin outcome. According to one of them:

> "NGOs had a very good performance, both in terms of their participation and parallel forums. Their presence has in a way made sure that the COP actually achieved something. Before Berlin it was hard to resolve the issue on the enhancement of the commitment of Article 4.2. But NGO presence, papers, documents, and media exposure was of some help to the process."

However, most of these respondents also refer to Germany playing a similar role. After all, the host of the Berlin conference did not want it to be a failure and put a lot of pressure on those dragging their feet in the negotiations. Moreover, the role of NGOs was limited, according to some of the government respondents, as the final negotiations on the Berlin Mandate took place behind closed doors. One even denies that NGOs could play any substantial role in Berlin, due to this 'closed-door diplomacy'. But all in all government respondents are positive on the indirect political influence of NGOs. A direct impact on texts which were adopted by governments in Berlin is, however, not referred to.

D *Causal analysis*

GOAL-ACHIEVEMENT — Most accredited NGOs at COP1 invested nearly all their resources in getting at least a Berlin Mandate adopted. Obviously, they achieved their goal with respect to that. What is more, they were extremely happy and congratulated each other when the agreement was finally reached, as the researcher could observe. This was however behind closed doors, as they publicly criticized the substance of the mandate as being far too weak. After all, NGOs did *not* get what they had preferred at an earlier stage: the adoption of a reduction protocol at COP1 (Doc 1.50). They failed to get an early protocol, but may have – again – prevented worse, such as an 'empty' mandate or no mandate at all.

NGO INTERVENTIONS — NGOs claim they co-determined this outcome by pressurizing both the North and the South to compromise and, secondly, by delivering text elements for the Green Paper as well as the Berlin Mandate. To start with the second claim, it could not be confirmed by any source available that NGO representatives were involved – as co-drafters – in the formulation of these documents, either as official delegates or lobbyists. Therefore it remained unclear whether NGOs *directly* affected the substance of these documents. *Indirect* NGO influence on the mandate seems on the other hand more likely. It is acknowledged by both sides – given ego-perception, alter-perception and additional interviews – that NGOs played a part in the setting up of the compromise, mainly as mediators between the 'progressive' delegates of the North and the South. As NGO representatives come from both blocs and operate in one network at the same time, they know the political arguments and cultural sensitivities on both sides and can explain them to the other party. To that end, Northern NGO representatives pushed the key 'progressive' countries of the North to accept the fact that additional measures for developing countries were unacceptable at this stage. At the same time Southern representatives of NGOs had to convince the key 'progressive' countries of the South that industrialized countries were indeed willing to exclude additional commitments for developing countries, but that a reference in the mandate to general commitments already adopted under Article 4.1 of the FCCC was a prerequisite for getting a compromise. In this process of trust-building among 'progressive' countries on both sides NGOs indeed played, according to observers, a useful role, although other players such as (German) country delegates and UN officials were involved as well. Consequently, the 'conservatives' became ever more isolated, which made the final compromise possible.

NGOs also mobilized general pressure on the delegations through the media. As COP 1 in Berlin was a huge media event and portrayed – at least in Western Europe – as the follow-up to Rio, the exposure of the entire event was enormous. The NGOs' message that COP1 had to produce a mandate at least was therefore heard not only in the conference building, but also in the outside world. It soon became obvious that countries had to come up with something, otherwise the summit would be perceived as a failure by many, if not all, activists and journalists. In addition, large groups of more radical NGOs organized parallel meetings, protest marches and sit-ins in Berlin (Doc 1.40). The effects of these initiatives on the formal negotiations and the decision-making were probably limited, although they may have contributed to the general feeling among country delegates that "the world is watching you".

E *Extent of political influence*

Without the intervention of the NGOs, the outcome would probably have been less, maybe even no Berlin Mandate at all. Therefore the counterfactual is considered different. Again, one should be careful not to exaggerate, as the main pressure on the delegations to proceed and adopt a mandate came from Germany. The host country could not afford to have a failure. The NGOs' role was therefore additional, mainly focusing on getting the South on board. Besides, the impact of NGOs was indirect at best. A number of their goals could not be achieved, such as clear references to targets and timetables in the mandate, or their initial wish to adopt a binding protocol in Berlin. The political relevance of this topic was definitely high. Again, we deal with possible obligations for industrialized countries to reduce their GHG emissions – and, hence, restructure their economies – here. Huge interests were at stake.

TABLE 4.10 *Scores on political influence in case 8 (Berlin Mandate)*

Counterfactual – policy outcome without NGO interventions (C)	different (1)
Goal-achievement by NGOs (GA)	some (1)
GA due to NGO interventions (OI)	some (1)
Political relevance of topic (PR)	high (3)
Directness of impact (D)	indirect (1)
Extent of political influence: PI = C [(GA x OI) + PR + D]	some (5)

F Final conclusion

The main outcome of the first COP was the Berlin Mandate. This mandate enabled parties to immediately start negotiations on additional instruments to reduce GHG emissions after the year 2000. The role of NGOs in Berlin was quite similar to the one played at INC5 just before Rio (see case 3). Due to media exposure and protest activities, NGOs could mobilize additional pressure on the delegations to come up with a substantive result. In addition, they mediated between those rich and poor countries which were willing to negotiate on such a mandate, thus isolating the blockers (OPEC countries, Russian Federation, coal lobby). Of course, other players contributed to this process of mediation as well, and probably more than NGOs. A good example is the host Germany. Finally, NGOs claimed to have impacted the wording of the Berlin Mandate, but such a direct influence could not be verified. Hence, NGOs exerted some indirect political influence.

— 4.5 An overview of the eight cases —

Table 4.11 summarizes the conclusions on the eight cases concerning the FCCC. Obviously, most non-government respondents believe NGOs were influential on the AOSIS Protocol. Most government respondents, however, consider NGOs successful with regard to the Berlin Mandate (see shaded cells in table 4.11). Overall, ego- and alter-perceptions yield relatively higher scores in the implementation phase (cases 5-8) than in the formation phase (cases 1-4). However, this difference may be an artificial. First of all, 6 out of 13 NGO respondents were *not* involved in the INCs before UNCED in 1992, so they could not answer questions on the formation phase. Therefore it is no surprise that ego-perceptions on the implementation phase yield higher scores, as more respondents were able to reflect on this phase. But this is not the case for alter-perception, as all government respondents had been involved in the climate arena since 1990, and yet we see a similar pattern of higher scores in the implementation phase there. Secondly, the mechanism of temporal distortion may be operating. The formation phase covers the period 1990-1992, whereas the implementation phase refers to the period 1992-1995. The latter is therefore more easily remembered and reflected upon by respondents who were interviewed in the period 1994-1996. Thirdly, a specific question on the role of NGOs in COP1 was part of the interview guide. That given fact may explain the relatively high scores in case 8. To sum up, one should be careful in drawing conclusions based on the different scores of ego- and alter-perception in table 4.11.

Given the outcome of the causal analysis as shown in table 4.11, half of the NGOs' claims on political influence could be confirmed. Obviously, NGOs overestimated their own impact. The same applies to government respondents, who also perceived NGOs as being influential in cases 1 and 2. This, though, could not be confirmed by the causal analysis. The cases which still yield positive outcomes after this analysis are (see shaded cells in table 4.11):

- the inclusion of the (diluted) stabilization target in Article 4.2 (case 3);
- the substance and tabling of the AOSIS Protocol (case 5);
- the restrictions on joint implementation (JI) to date (case 6);
- the adoption of the Berlin Mandate (case 8).

TABLE 4.11 *Overview of the eight FCCC cases*

FCCC	*Ego-perception (n respondents)*	*Alter-perception (n respondents)*	*Causal analysis (confirmation?)*	*Extent of political influence*
1 Objective	n = 3	n = 3	-	none
2 Principles	n = 3	n = 2	-	none
3 Target	n = 2	n = 2	+	some
4 Implementation	n = 3	n = 0	-	none
5 AOSIS protocol	n = 7	n = 4	+	substantial
6 Joint implementation	n = 5	n = 3	+	some
7 First review	n = 2	n = 0	-	none
8 Berlin Mandate	n = 5	n = 8	+	some

The first topic refers to the formation phase, the other three to the implementation phase. They were influenced by NGOs to some extent only, except for case 8 on the AOSIS Protocol which was substantially impacted. The reason for this is that this case was *directly* affected, as NGOs were co-drafters of the proposed legal text. In other cases, NGOs impacted the policy outcome indirectly only, for example by mobilizing pressure on delegations to compromise or by delivering arguments and views which changed those of others. No NGO proposal or text elements were, however, taken over literally and formally adopted by governments in these cases.

The fact that NGOs mainly exerted *some* influence on these topics implies that other political players have been (more) influential. This is no surprise, as several of those players have better positions and more resources at their disposal to determine policy outcomes. For example governments are the participants who formally decide on international policies, not NGOs. Although the influence of the former is not precisely assessed in this chapter, it is obvious that among others the USA (case 3), the AOSIS countries (case 5), the G77 and China (case 6), and Germany (case 8) impacted the above topics far more than the NGOs did. In addition, other players such as international organizations, the business lobby and scientists may have been (more) influential than NGOs. All were for example involved in case 3 (target of the FCCC). In particular the coal lobby is generally perceived as having blocked the literal adoption of the stabilization target. This lobby exerted its influence mainly through the USA delegation and OPEC countries. All in all, other political players were generally more influential than NGOs, although the division of political influence differs on a case-by-case basis.

Half of the NGOs claims could, however, not be confirmed. With regard to the objective of the FCCC (Article 2), the principles (Article 3), the provisions on implementation (Articles 7, 10, 12 and 13), and the first review of national communications, the NGO interventions were not effective in terms of policy outcomes. Nonetheless, they were generally able to contribute to the debates on these topics, but not in such a way that their views made a difference in decision-making (process influence). But NGO arguments on implementation mechanisms seem not to have played a role in the negotiations on the Climate Convention (case 4). Here NGOs lacked any impact, either on the process or on the product.

Apart from the NGO claims which could not be confirmed, there are probably several other topics NGOs did not impact either. But as respondents were aware of that, they did not claim success. An example is the financial debate which was so dominant in the climate arena. In particular North and South clashed on the amount of extra money which should be transferred from the rich to the poor, as well as on the financial mechanism through which this assistance should be provided (see subsection 4.2.3). Many NGOs participated in this debate, but their arguments were not heard or coincided with those of the G77 and China. Therefore NGOs could hardly play any additional and useful role in this debate. In other words, there are probably more NGO 'failures' to point to than the 4 cases referred to above.

— 4.6 Conclusions —

The assessment of the NGOs' political influence, which was based on the so-called EAC Method, produced the following conclusions:

1 NGOs had no *direct* impact on the FCCC text, despite their claims about for example Article 2 (objective). In other words, none of their text proposals were adopted by governments and included in the treaty, at least not proposals which are covered by this study.

2 NGOs had some *indirect* influence on the specific commitment in Article 4.2. Without the interventions and pressure of NGOs, the outcome would probably have been weaker. This was also claimed with regard to the principles of the FCCC (Article 3) and provisions on implementation (Articles 7, 10, 12, 13). But these NGO claims turned out to be invalid.

3 With regard to the period after the adoption of the treaty text (1992-1995), NGOs impacted the AOSIS Protocol, the Joint Implementation Pilot Phase, and the Berlin Mandate. They however failed to have an impact on the First Review of National Communications of industrialized countries.
4 In general, NGOs exerted *some* political influence. An exception is the AOSIS Protocol. Here the NGOs are credited with *substantial* political influence.
5 NGO representatives definitely overestimated their own political influence. Only 50% of their claims could be confirmed on the basis of further assessments (4 out of 8 cases).
6 NGOs nonetheless contributed to the debates in the climate arena on all the above topics, except for one. However, such *process* influence should be sharply distinguished from *political* influence, as a (temporary) contribution to a debate is no guarantee that the policy outcome will be more in line with one's goals.

CHAPTER 5

NGOs and the UNEP Convention on Biological Diversity

— 5.1 Introduction —

This chapter is similar in arrangement to chapter 4. The objective is to assess the political influence of NGOs, this time however on the *Convention on Biological Diversity* (CBD). Again, the content and policy cycle of the CBD will be described, and the EAC Method will function as a format to assess the various cases. The number of cases, though, is ten compared to the eight of the previous chapter. This difference is due to the selection procedure based on ego-perception (see chapter 1). To repeat, NGOs in the biodiversity arena simply claimed more topics as influenced by them than their colleagues in the climate arena. The government, intergovernmental and non-government policy documents which will be referred to in the text below are listed in Appendix II of this book (page 346). These references will – in accordance with the system used in the previous chapter – be shown as (Doc II.6) or (Doc II.83, p. 45).

— 5.2 The UNEP Convention on Biological Diversity (CBD) —

5.2.1 The issue of biological diversity

The notion of *biodiversity* became popular only recently. The CBD defines it as follows:

> "*Biological diversity* means the variability among living organisms from all sources, including, *inter alia*, terrestrial, marine and other aquatic ecosystems and the ecological complexes of which they are part; this includes diversity within species, between species and of ecosystems."

In short, biological diversity refers to the richness of genes, species and ecosystems. Therefore the notion brings together different areas of concern: 1. the problems relating to genetic erosion and genetic engineering; 2. the endangering and extinction of species; 3. the destruction and loss of ecosystems. The problem of genetic erosion was formulated in the area of agriculture in the 1970s (Crucible Group, 1994; Pistorius and Van Wijk, 1997; Tolba and El-Kholy, 1992). The use of only a few high-yielding varieties (HYVs) of food crops all around the globe raised concern about the small genetic base of modern agriculture. Such a small basis might, according to scientists, decrease the ability of species to adapt to changing circumstances and, consequently, increase the vulnerability of seeds to environmental changes (including new diseases). Scientists therefore continuously warned that food security might be threatened if agriculture continued to rely on only a few HYVs. Modern agriculture should modify its practices and use a greater diversity of seeds including traditional varieties, which are generally less productive but more resistant to environmental change and disease. This message – initially formulated by concerned scientists between 1968 and 1970 – was also heard by the Food and Agriculture Organization (FAO) of the United Nations (UN) and this organization initiated programmes to address the concern. The environmental movement, in turn, added the threat of genetic erosion to their critique on the Green Revolution, the modernization of agriculture in the Third World (Pereira, 1993). This 'revolution' was considered to favour the strong – the rich land owners – and constrain the poor – the small farmers. Moreover, it was considered to have a negative impact on the environment, given the huge quantities of pesticides and fertilizers used. Now it was also considered to be harmful from a genetic point of view.

Another problem related to genetic diversity is genetic engineering. Genes are the raw materials for biotechnological innovations nowadays, for example in agriculture and the pharmaceutical industry (Juma, 1989; Crucible Group, 1994; Sanchez and Juma, 1994). Prophets of this new technology believe that it will yield many new and useful products for humanity, such as better food crops and medicines, and make billions of dollars in the coming decades. On the other hand, the potential disadvantages of biotechnology are also recognized. For example, many environmental organizations and concerned scientists believe that genetically modified organisms (GMOs) might threaten other species or human health (Regal, 1994). Some even hold that society as a whole might be in danger, given the possibility of modifying *human* genes, and hence human beings, as well. In addition, the patenting of biomaterials such as genes, as well as the establishment of test fields for modified plants around the world by (mainly) Western industry, have become hotly debated issues.

The second area of concern – the extinction of species – was already highlighted by conservation groups at the end of the previous century (Caldwell, 1990; McCormick, 1989). In particular the disappearance of birds was emphasized. This concern was extended to other species during the course of this century – elephant, rhinoceros, tiger, bear – and many international organizations were founded, for example the World Conservation Union (IUCN) in 1948 and the World Wide

Fund for Nature (WWF) in 1968. The former organization is a so-called *hybrid* NGO in which both conservation groups and governments from all around the world co-operate. Its aim is to design multilateral programmes and laws as well as guidelines for national policies, to promote the conservation of nature. A worldwide IUCN initiative which just preceded the Biodiversity Convention was the World Conservation Strategy in 1980 (Dankelman *et al.*, 1981; McCormick, 1989). This strategy urged individual governments to increase conservation objectives and start programmes on the protection of nature. A successor of this strategy was IUCN's (and others') Global Biodiversity Strategy in 1992. This new strategy put more emphasis on sustainable use of biological resources within the context of conservation objectives (Glowka *et al.*, 1994). The WWF is known for its efforts to save specific species, such as the panda and the elephant, although this organization recently has started to focus on environmental and development issues as well. WWF raises money to finance conservation projects (Doc II.79). Furthermore, it tries to raise awareness on the subject among the general public.

Although most conservation groups focused on individual species in their early years, they extended their policy and programmes to entire ecosystems. After all, it is of no use saving a few samples of a species, when the ecological preconditions for their survival disappear at the same time. In other words, the focus shifted from species to habitats and ecosystems. This shift was also due to the growing popularity in the 1960s of the ecological sciences. Nature was considered to be an ecological web on which all species – including man – are dependent. Furthermore, the diversity of species was considered to be a good in itself and related to the stability of ecosystems (Odum, 1971). Although this 'diversity-is-stability' thesis was criticized in the 1980s, nature should, according to many conservationists, be as diverse as possible. Finally, the shift in orientation – from species to ecosystem – can also be linked to the fact that the disappearance of entire ecosystems, such as the tropical rainforests, was addressed by ever more NGOs (Humphreys, 1994; Kolk, 1996).

A second shift in focus among the conservation groups concerns the *integrated* approach. Whereas they tended to focus entirely on species other than man in the past, the link between nature and society has become emphasized recently (Adams, 1991; Redclift, 1984, 1987). Groups found that it was difficult to conserve nature without addressing human issues, in particular in densely populated areas such as the Netherlands, or in Third World countries where local communities may be very dependent on protected areas. Leaving man out of the reckoning resulted in bad outcomes, for example in small and marginalized areas of nature or in non-co-operation by locals. Therefore conservation programmes had to be replaced by integrated policies which focus on the sustainable use of biodiversity as well. For example agriculture and nature conservation were linked.

All the above aspects – genetic erosion, genetic engineering, extinction of species, destruction of ecosystem, and man-nature linkages – were fused into one concept: biodiversity. And into one treaty: the Convention on Biological Diversity (CBD).

5.2.2 The content of the CBD

The CBD is a comprehensive framework convention, comprehensive because it addresses all aspects of biodiversity. This will become obvious if one reads the objective of the treaty in Article 1:

> "The objectives of this Convention, to be pursued in accordance with its relevant provisions, are the conservation of biological diversity, the sustainable use of its components and the fair and equitable sharing of the benefits arising out of the utilization of genetic resources, including by appropriate access to genetic resources and by appropriate transfer of relevant technologies, taking into account all rights over those resources and to technologies, and by appropriate funding."

Given this comprehensiveness, the treaty formulates measures on many items in several articles. For example parties to the convention are urged to:

- establish a system of protected areas and promote the recovery of threatened species (so-called *in-situ* conservation);
- establish facilities such as herbaria and gene banks (*ex-situ* conservation);
- use components of biological diversity in a sustainable way;
- integrate biodiversity considerations in national decision-making;
- adopt economically and socially sound incentives for the conservation and sustainable use of biological resources;
- promote research and public education;
- assess and minimize adverse impacts on biodiversity;
- facilitate access to genetic resources and (bio)technology for other parties;
- share with countries, indigenous groups and local communities which provide genetic resources the benefits which arise from the utilization of these resources.

Furthermore, developed countries will contribute to a fund through which the poor countries are enabled to implement the convention.

The CBD is also a *framework* convention. This implies two things (Glowka *et al.*, 1994; Schermers, 1985). First, the provisions of the treaty are not so much precise obligations as well overall policy goals. Specific targets are therefore not set (e.g. a quantified norm on the extent of protected areas in each country). Secondly there is a possibility for the parties to the framework treaty to elaborate more specific rules and targets at a later stage, for example by amending the treaty or by adding protocols to it. Hence, the obligations in the CBD are rather broad and need to become more precise during the implementation phase. However, one may question whether the CBD indeed formulates *obligations*, as most articles are introduced by the qualifier: 'Parties shall, as far as possible and as appropriate'. In other words, parties need to implement the convention for as far as they are able (or willing) to do so. This implies that there is a great deal of flexibility with regard to implementation. Some believe this is an advantage, as the parties to the convention are far from equal. But it might be a disadvantage as well, because one can always escape from any obligation with the help of such a qualifier. What is more, most provisions in the CBD focus on the national level. This makes intergovernmental control and review of implementation rather difficult, as states are sovereign entities within the world community. It is no coincidence that this sovereignty principle is confirmed in the CBD several times.

The CBD also contains a number of provisions on institutional matters. It establishes a Conference of the Parties (COP), which is the governing and review body of the biodiversity treaty and which will meet regularly. It also establishes a secretariat to deal with the daily activities and a Subsidiary Body on Scientific, Technical, and Technological Advice (SBSTTA) which provides all other relevant bodies with advice. Furthermore, the CBD contains provisions on reporting (parties have to report their domestic implementation of the CBD to the Conference of the Parties), on the settlement of disputes, on the adoption and amendment of protocols and annexes, on voting procedures and on ratification.

Although both the Biodiversity and Climate Conventions are framework conventions, some differences are striking. In general, the CBD contains more detailed measures on several topics than the FCCC. Therefore the number of articles of the former exceeds those of the latter (42 and 26 respectively). However, there is one exception: the target on greenhouse gas emission reductions in the FCCC is more detailed than any of the measures in the CBD. Furthermore, similar articles, for example on principles, are very differently elaborated on in terms of substance and length in the two treaties. This may have been caused by the fact that different country delegates have negotiated these instruments in different UN bodies, although there was still some overlap in personnel, but it is probably caused by the different nature of the two issues as well. Three aspects seem important here. Firstly, climate change has to do with air pollution, whereas biodiversity loss has to do with (over)exploitation of biomaterials and damage to ecosystems (Boersema *et al.*, 1991; Bouwer and Leroy, 1995; Tolba and El-Kholy, 1992). These terms cover different kinds of environmental problems and urge countries to design different policies. Secondly, climate change is a so-called *sink*-related issue, whereas biodiversity loss is a so-called *resource*-related issue (Weale, 1992). The atmosphere is used as a common sink to dispose of our greenhouse gases, whereas biodiversity is used as a common resource from which useful materials can be extracted. The former leads to a struggle between the (main) polluters of the atmosphere on the one hand and the potential victims of global warming on the other; the latter to a struggle between the owners and the users of biological diversity. The division of (potential) costs and benefits among the parties is therefore not the same in these arenas. Thirdly, global warming refers to processes in the higher atmosphere, which is one of our *global common resources*, whereas biodiversity loss is mainly experienced within the *national jurisdiction* of states (Glowka *et al.*, 1994; Mintzer and Leonard, 1994; Sanchez and Juma, 1994). Separate legal questions therefore come into play in the climate and biodiversity arenas. To conclude, the above three dissimilar characteristics of the two issues have probably led to different negotiation processes and, consequently, to different outcomes.

5.2.3 The policy cycle of the CBD (1980-1997)

A Issue-raising and agenda-setting (1980-1990)

The origin of the biodiversity treaty goes back to recommendations of the 1984 meeting of the General Assembly of the IUCN (Burhenne-Guilmin and Casey-Lefkowitz, 1992; Glowka *et al.*, 1994; Sanchez and Juma, 1994). It was recognized that a global instrument was needed, irrespective of the fact that quite a number of international agreements on the conservation of biological diversity had already existed for many years. Examples of these are treaties on specific natural sites (World Heritage Convention), on trade in endangered species (CITES), on specific ecosystems (Wetlands Convention), on specific groups of species (Migratory Species Convention), and on specific regions (several regional programmes). But even together, these instruments could, according to the IUCN, not guarantee the conservation of global biodiversity. Therefore an umbrella instrument was needed. Subsequently, some of IUCN's commissions and experts prepared draft articles for a global convention. This draft convention, though, was still very much focused on *in-situ* conservation, although it was recognized that the developing countries should be compensated for their costs in implementing such an instrument (Doc II.52). This 'deal' was therefore already being negotiated within the IUCN.

Parallel to the discussions in the IUCN the FAO had extensively debated the need to conserve plant genetic resources (PGRs) in the 1980s (Fowler, 1993). An FAO Commission on PGRs was established in 1983. Furthermore, the (non-binding) FAO Undertaking on PGRs was adopted in the same year. These initiatives aimed at discussing and regulating the conservation and exchange of genetic resources, resources which tended to 'erode' and which had become so important to modern agriculture in the development of new varieties. However, views on the content of such a regime were extremely varied, as the rules on the access to the raw materials (genes) as well as to the end-products (new varieties) were heavily disputed. Not surprisingly, the suppliers of the raw materials – notably farmers in developing countries – emphasized the need to protect their rights regarding compensation and reward, whereas the suppliers of the end-products – industrial breeders in the North – stressed the need to protect the rights of the inventors (intellectual property rights). In consequence, an international dialogue was organized by the Keystone Centre (USA) to foster the making of compromises between donors and receivers of genetic resources. Although these debates were very much focused on the use of genetic biodiversity, the issue of conservation was addressed as well, especially because NGOs were involved. In the 1990s, the debates were continued by the so-called Crucible Group. Furthermore, FAO is now reviewing its system on plant genetic resources (PGRs). Linkages with the CBD are sought, as it is increasingly recognized that the issues discussed in both arenas are interdependent. Therefore the FAO also contributed to the agenda of the CBD negotiations. For example, it presented – just as the IUCN did – a draft for a Biodiversity Convention just before the formal negotiations started (Doc II.1).

In 1987 the United Nations Environment Programme (UNEP) also recognized the possible relevance of a global instrument on biological and genetic diversity. It therefore established an ad hoc working group of experts to consider the need for a new convention. This group reached consensus in 1990: a new global framework convention should be negotiated. Subsequently, multilateral discussions on the scope of the new instrument were held and it appeared that there would be no general basis for an IUCN-like treaty, as many states did not want to focus on conservation aspects only. Therefore elements such as *ex-situ* conservation, sustainable use and access to genetic resources and biotechnology were added to the agenda. Finally, the UNEP wrote a so-called element paper to frame the negotiations (Doc II.8), which distinguished many provisions for possible inclusion in a biodiversity treaty. This paper was discussed and UNEP formulated the first draft convention on the basis of these debates (Doc II.10). Finally, the formal negotiation process started in February 1991, when the ad hoc group was renamed the Intergovernmental Negotiating Committee (INC). The first meeting of the INC had the UNEP draft convention before it. Real negotiations could now start.

As far as the start of this negotiation process is concerned, one difference between the CBD and FCCC is striking: the institutional and organizational context. Whereas the CBD was prepared and negotiated within UNEP, the FCCC was dealt with within the UN General Assembly. The developing countries in particular were responsible for this difference. They primarily perceived biological diversity as a technical and scientific issue, but climate change as a political and economic one (Mintzer and Leonard, 1994). Therefore they preferred different arenas to negotiate the two conventions. This may have impacted the nature and outcome of the negotiations, as the formal and informal rules of the game differ in these two arenas. To mention an example: the UNEP is probably more responsive to (technical) NGOs and scientists than the UN General Assembly is, as the former is an issue-oriented and minor body in the UN system, whereas the latter is the supreme political one.

B Treaty formation (1990-1992)

UNEP established – as was noted above – an intergovernmental expert group to prepare for the formal negotiations of a biodiversity treaty in 1990. This group met twice before the Intergovernmental Negotiation Committee (INC) was established and negotiations could formally begin. This INC met five times before the text could be agreed upon and adopted during the final diplomatic conference in Nairobi on 22 May 1992 (the so-called Nairobi Final Act). As well as the text of the CBD, the diplomatic conference adopted three resolutions: one on interim arrangements (the interim secretariat and the interim financial mechanism to the treaty), one on a prompt start (the establishment of intergovernmental meetings of signatories of the convention) and one on agricultural biodiversity (linkage with the FAO). Subsequently, more than 150 countries signed the UNEP Convention on Biological Diversity (CBD) at the UNCED in Rio de Janeiro (June 1992). One exception was the USA. Although this country played a significant and influential role in the negotiations, President Bush refused to sign the result. He believed that the economic interests of the USA were not sufficiently addressed by the CBD. In particular the provisions on biotechnology, intellectual property rights and finance were, according to Bush, inappropriate. His successor Clinton has now signed the CBD, but the treaty has still not been ratified by the USA. Consequently, the USA is not a party to the CBD and can therefore only participate as an observer in the Conference of the Parties (COP).

The expert meetings did not envisage formal negotiations and were therefore aimed at the exchange of views and national positions. The above mentioned UNEP element paper functioned as a catalyst in this respect. In particular the different perspectives on biodiversity of the industrialized and developing countries determined the debates. The South tended to focus on the development aspects of biodiversity (agriculture, sustainable use of biological resources, gene banks, biotechnology, funds for technology transfer), whereas the North tended to emphasize the conservation side (nature parks, protected areas and species, ecological corridors, funds for nature). Several countries were in a middle position, as for example the Nordic – Sweden, Norway, Finland, Denmark – and several African and Latin-American countries emphasized both cases: biodiversity should indeed be preserved, but man should be able to utilize it in a sustainable way as well (Sanchez and Juma, 1994).

INC1 in Madrid (June/July 1991) witnessed very little progress, especially since the organization of the INC – the bureau and working groups – could not be agreed upon. The session consisted therefore of a reiteration of national positions.

Hotly debated issues were the financial assistance given by the rich countries to the poor in the context of the CBD, as well as the question of technology transfer from North to South. The G77 and China – the block of developing countries – asked for substantial commitments from the rich countries, but the latter were very reluctant. INC2 in Nairobi (October 1991) lost one week on procedural issues. The second week, however, dealt with substance. The discussions on finance and technology were continued, again with little progress. In addition, other issues were (briefly) dealt with, such as in-situ conservation and the role of indigenous people in the sustainable use of biodiversity. INC3 in Geneva (December 1991) saw some substantive progress, although some new controversies were born. For example, the safety of genetically modified organisms (GMOs) was introduced into the biodiversity arena by the G77, a move which was very much opposed by some industrialized countries, which believed that this item should be covered elsewhere and not in a biodiversity treaty.

INC4 in Nairobi (February 1991) faced yet more deadlock. The G77-OECD struggle over the inclusion of principles in a treaty was repeated in the biodiversity arena (compare case 2 of the FCCC, subsection 4.3.3). In addition, the topic of intellectual property rights played an ever more important role. These rights grant compensation to the inventors of new knowledge or technologies where others want to utilize them. But IPRs may hamper technology transfer, as they make the access to innovations very expensive. Developing countries therefore favour the abolition or softening of IPRs in the case of technology transfer, whereas the industrialized countries defend these rights on behalf of their private sectors. However, INC4 also saw some progress, for example in the field of a mechanism for access to genetic resources in other countries and mutual benefit-sharing. INC5 in – again – Nairobi (May 1992) had to produce the final compromises just before UNCED, a difficult task. The Global Environment Facility (GEF) was, just as in the FCCC case, accepted as an interim financial mechanism to the CBD. The question of technology transfer and IPRs was also 'solved' through an ambiguous compromise, and the USA had to accept that the issues of biotechnology and biosafety were indeed addressed in the CBD.

The nature and content of the CBD changed drastically during the negotiations. If one compares the drafts which were negotiated in the INCs with the final outcome, substantial differences will be found (compare Doc II.4, II.8, II.10 and II.15). To name a few: the convention became, first of all, less 'green'. Ever more development aspects were added. Secondly, the CBD became less multilateral. The possibility of co-operating through international organizations was less emphasized during the negotiations. In addition, emphasis on national priorities increased. Thirdly, the convention became vaguer. For example, specific categories of adverse impact on biological biodiversity were deleted (e.g. trade politics, economic incentives, pollution, population growth, ozone depletion, climate change).

Last but not least, the language of the convention changed drastically. The earlier drafts generally contained quite obligatory language, whereas the qualifier 'as far as possible and as appropriate' was added to most of the provisions in the final convention.

It is striking that a number of topics led to similar controversies in both the biodiversity and the climate arenas, for example financial assistance, GEF and technology transfer. At the same time, other topics were quite easily determined in both arenas, such as provisions on technical and scientific co-operation and on public education and awareness. But issue-specific confrontations occurred as well. Examples of the latter are clashes on emission reduction targets in the climate arena and on biotechnology and biosafety in the biodiversity arena.

C *Treaty implementation (1992-1997)*

It was decided that the CBD would enter into force after the thirtieth country had ratified it (Article 36). This happened on 29 December 1993, eighteen months after UNCED. Consequently, the first Conference of the Parties (COP 1) would be convened not later that one year after this date (Article 23.1). COP 1 was held in Nassau, Bahamas, 28 November – 9 December 1994. In the meantime, the signatories to the Biodiversity Convention had decided to continue to meet in order to realize a prompt start to the convention. The most important sessions were those of the Intergovernmental Committee on the Convention on Biological Diversity (ICCBD1 and 2). This committee was established by Resolution 2 of the diplomatic conference in Nairobi when the CBD text was adopted in May 1992. The most important task of the ICCBD was to prepare for the debates and decisions of the first Conference of the Parties (COP 1). Additionally the UNEP, which functioned as the interim secretariat to the CBD, immediately after UNCED established four panels and additional conferences on the convention were also organized in 1993 and 1994 (for example in Trondheim, Norway and Mexico City).

The four UNEP panels covered the following items: 1. priorities on action and research; 2. economic implications of conservation measures and valuation of biodiversity; 3. technology transfer and financial issues; 4. biosafety (Doc II.5). A group concerned with guidelines for country studies on the conservation and sustainable use of biodiversity (sometimes referred to as Panel 5) was also established. The panels met three times. Their aim was to prepare the substance for the meetings of the Intergovernmental Committee on the CBD (ICCBDs). Also, other sessions than those of the UNEP Panels and ICCBDs were held to interpret and elaborate on the Biodiversity Treaty. The meeting in Trondheim (Norway, May 1993) was a gathering of governments and NGOs involved and had a lot of workshops on several issues. There were also special working groups on agricultural biodiversity and on the question of whether an interim Scientific Advisory Committee (STAC) should be established to cover the period before the first Conference of the Parties (COP 1). Countries decided against this, however, as they disagreed on the need for and the modalities of a scientific committee, and instead a meeting of experts on biodiversity was organized in Mexico City (April 1994). This session produced, among other things, a research agenda on the conservation and sustainable use of biological diversity (Doc II.22).

The main sessions before the first COP, though, were the two meetings of the Intergovernmental Committee on the Convention on Biological Diversity (ICCBDs). These were directly aimed at a prompt start of the CBD, as well as at preparing the agenda and decisions for COP 1. The first ICCBD, however, was a failure. A lot of time was lost to procedural issues, and those substantive issues which were dealt with tended to result in deadlock. The second meeting was more successful, although a lot of topics were postponed to COP 1. Nonetheless, the debates were more constructive and substantive. Among the items discussed were mechanisms for regulating the access to genetic resources, the role of indigenous people in the conservation of biological diversity, the financial mechanism, and biosafety.

The first meeting of the Conference of the Parties in Nassau was quite technical and procedural in nature, as the rules of procedure for the COP, the secretariat, its budget, the Subsidiary Body for Scientific, Technical, and Technological Advice (SBSTTA), a clearing-house mechanism for exchanging information among the parties and the financial mechanism had to be considered and, if possible, established. The most crucial decisions in this field were the maintenance of the interim status of the GEF as the financial mechanism to the CBD and the permanent designation of the secretariat to the UNEP. Substantive debates, though, focused on criteria with regard to the use of financial resources by the developing countries, as well as on the medium-term programme of work for the Conference of the Parties (COP). In other words, countries pondered which biodiversity projects should be financed and which issues should be addressed by the COP first. Obviously, different countries had different priorities. Whereas the developing countries prioritized issues such as technology transfer, eradication of poverty and biosafety, most industrialized countries emphasized issues such as the conservation of marine and land biodiversity and sustainable development.

This study focuses on the period between UNEP expert meeting 1 (November 1990) and the closure of COP 1 (December 1994). Therefore the COPs which have been held since then – between Nassau and the publication of this thesis – are excluded. Nevertheless, some information on more recent sessions follows here. The second meeting of the Conference of the Parties (COP 2) convened in Djakarta, Indonesia, from 6 to 17 November 1995 (Doc II.33, II.86). Its agenda contained more substantive issues than the first one. The parties addressed, among other things, the following items: biosafety, biological diversity under threat, coastal and marine biodiversity, access to genetic resources and mechanisms for benefit-sharing. The most remarkable and difficult decision of COP 2 was the establishment of an open-ended ad hoc working group to negotiate a draft protocol on biosafety. This protocol was to establish regulations in the field of the safe transfer, handling and use of so-called living modified organisms (LMOs, for example bacteria or animals with built-in human genes). The third Conference of the Parties (COP 3) met in Buenos Aires, Argentina, from 4 to 15 November 1996 (Doc II.87). It suffered from an agenda overload, and delegates discussed how to properly manage the many topics of the CBD. Several delegates proposed a realistic focus on sectoral topics at each COP. Furthermore, the functioning of the GEF and the secretariat were – again – assessed. The substantive items which were more thoroughly addressed were indigenous and local communities, agricultural biodiversity, forest biodiversity and intellectual property rights. Much emphasis was placed on the interfacing of the COP with other international forums such as the World Trade Organization (WTO), World Intellectual Property Organization (WIPO), Commission on Sustainable Development (CSD) and FAO. The progress of the open-ended ad hoc working group on a biosafety protocol was also briefly addressed. So far no protocol has been adopted and negotiations will continue during the course of 1997 and 1998. Finally, it was decided in Buenos Aires to convene COP 4 in Bratislava, Slovakia, in May 1998.

If one compares the implementation of the CBD and FCCC so far, some differences are – again – striking. First of all, the number of national ratifications to ensure the entry-into-force of the two treaties differed, 30 and 50 respectively. Such a number is generally based on the negotiators' assessment of how many parties are needed as a minimum to make a legal instrument effective. As global warming can only be prevented if many countries address their GHG emissions, the number of ratifications was set at 50. The loss of biodiversity, though, can be substantially reversed even if only a few, biodiversity-rich countries undertake conservation measures. It is therefore no surprise that the number of ratifications to trigger the entry-into-force of the CBD was set at a lower rate (n = 30). Secondly, in the climate arena the INC continued to meet, whereas a new body – the ICCBD – was established in the biodiversity arena. This was not only due to the different institutional contexts of the two arenas, UNGA and UNEP respectively, but to the substance as well. The prompt start of the FCCC was less procedural in nature and more focused than that of the CBD. A strengthening of the Climate Convention was more or less immediately addressed by the INC, whereas the ICCBD witnessed a rather slow start in terms of substance. For example, formal negotiations on a reduction protocol were started immediately after the first COP of the FCCC. The negotiations on a biosafety protocol, however, were initiated only a year after the first CBD COP. And whereas country reports on the domestic implementation of the FCCC in OECD countries had already been reviewed *before* COP 1 in Berlin in 1995, national reports on the CBD will not be reviewed before COP 4 in 1998.

Finally, the follow-up of the FCCC was more visible to the general public than that of the CBD. Whereas COP I in Berlin was visited by hundreds of journalists and portrayed as 'the' follow-up of the UNCED, the CBD's COP I in Nassau was entirely without international press coverage.

5.2.4 NGOs IN THE BIODIVERSITY ARENA

The participation of NGOs in the formation phase of the CBD was limited. The World Wide Fund for Nature (WWF) had to report that it was the *only* NGO participant at the first expert meeting (Nairobi, November 1990) (Doc II.78). Moreover, several respondents – government as well as non-government – complained about the lack of NGO presence. One country delegate said:

> "NGOs were very disappointing to me. I did not get much interest, pressure and support. That was very disappointing indeed."

Whereas the INCs of the Climate Convention were in general visited by approximately 50 to 75 NGO observers, the INCs on the Biodiversity Convention could usually welcome some 5 to 10 NGOs only (Doc II.11, II.18, doc II.19). In other words, the NGO participation in the formation sessions of the FCCC was about ten times as high (note that these figures refer to NGOs *in general* – scientific, business, development, conservation and environmental NGOs). Examples of relevant NGOs which did indeed participate in the formation phase are: the World Wide Fund for Nature (WWF), World Conservation Union (IUCN) and Genetic Resources Action International (GRAIN).

The low participation of NGOs in the CBD meetings may be explained as follows:

1 The CBD seemed in the first instance to become another conservation treaty. Therefore the large group of environmental and development NGOs was not interested.

2 NGOs in general favoured involvement in the climate treaty, the forest treaty and the UNCED process, negotiations which took place simultaneously.

3 Those conservation NGOs which *were* interested – IUCN, World Resources Institute (WRI), WWF – generally had good access to UNEP and governments. They may have spread their papers and ideas among the delegates through domestic and international channels in advance of international negotiation sessions. IUCN, WRI and UNEP also convened pre-INC meetings for NGOs and government delegates where specific issues were addressed. In other words, some major NGOs probably fed the CBD negotiations not through an active lobby at the meetings themselves, but through spreading ideas in advance.

The last two INCs before UNCED, though, were visited by some additional NGOs, partly because the NGO-Task Group on the Convention on Biological Diversity had been established and was able to mobilize additional resources to enable some Southern organizations to participate as well (Doc II.63).

The participation of NGOs grew immensely in the implementation phase of the CBD. Whereas 5 to 10 NGOs observed the formation sessions, 70 to 100 do so in the implementation phase. The NGOs 'discovered' the Biodiversity Convention after Rio. One reason may be the refusal of the USA to sign the convention. As one government respondent noted:

> "The fact that Bush refused to sign the CBD woke up the NGOs. They argued as follows: there must be something in it if America is so frightened of it."

Another reason might have been the substantive change which the CBD underwent during the negotiations. As it became less conservation oriented and more environmental and developmental in nature, other NGOs became interested. Furthermore, the end of UNCED gave the main environmental and development NGOs a lot more time. They could re-orientate themselves and the CBD was an interesting 'target'. Important NGOs which were not involved before Rio but started to participate in the biodiversity arena after June 1992 are: Greenpeace International, Friends of the Earth International (FOEI), Environment Liaison Centre International (ELCI), Third World Network (TWN), Birdlife International, Rural Advancement Foundation International (Rafi), Cultural Survival and others. Furthermore, new international coalitions and networks have been established since then, for example the US-based Bionet and the Indigenous Peoples' Network on Biodiversity. Additionally, NGOs also co-ordinated their work through the Interim International Liaison Group on Biological Diversity (IILG) between UNCED (June 1992) and COP 1 (November 1994) (Doc II.90).

As far as the interventions of NGOs in the biodiversity arena were concerned in the period 1990-1995, these have a lot in common with those of the climate NGOs (see table 4.1, page 109). All activities distinguished there could be observed or reconstructed in this arena as well. There is one difference though. Hardly any protest activities were organized around any of the formal negotiation sessions on the CBD. Hence, there were two main differences regarding NGO participation in the two arenas: *number* (hardly any organizations in the CBD formation phase) and *protest* (hardly any activists aiming at the Biodiversity Convention).

— 5.3 NGOs in the formation phase (1990-1992) —

5.3.1 The 'counterfactual'

One way to obtain a general impression of the political influence of NGOs is – as was outlined before – the employment of 'counterfactual reasoning' by the respondents. They were therefore asked whether the treaty text would, according to them, have been the same in case the NGOs had not intervened in the negotiation process of the CBD. Most respondents believe not. One government representative puts it as follows: "Yes, we would have had this convention if we did not have the NGOs, but it would not have been as strong as the CBD is now." But some non-government respondents go even further:

> "There would have been no convention, as the idea for the CBD came out of the concerns of nature conservation scientists and NGOs in the late 1970s. And it is still very much our baby. We regard ourselves as having had a profound influence on what the governments talk about now."

Others, though, refer to the minimal presence of NGOs during the negotiation sessions and therefore believe that these organizations' impact on the text is negligible. One country delegate said:

> "The CBD would have been pretty much the same, because the input of the NGOs was disappointingly small in the beginning. Most of the traditional conservation NGOs were absent at the negotiation meetings."

TABLE 5.1 *The counterfactual of the CBD as perceived by the respondents from the biodiversity arena*

	NGO respondents	*Government respondents*
convention the same	1	5
convention *not* the same	5	5
no convention *at all*	4	-
do not know	2	1
number of respondents	12	11

The figures in table 5.1 do not differ substantially from those of the FCCC (table 4.2, p. 111). Again, most respondents believe that NGOs indeed made a difference (categories 'convention not the same' and 'no convention at all'). But NGO respondents are more optimistic than government respondents with regard to their impact. What is more, four of them believe that NGOs were not only *a* causal factor in treaty formation, but *the* causal factor, a view which is not shared by any of the government representatives. Furthermore, three respondents were unable to come up with an answer.

These outcomes correspond to those of the FCCC (table 4.2). However, there may be one exception. More government respondents in the biodiversity arena deny the influence of NGOs than their colleagues of the FCCC do (respectively 5 out of 11 and 2 out of 10). This observation may suggest that NGOs had more impact on the FCCC than on the CBD text, but this conclusion is, of course, premature. After all, it is only based on one interview question and on a few perceptions. The ten cases below are assessed in accordance with the EAC Method. Only after such detailed assessments are employed can a more reliable comparison of the FCCC and CBD be made.

5.3.2 Case 9: the preamble

A Introduction

Preambles of international agreements generally refer to the substance, background and context of the treaty as well as to the concerns and motives of those who agreed upon it (Glowka *et al.*, 1994). In terms of the regime theory, preambles generally include the *principles* of a treaty (see subsection 2.4.2). It is important to note that preambles themselves do not establish binding obligations. Therefore the provisions of a preamble may go beyond the binding commitments of a treaty. Reluctant states may accept such more far-reaching provisions *because* they are part of the non-binding introduction to the treaty. On the other hand, the preamble offers an opportunity to the parties to include more 'progressive' (or 'conservative') provisions compared to the convention's binding parts.

Lawyers differ in opinion on the legal status of preambles (Bodansky, 1994). Some believe they have no real meaning or any consequences whatsoever. States can proclaim the most beautiful principles in a preamble, while binding themselves to the opposite in the rest of the treaty. Others, though, believe that more 'progressive' provisions in a preamble may become important steps in the development of customary international law, or they may become building stones for new and binding agreements in the future. Hence, principles of preambles may evolve into binding law at some time, and for example NGOs may play a role in this evolution by mobilizing these agreed principles in other policy arenas or new negotiation rounds.

The Preamble of the CBD contains paragraphs on among other things: the value of biodiversity, the problems regarding its loss, the sovereignty and responsibility of states, the need to preserve biodiversity, the precautionary principle, the position and role of indigenous groups as well as women, the need for international co-operation, and on the transfer of finance and technology to the developing world. In earlier drafts of the Biodiversity Convention, the substance of this Preamble was partly attached to Article 3 on Fundamental Principles (Doc II.10, II.12, II.14). However, delegations disagreed on both the content and length of this article. Furthermore, the overlap with the Preamble was extensive and therefore it was decided to fuse this with Article 3. The section on principles was thus deleted in the final CBD, except for one paragraph which reiterates Principle 21 of the Stockholm Declaration on the sovereign rights and transboundary responsibilities of states with regard to the environment. In doing so this principle became binding law.

B Ego-perception

Two of the twelve NGO respondents believe they impacted the content of the Preamble and they particularly refer to two different provisions: the paragraph on the value of biodiversity and the one on the role of women. These paragraphs run as follows:

> "*Conscious* of the intrinsic value of biological diversity and of the ecological, genetic, social, economic, scientific, educational, cultural, recreational and aesthetic values of biological diversity and its components." (Paragraph 1)

And:

> "*Recognizing* also the vital role that women play in the conservation and sustainable use of biological diversity and affirming the need for the full participation of women at all levels of policy-making and implementation for biological diversity conservation." (Paragraph 13)

One NGO respondent said with regard to the first paragraph:

> "NGOs influenced a lot of the philosophical substance at the beginning of the text. That was a way to absorb many of the concerns that did not enter into the negotiations as articles. NGOs continued to raise awareness in the delegates that these were vital components to be included in the conceptual framework. It emerged in that form as preamble. Intrinsic value is one example; the range of values another."

Another respondent claims to have been responsible for the inclusion of the paragraph on women, as her text proposal entered into the negotiations through a friendly country delegation. Finally, her words were re-written and adopted by all governments and included in the CBD.

C Alter-perception

This NGO claim on biodiversity values was spontaneously confirmed by only one of the eleven government respondents. He remarked: "In the Preamble the NGOs played a key role in keeping on reminding the delegations of the many roles of biodiversity beyond those that just fit in the market place." Neither this respondent nor any other, however, confirmed the NGO impact on the paragraph on women.

D *Causal analysis*

GOAL-ACHIEVEMENT — One can safely assume that NGOs achieved their goals with regard to Paragraphs 1 and 13 of the Preamble, as several NGO documents stress the importance of the different values of biological diversity and of the role of women in the conservation and sustainable use of biodiversity. For example Paragraphs 2 and 3 of the preamble of the IUCN's draft convention reaffirm the fact that wild species have a right to exist independently from the benefits they may provide to humanity, but that these species may be of immense value to the biological and medical sciences at the same time (Doc II.52). The Citizens' Treaty on Biodiversity, which was designed by NGOs at the Global Forum in Rio de Janeiro, also emphasizes the intrinsic value of the diversity of life in its preamble (Doc II.51). IUCN lawyers commented on the CBD as follows (Glowka *et al.*, 1994, p. 9):

> "It is noteworthy that for the first time the intrinsic value of biological diversity is recognized in a binding international instrument. This is a very important innovation."

In addition, a small survey among key NGOs in the biodiversity arena, such as WWF, Third World Network, World Resources Institute and IUCN, conducted by the NGO-Task Group in 1991 not only revealed that all of them think that the intrinsic value of biological diversity should be emphasized in the treaty, but also that the role and interests of women in conservation practices should be emphasized throughout the legal instrument (Doc II.65). This latter topic is also underlined in the Citizen's Treaty mentioned above. Hence, NGOs achieved their policy goals in the CBD's Preamble, although they may have used other language and put emphasis on different aspects than the governments.

It should be stressed, though, that this conclusion on goal-achievement only refers to the two paragraphs quoted above and does not cover the entire Preamble. There are definitely several provisions which are less favoured by many NGOs, such as Paragraph 4 which reaffirms the states' sovereign rights over their own biological resources. But this paragraph was not claimed as impacted by NGOs and therefore has not been included in the analysis.

NGO INTERVENTIONS — As the value of biodiversity was recognized even in the earliest UNEP drafts of the convention (Doc II.8), it seems unlikely that governments would not have recognized this value without NGO interventions. The simple fact that they had decided to negotiate a Biodiversity Convention is already an indicator for such a recognition. However, one might argue that the activities of the conservation movements in the 1970s and 1980s were one of the main prerequisites for the governmental appreciation of biodiversity and, hence, for the CBD, as some NGO respondents do. True, that may have been the case, but the role of NGOs in issue-raising is not assessed here. What we want to know is the impact of NGOs on policy formation and implementation in the biodiversity arena in the period 1990-1995, and it is unlikely that NGOs caused governments to recognize the value of biodiversity during that period. The latter already recognized this, at least many of them did.

What values are underlined by governments and in what language is, however, another question. It is striking that ever more functions of biodiversity were stressed in the documents during the negotiations (ecological, genetic, social, economic etc.), but that some were deleted as well (ethical and environmental values). In addition, the sequence of the different values changed in the paragraph, and the paragraph as a whole 'moved' through the Preamble in the different draft conventions. Whereas it was Paragraph 17 in the Preamble of the first draft, it became the first paragraph in the final CBD. It cannot be excluded with certainty that NGOs played a role in these changes of language and position, but it was not possible to confirm that by causal analysis either. One exception, though, may be the inclusion of the *intrinsic value* of biodiversity. This notion was introduced in Article 2 on the fundamental principles in the January 1991 draft convention (Doc II.10). It was deleted again in subsequent drafts, but reappeared in the Preamble of the final convention. This sequence of inclusion, exclusion and re-inclusion refers to disagreement among governments about accepting the intrinsic value notion. But they finally did so. This fact may have been co-determined by external pressure from NGOs which so fiercely promoted this principle, although it should be recognized that some countries defended it as well. One example is the Netherlands (Doc II.5).

Besides, all government drafts for a preamble included a paragraph on the recognition of mankind's co-existence on earth with a wide range of other living organisms *which exist independently from the benefits they may provide to humanity* (e.g. Doc II.10, p. 7). These words in italics, though, stem from the preamble of the non-government IUCN draft convention (June 1979) and refer to the fact that non-human organisms have a value in their own right (Doc II.52, p. 1). Although the wording is not a literal version of the intrinsic value concept, it has a lot in common with it. It seems that this distinct paragraph in the UNEP draft convention – including the IUCN wording quoted – was finally fused in the CBD with the paragraph on the range of other values of biodiversity. Consequently, the former paragraph was deleted, whereas the notion of intrinsic value was added to the latter. Therefore it is believed that through – among other thins – NGO pressure as well as some IUCN language the notion of intrinsic value became part of the final CBD's

Preamble. This NGO influence is, however, indirect. No NGO text proposal was taken over literally.

Paragraph 13 provides a much stronger case. The provision on the role of women in the conservation and sustainable use of biodiversity was proposed by the Netherlands at INC4 and subsequently included in the final treaty (Doc II.16). This initiative, though, came from an NGO, which delivered a text proposal to the Dutch delegation. Subsequently, it was rewritten by this delegation, formally tabled in the INC and finally adopted by all governments. This sequence of events was confirmed by several players both from the non-government and government sides. Hence, the NGO concerned was able to influence Paragraph 13 of the Preamble. This influence was indirect in nature, as the final legal text was written by Dutch delegates.

It should be emphasized that there may be more paragraphs in the Preamble which were impacted by NGOs. Possible examples are Paragraph 7 on information and knowledge, and Paragraph 12 on indigenous and local communities. The latter will be dealt with in the next case, as NGO respondents claim to have influenced a wider range of provisions on indigenous and local communities in the CBD (case 10). The former, though, was not claimed by any NGO respondent, but brought forward by a government representative as an example of NGO influence:

> "The paragraph in the Preamble relating to the lack of information and knowledge regarding biological diversity and the need for capacity-building in this area was an input from an NGO during a meeting our delegation had arranged with NGOs over lunch. As we liked it, we re-drafted and proposed it."

Given the EAC Method, to build exclusively on ego-perception, this example was not thoroughly assessed and, hence, verified or otherwise.

E *Extent of political influence*

Obviously, the counterfactual of the CBD's Preamble would have been different without the NGOs (probably no inclusion of the intrinsic value notion and certainly no inclusion of the paragraph on women). Although NGOs achieved all their policy goals with regard to these two topics, they were not equally responsible for this themselves. The intrinsic value principle was also promoted by many states, so that NGOs were only to some extent responsible for its inclusion. But the paragraph on women was entirely due to the interventions of NGOs, therefore the variable OI (own intervention) is 'much' here. Taking both topics together, OI is considered 'substantial'. However, the topics are not seen as of high or even moderate relevance. To repeat, preambles are non-binding parts of treaties, and governments can locate all their 'nice intentions and language' there. As the direct consequences of these provisions are negligible for the parties, they are of minor relevance. Finally, NGOs influenced these provisions indirectly, so that the overall extent of political influence is considered 'substantial'.

TABLE 5.2 *Scores on political influence in case 9 (Preamble CBD)*

Counterfactual (C)	different (1)
Goal-achievement by NGOs (GA)	much (3)
GA due to NGOs' own interventions (OI)	substantial (2)
Political relevance of topic (PR)	low (1)
Directness of impact (D)	indirect (1)
Extent of political influence: PI = C [(GA x OI) + PR + D]	substantial (8)

Legends
C : similar = 0; different = 1
GA: no = 0; some = 1; substantial = 2; much = 3
OI: no = 0; some = 1; substantial = 2; much = 3
PR: no = 0; low = 1; moderate = 2; high = 3
D : no impact = 0; indirect impact = 1; direct impact = 2
PI: no PI = 0; some PI = 3 to 6; substantial PI = 7 to 10; much PI = 11 to 14

F *Final conclusion*

NGOs in the biodiversity arena – in particular organizations such as IUCN, WRI and WWF – stressed the need to recognize the many values of biodiversity as well as the role of women in its conservation and sustainable use. Given the substance of the CBD's Preamble, governments agreed. The fact they did so was partly due to these NGOs. Mainly through the reformulation of some of their text proposals by governments, NGOs were able to substantially impact some parts of the Preamble.

5.3.3 CASE 10: INDIGENOUS AND LOCAL COMMUNITIES

A *Introduction*

The position and role of indigenous people in the conservation and sustainable use of biological diversity have become important themes in the 1980s. There are a number of reasons for this:

1 It is obvious that most traditional cultures are directly dependent on nature (Glowka *et al.*, 1994). If their natural habitats disappear, their livelihoods will be severely threatened. Good examples are the indigenous people living in the tropical forests. Their needs are arguments in favour of protecting the surrounding ecosystems.

2 Because many of these cultures live in harmony with the carrying capacity of their natural surroundings, their livelihood systems are considered to be sustainable. In other words, their way of life includes examples of how to achieve sustainability. These examples may be valuable to others as well, and some even go so far as to extract *blueprints* for survival from traditional practices (Bandyopadhyay and Shiva, 1989; Pereira, 1993).

3 Many tropical ecosystems in which indigenous people live have become a storehouse of profitable biological resources in the eyes of Western industry (Juma, 1989; Sanchez and Juma, 1994). Ever more pharmaceutical and agricultural companies are looking for new species, traditional crops, genes and even human cells, which they can transform into marketable products: for example medicines, high yielding varieties and biotechnology. However, these biomaterials are generally taken away free, without any notice in advance or payment afterwards to the holders of the material, and the benefits coming from the products after their introduction into the market are not shared either. What is more, the knowledge on these biomaterials and the resulting products are generally protected by intellectual property rights (IPRs). This implies that the original holders of the biological resources may be denied access, and that they have to pay high prices or obtain expensive licenses in order to use the knowledge or products. NGOs label these practices 'exploitative' and demand that countries of origin, indigenous people, farmers and local communities shall have the right to decide on access to their biomaterials by third parties and to share benefits from the commercial use of this material elsewhere (Doc II.41, II.67, II.72).

The CBD covers these topics. Paragraph 12 of the Preamble was already mentioned above (case 9). It recognizes the dependence of indigenous and local communities on biological resources and stresses the desirability of equitable sharing of benefits arising from the utilization of traditional knowledge, innovations and practices. The aspect of benefit-sharing is also formulated in Article 8(j) on in-situ conservation, although the treaty lacks any mechanism to implement such benefit-sharing. This is up to the individual countries. Furthermore, states should, according to this Paragraph 8(j), respect, preserve and promote traditional knowledge, innovations and practices relevant to the conservation and sustainable use of biodiversity. This is reiterated in Article 10(c). Parties shall, according to this paragraph, protect and encourage customary use of biological resources in accordance with traditional cultural practices that are compatible with conservation or sustainable use requirements. Finally, references to indigenous and traditional knowledge and technologies are also made in Article 17 (exchange of information) and Article 18 (technical and scientific co-operation).

If one takes the evolution of the CBD into account, one has to acknowledge the fact that the topic was addressed in all drafts of the treaty. Obviously, there was a consensus among governments to deal with it. However, conflicts arose over the specific wording of the provisions. The words 'indigenous people', for example, turned out to be unacceptable to some Asian and Latin American countries, as these do not officially recognize such distinct groups within their territories

(Doc II.5). Therefore some preferred other terms, such as 'local communities' or 'traditional cultures'. Finally the wording 'indigenous and local communities' was agreed upon. Additionally, some of these governments had already made arrangements with such groups regarding the public and resource management of their areas, and did not want the CBD to interfere in internal and sensitive affairs. Consequently, a clause was added to Article 8(j) which says that the provision is subject to national legislation. This is rather unusual language for international law and may even undermine the entire paragraph. Another change which resulted from the negotiations was a decrease in the number of references to indigenous and local communities in the treaty text. Some provisions were omitted. For example, drafts of the treaty included a separate – but entirely bracketed – Article 14*bis* on the issue of access to and reward of traditional knowledge, innovations and practices (Doc II.15), but this article was deleted and some elements were fused with Article 8(j). Moreover, nearly the whole section on principles (Article 3) was omitted in the CBD, including one reference to indigenous people. As a result the number of provisions obviously decreased. Finally, it should be stressed that the obligatory language on indigenous and local communities in Articles 8 and 10 was weakened by the qualifier 'States shall, as far as possible and as appropriate (...)', but such clauses were added to many articles of the convention.

B *Ego-perception*

Three NGO respondents claim they were jointly responsible – together with other policy players – for references to the role and interests of indigenous people and local communities in the CBD. However, their claims are quite unspecific. One for instance said: "I think that the provisions on indigenous people and traditional knowledge are good examples of NGO influence." When asked to elaborate on this example, the respondent remained as vague as he was before. For example, no references were made to specific wording in the CBD. He nonetheless re-affirmed his claim that the emphasis NGOs put on this topic led to the inclusion of the theme in several articles of the CBD. One of his colleagues agreed, but was nonetheless a bit more pessimistic about their impact. After all, a lot of good and strong language was according to him lost:

> "We lost nearly everything with respect to rights and interests of indigenous people and local communities, except for Article 8(j). At the beginning of the Nairobi meeting in 1990 we had 3-4 references. Finally, we had one."

Although this view is rather too pessimistic – it is obvious from the above that more references remained in the CBD – some draft provisions on indigenous people were indeed deleted. In short, NGOs claim that they indirectly influenced provisions on this topic, but implicitly add that they could have done a better job.

C Alter-perception

Seven (!) of the eleven government respondents believe that NGOs indeed co-determined the inclusion of forgoing provisions. To quote one of them:

> "I do remember one issue that stands out in my mind where NGOs definitely had a very important influence. Any time 'indigenous people' is mentioned in the convention – and I remember at least two times – that was really the work of NGOs. If it had not been for NGOs and particularly indigenous NGOs, they would have not been mentioned at all."

However, not all delegates were as positive as this respondent and one of them even reiterates the view that NGOs could have performed better:

> "One area which I think draws a bit from the role of NGOs was the whole issue of indigenous people. I think if the NGOs had not been involved at all, that issue would have not been articulated as well as it is now. But I go even further: if NGOs had been more active at the last negotiation session, that issue would have been better articulated than it is now, because in earlier drafts there was stronger language on indigenous people. However, this was downplayed subsequently at the final negotiation session where NGOs were almost non-existent."

One aspect is particularly highlighted by government respondents, namely benefit-sharing. They perceive NGOs to have been most influential in this area. NGOs promoted the idea that indigenous people should be rewarded for the use of their biological resources, knowledge and techniques by others and share in the possible benefits. This message was indeed heard and taken up by governments, according to these respondents.

It is remarkable that so many *government* representatives confirm the above NGO claims (7 out of 11). The number is about twice as many as in other cases and they even outnumber the NGO respondents who had claimed success on this topic. Obviously, governments believe that the input, importance and impact of NGOs mainly refer to this item. NGOs themselves however, given their own perception, disagree.

D Causal analysis

GOAL-ACHIEVEMENT — The NGOs in the biodiversity arena, and in particular GRAIN, stressed the need to recognize the importance of cultural diversity in the CBD, to respect and protect the interests of indigenous people and local farmers, and to compensate for the utilization of customary and traditional knowledge, innovations and practices by third parties (Doc 11.41, 11.65). Given the substance in the Preamble and Articles 8, 10, 17 and 18, NGOs seem to have achieved their policy goals to a very large extent. However, the notion of cultural diversity did not reappear in the Biodiversity Convention. Moreover, NGOs had campaigned for even more and much stronger language on indigenous people than was finally adopted. For example, the establishment of community property rights on local biodiversity for indigenous people was campaigned for, but not taken up by governments (see case 12 on genetic resources in subsection 5.3.5 below). In addition, references to increased participation of indigenous groups in decision-making, the need for national legislation, more control over local resources, stronger provisions on benefit-sharing and special references to indigenous women were proposed (Docs II.60, II.61, II.67), and provisions for indigenous people in the articles on access to genetic resources (Article 15), technology (Article 16) and biotechnology (Article 19) were also lobbied for but not accepted. Finally, NGOs strongly supported the separate Article 14*bis* which was – as outlined above – deleted by governments in the final stages of the negotiations. In short, NGOs only got a small proportion of all their wishes, and those they realized were written in often soft and ambiguous language. Therefore Posey (1994) notes that indigenous groups are rather sceptical about whether the CBD will adequately recognize and protect their knowledge and practices. To sum up, NGOs achieved *some* of their policy goals on this topic at best.

NGO INTERVENTIONS — Although NGOs made many interventions on this topic in the biodiversity arena in 1991-1992 and although so many government respondents assume that NGOs impacted the provisions on indigenous and local communities in the CBD, such political influence could not be confirmed by the causal analysis. The following two arguments should underpin this conclusion. Firstly, the UNEP element paper for a Biodiversity Convention already contained most of the provisions on indigenous people in the CBD (though in different language; Doc II.8). The most important exception was Article 14*bis*, which was proposed by Canada and Peru at INC3 and temporarily included in the draft convention (Doc II.15). Maybe NGOs played a role in the tabling of this provision, but as it was deleted later on by the INC the impact on the final policy outcome should be considered negligible. Hence, the substance of the provisions which were finally adopted at the Diplomatic Conference in May 1992 were already there right from the start of the negotiations in November 1990. Secondly, NGOs seemed not to be the source of the idea to include indigenous and local communities in the first negotiation documents of UNEP. Potential sources might have been GRAIN, WWF and IUCN (as indigenous groups themselves were absent during the early negotiation sessions). However, WWF and IUCN only paid attention to this topic in the biodiversity arena more recently, after being pushed by indigenous NGOs to do so (Posey, 1994). It is no coincidence that the IUCN draft does not include a single provision on the roles, rights and interests of indigenous and local communities in biodiversity conservation (Doc II.52). GRAIN on the other hand *did* deliver some written material on this topic to the delegates in October 1990, but so did the FAO and UNEP in the same period (Doc II.1, II.8, II.41). Besides, the (only) representative of GRAIN to observe the negotiations at that time did not claim to be the source of the idea to put indigenous people on the negotiation agenda, particularly as he focused on other topics (mainly genetic resources).

One may still wonder where the inclusion of this topic in the draft treaty stems from. Although the available policy documents and some additional interviews did not produce an answer – the respondents concerned simply did not know – it is likely that this inclusion came from the experts meetings which the UNEP had organized in the period 1988-1990 to prepare for the treaty. Additionally, the International Society on Ethnobiology (ISE) might, according to some respondents, have played a role as well (see also: Posey, 1994). The ISE had already in 1988 adopted its Declaration of Belem, which was one of the first international documents to call attention to the link between the conservation of biodiversity and the preservation of cultural diversity and rights of native people. In addition, protest and pressure groups may also have played a role in the background. Since the second half of the 1980s, activists have condemned practices of what they call *patent piracy* (Doc II.44). This notion refers to the fact that Western government departments, research institutes and transnational corporations have started to apply for patents on biomaterials from wild species and indigenous peoples in tropical regions. For example the US Department of Health applied for a patent on cell lines from Hagahai people in Papua New Guinea in 1989. These cells may contain viral antigens which are associated with the cure of leukaemia. This led to (local) protests by NGOs. The UNEP experts meetings may have been responsive to these protests (although this could no longer be checked). Moreover, NGOs may have impacted the debates on indigenous people in the biodiversity arena through the UNCED process, where many indigenous groups visited the PrepComs (Preparatory Committee of the UNCED, which met four times in the period 1990-1992). Finally, NGOs which actually observed the biodiversity treaty negotiations definitely contributed to the debates at the INCs (process influence). But all in all, these NGOs did not impact the policy outcomes on indigenous and local communities in the formation phase.

It should be reflected upon, though, why such a relatively high, positive alter-perception score (n = 7) can be ignored here. Although seven government respondents obviously believe that NGOs *did* influence this topic, the final conclusion is different. Two arguments may explain this: 1. the government respondents involved may have mixed up process influence (affecting a debate) and political influence (affecting policy outcomes) of NGOs, although they were instructed to focus on the latter before any interview started; 2. these respondents may have mixed up the NGOs' influence on this topic in the formation and the implementation phase (see case 14 in subsection 5.4.1 below). True, ever more indigenous groups have entered the biodiversity arena, but mainly in the period *after* UNCED. In addition, there is some *political correctness* to be found in this field, as one respondent noted. Indigenous people are a 'nice subject' to show one's responsiveness to societal demands. This may have led to strategic answers.

E *The extent of political influence*

According to the above analysis, the counterfactual would have been the same without NGOs. Maybe they (temporarily) affected the debate on indigenous people and maybe they had some background effects through UNCED, but they did not impact the policy outcomes themselves (e.g. Articles 8(j) and 10(c)). These were determined by governments and professional experts, irrespective of some goal-achievement on the NGOs' side. The relevance of this topic is considered moderate. It indeed provoked serious conflict among some country delegations and was an important aspect of the convention, but the topic was not at its core. Besides, the provisions on indigenous people are rather weak, even weaker than most of the other provisions in the CBD.

TABLE 5.3 *Scores on political influence in case 10 (Indigenous people)*

Counterfactual (C)	similar (0)
Goal-achievement by NGOs (GA)	some (1)
GA due to NGOs' own interventions (OI)	no (0)
Political relevance of topic (PR)	moderate (2)
Directness of impact (D)	no impact (0)
Extent of political influence: PI = C [(GA x OI) + PR + D]	none (0)

F *Final conclusion*

The topic of the role of indigenous people and local communities in the conservation and sustainable use of biological diversity had already been addressed by early UNEP documents preparing for the negotiations of the CBD. This was probably due to experts meetings and organizations such as the International Society of Ethnobiology. Since then, it seems that the strength of these provisions was only weakened during the negotiation process, irrespective of the fact that some very strong proposals were considered by the INC. Given this course of the decision-making process, the absence of indigenous NGOs in the biodiversity arena in 1990/1991, the initial focus on other topics by the NGOs which *were* present, and their possible but certainly limited impact on the debates at a later stage, it was concluded that the biodiversity NGOs did not exert political influence in this specific case. It is likely that some colleagues in other arenas (UNCED) and some local protests against patent piracy may have influenced the debates in the biodiversity arena through the back door, but this was not and could not be assessed.

5.3.4 Case II: in-situ conservation

A Introduction

The CBD defines 'in-situ conservation' as follows:

> "The conservation of ecosystems and natural habitats and the maintenance and recovery of viable populations of species in their natural surroundings and, in the case of domesticated or cultivated species, in the surroundings where they have developed their distinctive properties."

This concept is the opposite to *ex-situ* conservation, which refers to the conservation of components of biological diversity outside their natural habitats, for example in gene banks, laboratories and zoos. These two approaches to biodiversity conservation are dealt with in Articles 8 and 9 respectively. Originally, the emphasis was on the former, as conservation groups tend to focus on natural systems. But due to the intervention of developing countries and the FAO, ever more attention was given to ex-situ conservation. According to these players, both should be of equal weight in the CBD. However, most Northern countries and NGOs opposed this idea of equality. For them, the emphasis should be on the conservation and utilization of biodiversity within its natural habitats. Ex-situ conservation should therefore be complementary at best and this is also recognized in the CBD.

Article 8 on in-situ conservation consists of 13 paragraphs and is the largest in the CBD. The provisions deal with among other things the establishment of a system of protected areas in each country; management of biological resources; protection of ecosystems; promotion of sustainable development near protected areas; restoration of degraded ecosystems and recovery of threatened species; control of living modified organisms and alien species in natural surroundings; compatibility between present use of biodiversity and conservation and sustainable use objectives; regulation of adverse effects on biodiversity; development of national legislation and policies; and financial and other support to developing countries in order to improve in-situ conservation.

If one looks at proposals for in-situ conservation in older drafts, however, some differences are striking. The most important is the disappearance of the notion of global lists. Conservationists had proposed the adding of lists of valuable areas and species to the CBD in order to include their protection in binding law. However, this idea was opposed by many, although such lists were already broadly accepted instruments in other conservation treaties (for example CITES, Bern Convention on the Conservation of European Wildlife, Bonn Convention on Migratory Species, Ramsar Convention on Wetlands, and World Heritage Convention; Doc II.7). But several delegates feared overlap and duplication. Besides, most developing countries were reluctant to make detailed commitments with regard to nature conservation policies, in particular regarding their forests; many rich countries doubted whether global lists could be made effective and believed that lists unjustly excludes biodiversity, which is not mentioned, from conservation; and most delegations were not very keen on *global* action programmes, favouring national or regional approaches at best. As a result the global lists were omitted.

One could also argue that the 'status' of Article 8 changed during the course of negotiations. Whereas many of its provisions were at the heart of the IUCN draft, it became less important *vis-à-vis* other articles, such as those on ex-situ conservation and sustainable use of biological resources. This was definitely due to the change in nature of the CBD: from a conservation treaty to a sustainable development convention.

B *Ego-perception*

Four of the twelve NGO respondents refer to Article 8 as being influenced by them. One of them said:

> "Yes, if you look at Article 8, there are phrases there we drafted back in 1984. There are many elements in the convention that we put into framework outlines and that IUCN put into initial drafts."

Obviously, this respondent claims direct political influence through NGO proposals and the IUCN draft. Another refers to more indirect processes and believes that NGOs kept the conservation approach on board against the pressure of those who wanted to marginalize it:

> "If it were not for NGOs, I think the concept of sustainable use of biodiversity would have completely overshadowed everything else. Once most of the governments became alerted to the self-interest argument, and started to use it, and once business began to advise governments of its views on biodiversity, then the convention rapidly moved to a development and financial focus. One of the things the NGOs clearly did, was to retain within the CBD the conservation element."

C *Alter-perception*

No government respondent spontaneously confirmed these NGO claims. On the contrary, one even considered Article 8 an example of NGO failure:

> "The NGOs, participating from the nature conservation side, they did not have success with their ideas, because it was clear that the NGOs favoured a stronger protection philosophy than what was there at the end of the day. They also favoured the idea of the global lists and that failed."

D *Causal analysis*

GOAL-ACHIEVEMENT — With regard to conservation objectives and global lists there was a split in the NGO community. On the one hand, organizations such as

WWF and IUCN favoured strong in-situ provisions and extended lists. This was also labelled the *scientific-technocratic* approach. Others such as GRAIN and some Southern NGOs opposed this one-sided emphasis on in-situ conservation and special areas. They believed that sustainable use objectives were as important and this was referred to as the *socio-economic* approach. While negotiations proceeded, even some conservation groups became less convinced that global lists were indeed needed. As an IUCN staff member said: "We could live without them." Nonetheless, some continued to endorse the establishment of global lists, even at INC4 and INC5 (Doc II.63), but WWF was the most outspoken. Therefore this organization had a hard time, as one government delegate noted:

> "WWF was the only NGO which had the courage – and I would really call it courage – to speak up in favour of nature protection viewpoints and global lists, but it was a very lonely voice. I have a certain admiration for WWF, because they knew that their views were not met favourably by the developing countries, but in spite of that they did so."

WWF did obviously not achieve its goal, given the fact that the lists were rejected. The other NGOs, however, did to some extent. These organizations opposed the lists or were less concerned about them. In-situ conservation is still one of the central articles of the CBD – and the most lengthy – while it is also linked to sustainable development requirements. Of course, these NGOs did not succeed with regard to a number of details. For example proposals to include concepts such as 'wildlife corridors' and 'buffer zones around protected areas' in Article 8 were not adopted. The most significant difference, though, between the treaty text and NGO proposals is the level on which parties should take initiatives. Whereas most NGOs favoured a global approach to set common priorities and guarantee common policies in all countries, Article 8 is entirely nation-oriented. This implies that parties are relatively free to develop their own in-situ conservation policies at home. The CBD offers some general policy objectives, instruments and guidelines at best. To conclude, NGOs only achieved some of their objectives with regard to Article 8.

NGO INTERVENTIONS — It should immediately be recognized that since the first negotiation rounds a more or less similar split existed between governments as well. On the one hand, there were several countries in favour of a conservation approach based on common lists, notably France and the USA. On the other, developing and Nordic countries favoured a socio-economic approach based on sustainable development. As one government respondent put it:

> "We wanted a convention on the protection of biodiversity in the country landscape, in agriculture, forestry, fishery. We did not want to deal with lists and the traditional instruments to care for biodiversity in the old-fashioned style with reserves and national parks."

Furthermore, developing countries opposed a global approach. They emphasized their national sovereignty and demanded that they should be able to set their own priorities at home. The battle on how 'green' and 'global' the treaty should be was therefore fought between (blocks of) countries. The final result was a nice balance between all the wishes, although France mourned the loss of the global lists and threatened not to sign the convention (which they ultimately did anyway). This 'balance of wishes' was a political and sensitive compromise, which meant that different countries had to give way on their original policy positions. It seems therefore that the policy outcome – Article 8 – was entirely determined by governments.

Nonetheless, the NGO claim that Article 8 contains some elements which NGOs drafted earlier could be confirmed to some extent. This claim was checked by comparing the relevant provisions of the IUCN draft with those of the CBD (compare Doc II.4 and II.52). Some similarities and differences are striking. Obviously, both documents contain similar items: the establishment of protected areas, the recovery of threatened species, the control of man-made and alien species, the need to have compatibility between present use of biodiversity on the one hand and conservation and sustainability objectives on the other, and the regulation of adverse effects on biological diversity. But there are many differences as well: the IUCN's proposal to have a world list of special areas was not adopted, the article on the effects of pollution on biodiversity was omitted, the role of the COP in setting global priorities and guidelines disappeared, and the emphasis on wild species alone was weakened in the CBD. In addition, the language used in the IUCN draft is generally quite different from the CBD text. Furthermore, new elements were added to the latter during the negotiations: for example the paragraph on the management of biological resources, the reference to sustainable development, and the paragraph on the development of national legislation. Given this mixed picture, the IUCN draft had some indirect influence at best, and only on those items which are dealt with in both documents.

The question remains whether the IUCN draft could have functioned as a source of influence. Even an NGO representative noticed in 1991 that the draft had already become irrelevant, as it was not officially tabled as the formal document to build on during the negotiations (Doc II.78). But it was still used as a background file by the UNEP to formulate its first element paper for a Biodiversity Convention (Doc II.8). Some provisions were even taken over literally by UNEP, such as Paragraph (l) on compatible land use. In addition, the IUCN draft was distributed among delegates when the negotiations started. Some government respondents who were explicitly asked to assess the impact of the IUCN draft on the negotiations confirm that it had an impact on a number of delegations:

> "It is clear that many delegations were to some extent inspired by the IUCN draft and picked up ideas from that draft. This is unquestionable, because as

soon as you have provisions already formulated and you like these then you try to further them."

However, it is no surprise that a Biodiversity Convention would for example urge the protection of special areas, the recovery of threatened species, the control of alien species, and the management of adverse effects on biodiversity. Such provisions would probably have been included without the IUCN draft as well, especially as these are common items in many other nature conservation treaties (Bergesen and Parmann, 1993; Doc II.7). But some inspiration from this IUCN proposal is still possible as far as these items are concerned. On the other hand, an inclusion of paragraphs on compatible use and man-made species would probably have been less likely without the IUCN draft (see Paragraphs (g) and (i) in Article 8 of the CBD). Therefore it is believed that in particular these provisions were impacted by the IUCN.

It should be recognized, though, that the IUCN is a hybrid NGO. It is a worldwide conservation union of which both NGOs and governments are members. The fact that (mainly developing) countries refused to accept the IUCN draft as a formal negotiation document, however, highlights that this union is perceived more as a biased conservation organization than as a neutral body the products of which represent some general interest felt among both NGOs and governments. Therefore it is considered an NGO in this study. Besides, IUCN is officially defined as an NGO by the UN.

E The extent of political influence

It is believed that the counterfactual of Article 8 – in case the NGOs had not participated – would have been different. In particular ideas from the IUCN were fed into the negotiations and these indirectly impacted the outcome (protected areas, recovery of species, compatibility, man-made organisms, alien species, adverse effects). That is not to say that none of these topics would have been covered by the CBD without the NGOs, but probably not as explicitly and some items might even have been absent (compatibility, man-made organisms). Nonetheless, the extent of goal-achievement is limited, as many preferences of the NGOs were not adopted (e.g. the global lists). As far as objectives were attained, this was only partly due to the NGOs themselves, as several governments defended similar goals (in particular France and the USA). Maybe the inclusion of Paragraphs 8(g) and 8(i) was *substantially* due to the IUCN, but as this remained more of a real guess than an *informed* guess the 'OI score' is still considered 'some' (see table 5.4). The political relevance of this topic is assumed to be moderate. Although the interests at stake were high given the North-South clash on the global lists, and although some obligations result from Article 8, the most controversial items were deleted while the remaining language was severely weakened. It is therefore questionable whether the CBD has something substantive to add to current conservation policies in the countries which signed it. In short, the NGOs' political influence on Article 8 amounts to 'some'.

TABLE 5.4 *Scores on political influence in case 11 (In-situ conservation)*

Counterfactual (C)	different (1)
Goal-achievement by NGOs (GA)	some (1)
GA due to NGOs' own interventions (OI)	some (1)
Political relevance of topic (PR)	moderate (2)
Directness of impact (D)	indirect (1)
Extent of political influence: PI = C [(GA x OI) + PR + D]	some (4)

F Final conclusion

Conservation NGOs such as IUCN and WWF and socio-economic NGOs such as GRAIN disagreed on how to elaborate on in-situ conservation measures. Whereas the former defended the instrument of global lists to protect special areas around the world, the latter focused on sustainable development objectives. The green NGOs, however, did not achieve their most prominent policy goal, because the global lists were omitted during the negotiations. As governments turned out to be as divided as NGOs, they worked out a complex compromise without such lists. Therefore it seems as if the green NGOs did not influence Article 8. Nonetheless, some traces of a similar article in the IUCN draft for a Biodiversity Convention are still there in the CBD. Consequently, this NGO has influenced Article 8 to some extent.

5.3.5 Case 12: access to genetic resources

A Introduction

Biodiversity also refers to *genetic* diversity. Genes allow species to adapt to changing circumstances and they form the principal units of biological evolution. In addition, genes are now big business. Therefore genes have become a sensitive political issue given the *genetic world order* we live in (Heins, 1993). This order is characterized by a specific interdependence between North and South. To simplify this, the latter has the largest share of the genetic wealth in the world under its jurisdiction, whereas the former has the desire and capability to develop new (bio)technologies and products. For example, more than 95% of the genetic material on which the 20 most important food crops in the world are based originate from regions of the current Third World (Svarstadt, 1994). But these food crops were developed mainly by Western multinationals and research institutes. Given this situation, the 'deal' of the Biodiversity Convention seemed simple: the South places 'their' genes at the North's disposal, while the North grants financial compensation and/or easy access to (bio)technology, knowledge and end-products which are based on these resources to the South. However, things are not that easy. While the access to biodiversity and genes has indeed developed into a more or less 'free regime' since colonial times, the access to technology and knowledge is protected by intellectual property rights (IPRS). In other words, Northern institutes or companies may collect samples of biodiversity in tropical ecosystems quite freely, while Southern countries have to pay high prices for utilizing end-products or technologies which were based on these same resources. This implies that the developing countries have to pay *twice*: they do not receive compensation for the extraction of 'their' genes, while they – on the other hand – have to compensate for intellectual property rights and, moreover, do not share in the profits. It is no surprise that the developing countries since the early 1970s have wanted to fundamentally change this practice (Sanchez and Juma, 1994). The CBD was one mechanism to achieve this goal. At the same time, Northern countries were reluctant to deal with this topic in the biodiversity arena given their specific interests. According to them, commercial utilization of genes and biotechnology were distinct topics which should be dealt with elsewhere, for example in the FAO, while IPRS should be dealt with by the World Intellectual Property Organization (WIPO) and the World Trade Organization (GATT/WTO). Nonetheless, industrialized countries soon realized that they would have to face this topic in the biodiversity arena, at least if they wished to keep the G77 and China on board (Svarstadt, 1994).

Apart from the CBD, there already existed a regime with regard to *plant* genetic resources (PGRS): the FAO Global System for the Conservation and Utilization of PGRS. This system aims at co-ordinating the conservation and use of these resources in the realm of food and agricultural production and consists of – among other things – the legally non-binding International Undertaking on Plant Genetic Resources (Glowka *et al.*, 1994; Bergesen and Parmann, 1993; Pistorius and Van Wijk, 1997). This undertaking is based on the so-called common heritage of mankind principle. It states that PGRS are the world's common heritage and should be available to humanity without restriction. Although this principle seems morally and technically correct from the perspective of global agricultural progress, it nonetheless seems to legitimize the rule of free access which was referred to above (although it is put into perspective in one of the annexes of the undertaking, where it is stated that free access does not mean 'free of charge'). It is therefore no surprise that Southern countries opposed the inclusion of the common heritage of mankind principle in the CBD, all the more so since the CBD is legally binding and exceeds the undertaking's scope. For example, the common heritage principle in the CBD might have legitimized the free extraction of medicinal plants from the tropics by the Western pharmaceutical industry. In consequence, the premise on access in the CBD evolved in a quite different direction.

Paragraphs 1 and 2 of Article 15 define this topic. The authority to determine access to both in-situ and ex-situ genetic resources rests, according to the CBD, with national governments as states have sovereign rights over their natural resources. Hence, access is not based on *common heritage* any more, but on *national sovereignty*. Nonetheless, each country which is party to the convention should try to facilitate access to their genetic resources for other parties. It must therefore be guaranteed as far as possible. In addition, access by other parties should take place on the basis of 'mutual agreement' and 'prior informed consent'. This implies that countries – or industries and research institutions under their jurisdiction – which wish to utilize biological resources need to inform the country concerned and negotiate a contract. Subsequently, Paragraphs 6 and 7 deal with benefit-sharing. Parties which provide genetic resources to others should be able to participate in research by the latter which is based on the genetic resources provided, and they should be able to share in the benefits which arise from this research or the commercial use of these genes as well. Governments, though, disagreed on the question of who is a provider and who should consequently share benefits: so-called countries of origin – countries where the genes concerned can be found in the wild (mainly the South) – or countries which deliver ex-situ material originally stemming from other regions of the world as well (mainly the North). The latter was, under heavy pressure of the industrialized countries, decided. Furthermore, it is generally difficult to trace all the countries of origin where specific materials are used. Therefore it is technically easier to reward the direct provider, either a (poor) country of origin in case wild genes are provided or a (rich) country providing material from its gene banks.

Paragraph 3 of Article 15 indicates that genetic resources which were gathered before the entry into force of the CBD are *excluded* from its scope. Examples are gene bank collections which were established earlier, which will continue to be accessible as before and benefit-sharing is no prerequisite here. According to several observers, this exclusion poses a problem, as almost all the biomaterial that we know about today is not protected by the CBD (Crucible Group, 1994). Furthermore, the obligatory language used in Article 15 is, again, weak or subject to 'mutual agreement'. The first means that no party can be easily obliged to share benefits with others; the second that each party has to create its own mechanism in consultation with those who want access. Hence, no multilateral mechanism of benefit-sharing is established by the CBD. Finally, it should be mentioned here that access to technology as part of the benefit-sharing mechanism is dealt with in Article 16.

A much quoted example of how Article 15 could be used in practice is the Merck – INBio deal made in 1991 (Sanchez and Juma, 1994). The US-based pharmaceutical company Merck & Co and the National Biodiversity Institute of Costa Rica INBio agreed upon a contract by which the latter helps the former to collect genetic resources in Costa Rica's rainforests. Merck has received exclusive rights to develop pharmaceutical or agricultural applications of this material, in exchange for a US$1 million payment over two years and a 5% share of any royalties for INBio. In turn, INBio will spend a large percentage of these benefits on biodiversity conservation projects in Costa Rica. This deal seems a good precedent and may serve as a model for the parties to the CBD. However, the model is based on private and individual agreements, which might produce negative competition among developing countries. After all, Nicaragua possesses similar genetic material, but receives no payment. In addition, developing countries will need specific legislation, administrative and technical capabilities, and enforcement measures to benefit from such contracts and the provisions of Article 15. Therefore Svarstadt (1994) doubts whether developing countries are able to benefit soon, especially as gene collections and applications which were gathered and designed before the entry into force of the CBD are excluded from its scope.

B *Ego-perception*

Three of the twelve NGO respondents believe that NGOs positively impacted Article 15. They all refer to the fact that, according to them, pressure groups were one of the sources and pushers of the idea to link the access to biodiversity to some sort of compensation to the holders, either financially or in terms of technology or end-products. As one respondent noted, while claiming some indirect influence:

> "We did not propose Article 15, but we continuously pushed the issue of access and benefit-sharing, that it had to be dealt with by the CBD. The fact the convention did so is partly due to the NGOs."

C *Alter-perception*

Two of the eleven government respondents confirm this NGO claim. One refers to NGO influence on Article 15 in quite general terms, whereas the other is more specific:

> "The text of that whole Article 15, I brought it forward as a proposal of my country, but it was in fact drafted by an NGO representative – together with me – as he is a very good drafter. And the text is still there. (...) It was a question of facilitating access and at the same time giving the resource countries the right to set the value. There were long discussions and amendments but the basic architecture is still ours."

This government representative obviously credits the NGOs with some direct political influence on Article 15, although he could not point to specific phrases which were there due to the NGO concerned. The drafting process was, according to him, more or less a common exercise. Remarkably enough, the NGO respondent quoted in the above section on ego-perception was a representative of the same organization as the NGO drafter referred to here, but was obviously not aware of the close involvement of the latter.

D *Causal analysis*

GOAL-ACHIEVEMENT — Some NGO representatives themselves noticed a lack of goal-achievement in Article 15 during the interviews. The following arguments may underpin their observation. Firstly, the NGO Genetic Resources Action International (GRAIN) had opted for the so-called *negotiated agreement scenario* in Article 15 (Doc II.41). This scenario is based on a common heritage of biodiversity, with open access to genetic resources in return for payment and the recognition of the rights of small farmers and indigenous groups. Instead, a more or less *private property scenario* was followed by the country delegations and adopted in the CBD. This scenario is based on patent protection of biotechnology on the one hand and access to genetic resources restricted by states on the other. Therefore the negotiations went in a different direction from the one GRAIN had preferred. Secondly, NGOs had favoured a multilateral mechanism of benefit-sharing, through which countries of origin, small farmers, indigenous peoples and local communities would be rewarded if their resources and knowledge were utilized by others (Doc II.41, II.65, II.67, II.72). But the CBD defines a bilateral system of access and benefit-sharing based on the national sovereignty of countries. The community rights of local groups, though, are neither recognized nor dealt with in the CBD (although the role and rights of indigenous and local communities are recognized in Article 8(j), and the Diplomatic Conference that adopted the final treaty text recognized the fact that farmers' rights are 'an outstanding issue that seeks solution' in Resolution 3 of the Nairobi Final Act, May 1992). Thirdly, the existing ex-situ collections were excluded from the CBD's scope, something the NGOs had fiercely opposed (Doc II.43, II.76). At the same time, governments decided that benefit-sharing was no longer restricted to countries of origin, but extended to providers of ex-situ materials as well. Finally, some NGOs had opted for real targets of benefit-sharing, for example a 10% share of commercial profits for the holders of biodiversity, and had proposed a non-compliance mechanism based on non-access (Doc II.72), meaning that states violating the provisions of the CBD would be denied access to biodiversity of other countries. However, these NGO proposals were ignored.

Another NGO source of inspiration for the delegates could have been – again – the IUCN draft. Here a different and broader system of access was proposed. First of all, the export of and payment for all specimens and biomaterials are addressed, not genetic resources alone. Therefore the extraction and export of tropical hardwood for example is covered by the IUCN draft, but not by the CBD. Secondly, the draft reiterates the rules on access to the FAO Global System, as is outlined in the explanatory notes of the document (Doc II.52, p. 7 and 12). Access should be guaranteed, although it is not free of charge. After all, one of the basic premises of the IUCN draft was to raise money and create incentives to improve the conservation of biodiversity through an international tax system on the use of biomaterials. Therefore, thirdly, a multilateral fund was proposed. Any person who uses any biomaterial should pay a contribution to that fund, which in turn would compensate the state where the material comes from. Moreover, it would finance biodiversity projects, mainly in the South. These ideas, though, were definitely not reproduced in the CBD, as the Biodiversity Treaty's rules on access are more limited in scope (genetic materials), based on national sovereignty, and favour bilateral access and payment contracts.

Given these elaborations on the NGOs' ideas, the CBD does not address the preferences and demands of the agricultural and indigenous groups or of the IUCN. Therefore it seems justified to conclude that NGOs did not achieve their policy goals. Yet NGOs and developing countries did find some common ground on the question of regulated access. As the small survey among NGOs of the Task-Group revealed, most organizations, with the exception of IUCN, opposed the FAO system of free access (Doc II.65). They all wanted the Third World to benefit from the (commercial) use of genetic resources. Therefore most could live with the emphasis on national sovereignty – which restricts free access to resources and implies the right of benefit-sharing – at the end of the day. Hence, many NGOs were to some extent satisfied with Article 15, although their policy goals were different.

NGO INTERVENTIONS — In as far as NGOs can live with Article 15, the question is whether they were the ones which (co)determined that outcome. In other words: did NGOs prevent the adoption of a free access system? It seems not, as the developing countries introduced the emphasis on national sovereignty in the biodiversity arena and continued to fiercely oppose a system based on unrestricted and unregulated access. This influence of the G77 and China is confirmed by a large number of published and unpublished sources as well as respondents from both sides (Glowka *et al.*, 1994; Sanchez and Juma, 1994; Doc II.5). The G77 and China received some belated NGO support, but that support was definitely not needed by the developing countries given their strong position in the negotiations based on their genetic stock. Furthermore, the concept of benefit-sharing was elaborated on in a totally different way by the country delegations. Community rights are not recognized nor is a multilateral mechanism established by the CBD. Therefore one simply has to conclude that NGOs lacked political influence with regard to this topic.

Still, there is a claim of direct NGO influence on the wording of Article 15 (see alter-perception). Apparently contrary to the above conclusion, this NGO impact could be confirmed after consulting the key negotiators involved in the topic at that time. Specifically, the wording of Paragraph 15.2 on the facilitation of access to states' genetic resources by other parties was co-drafted by an NGO representative. However, this impact is considered to be legal advice, which did *not* modify the policy outcome in such a way that the NGOs achieved *more* of their goals. It was perfectly clear right from the start that the industrialized countries would demand a provision on the facilitation of access in Article 15 (Svarstadt, 1994). The general outcome was therefore more or less determined in advance, although the specific wording was not. Besides, even the NGO drafter concerned admitted that the same substance would probably have been adopted anyway, although expressed in different language, while the government respondent who confessed that he drafted the legal text for Article 15 jointly with this NGO representative said that the latter was "*instrumental* in assisting me to formulate the basic elements of that article" [italics by author]. He adds: "I would have been able to do it on my own." In other words, there was an NGO impact in terms of specific wording, not in terms of a reformulation of objectives. Even assuming that some 'NGO traces' are still left in Article 15.2, these do not compensate for the fact that the original policy goals on common heritage, community rights, multilateral mechanisms, taxes on biomaterials and existing ex-situ collections were not achieved. In fact, Article 15 was a North-South deal in which NGOs played a marginal role at best.

TABLE 5.5 *Scores on political influence in case 12 (Genetic resources)*

Counterfactual (C)	similar (0)
Goal-achievement by NGOs (GA)	no (0)
GA due to NGOs' own interventions (OI)	no (0)
Political relevance of topic (PR)	high (3)
Directness of impact (D)	direct (2)
Extent of political influence: PI = C [(GA x OI) + PR + D]	none (0)

E *Extent of political influence*

Given the foregoing elaborations, the extent of the NGOs' political influence is nil in this specific case. As NGOs did not achieve their policy goals nor prevented governments from adopting an even worse outcome, the counterfactual is assumed to be similar, irrespective of the fact that one NGO representative was able to impact some of the wording of Article 15 due to his legal expertise. Obviously, a small impact on wording and political influence do not equate. Given the interests of both industrialized and poor countries linked to this topic, it is considered to be of very high political relevance.

F Final conclusion

Many governments and NGOs in the biodiversity arena believed that developing countries and local holders of biodiversity should share in the benefits of the commercial use of genetic resources in the industrialized world (medicines, food crop varieties, biotechnology). However, these NGOs and governments disagreed on which regulations should be implemented. NGOs such as GRAIN and IUCN preferred a mechanism based on the common heritage principle to guarantee general access to genetic resources on the one hand, and on community property rights by which indigenous and local communities would benefit from the use of biological resources on the other. But most countries preferred a mechanism based on national sovereignty and bilateral contracts, so that individual governments would decide on access to genetic resources and the sharing of benefits. This option was therefore included in the CBD. As a consequence, NGOs were not able to really impact the policy outcome of Article 15, especially as their other policy goals were not attained either. This conclusion holds irrespective of the fact that: 1. NGOs supported the South's emphasis on national sovereignty in the second phase, and 2. one of their representatives was directly involved in the drafting of Article 15. In short, NGOs were definitely able to play a role in the negotiation process, but without positively affecting the substance of the policy outcome (process influence).

5.3.6 Case 13: biosafety

A Introduction

Article 19 of the CBD deals with the safe handling of biotechnology – which is also referred to as *biosafety* – and the distribution of its benefits. This topic, though, was only dealt with as a separate issue from INC3 (December 1991; Doc II.15). Before that it was debated in the context of other articles: on fundamental principles, general obligations and technology (Doc II.10). Both the provisions on principles and on general measures were, however, drastically cut during the negotiations and the elements on biotechnology were omitted. The technology provisions were, however, reshuffled during the last three INCs. Access to and transfer of technology from North to South – formerly two distinct articles – were fused into one (Article 16 of the final CBD). This move was mainly pushed for by industrialized countries. At the same time, biotechnology was not only included as an item in Article 16, but – moreover – separately dealt with in a new Article 19. This addition was, in turn, a preference of most G77 countries. To understand all the attention paid to biotechnology in what was originally meant to be a conservation treaty, one should not forget that this technology was considered one of the economically promising boom sectors in the 1980s (see subsection 5.2.1). In fact, biotechnology was already being considered by a special UNEP working sub-group of experts which prepared for the negotiations of a biodiversity treaty in 1990 (Doc II.9). However, the *biotech hype* decreased in the course of the 1990s. Progress in this field was not as fast as expected. Furthermore, consumers' concerns regarding safety and health slowed down its development.

Article 19 of the CBD contains two paragraphs on benefit-sharing and two on biosafety. The former aspect is interpreted in two ways: 1. benefit-sharing in terms of participation in biotechnology research by the providers of the genetic resources, and 2. benefit-sharing in terms of access to the resulting products, profits and knowledge. Paragraph 1 deals with participation in research. Its intention is to build a capability with regard to biotech development in all countries which are party to the CBD, but particularly in developing countries. This paragraph parallels provisions in Article 15.6 (see case 12, subsection 5.3.5), although the obligatory language of the latter is much weaker. Paragraph 2 concerns the sharing in results and benefits arising out of biotechnology. Similar provisions can be found in Articles 15.7 (genetic resources) and 16.3 (technology), where providers of genetic resources are also granted (some) access to the benefits of others, but in article 19 The emphasis is on biotechnology. Paragraphs 3 and 4 of Article 19 Deal with the safe handling of 'living modified organisms resulting from biotechnology' (LMOs). Paragraph 3 urges parties to *consider* the need for and modalities of a biosafety protocol to the CBD in the field of the safe transfer, handling and use of any LMO that may have an adverse effect on biodiversity (hence, it is not a *call* for a protocol). Legally-binding international standards on biosafety do not exist yet, although guidelines are being negotiated under the auspices of the Organization for Economic Cooperation and Development and the United Nations Environment Programme (UNEP) (Glowka *et al.*, 1994). Paragraph 4, finally, obliges parties to hand over all available information to countries where living modified organisms (LMOs) are introduced. This information should be given, irrespective of whether a protocol will be established or not, and should refer to the nature of these organisms, their safety and their potential adverse impacts.

This article was, according to Glowka *et al.* (1994, p. 96), one of the most difficult to negotiate. At one of the first experts meetings delegates already disagreed on whether biotechnology should be an integral part of the convention or whether it should be dealt with by the 1992 UNCED Conference (Doc II.11). The topic was nonetheless covered, but countries clashed on many of the details. The most furious debate was on the safety of genetically modified organisms (GMOs). Malaysia had introduced a proposal at INC3 to subject the introduction of any GMO to the prior informed consent of the country where the introduction would take place (Doc II.13, p. 26). Furthermore, all information on the safe use and handling of these organisms should be provided to the receiving country. Elements of these proposals, however, were fiercely objected to by others, in particular the USA (Doc

II.5), which believed that there was no need to worry about what they called 'products of biotechnology'. The USA particularly rejected the use of the term 'GMO', as it conjures up an image of danger. But 'products of biotechnology' are, according to them, not dangerous in the way chemical or nuclear wastes are. As both the USA and Malaysia stuck to their positions for a long time, it was difficult to find a compromise. Finally, the notion of 'living modified organism resulting from biotechnology' (LMO) – instead of 'genetically modified organism' (GMO) – was adopted. This term is, by the way, broader than the concept of GMO, and covers not only modified organisms based on genetic engineering, but those based on for example traditional breeding techniques as well. In addition, other aspects of the Malaysian proposals were mitigated too, including the obligatory language. Finally Paragraphs 19.3 and 19.4 were negotiated and a call to consider the need for a protocol to settle a kind of prior informed consent procedure in the near future became part of Article 19.

It should be noted that the CBD addresses the topic of biosafety in Paragraph 8(g) as well (see case II), where countries are urged to establish or maintain national means to manage the risks associated with LMOs likely to have adverse effects on biodiversity and/or human health. The difference, though, is that this paragraph deals with possible and appropriate means at the *domestic* level, whereas Paragraphs 19.3 and 19.4 refer to possible *international* regulations. Furthermore, the former includes human health, the latter do not.

B Ego-perception

Three of the twelve NGO respondents are convinced that they impacted the language on biosafety. One even believes that NGOs were the source of it:

> "The NGO influence, most significantly, was on biosafety. Biosafety was not in the convention at all."

Whereas this respondent seems to refer to biosafety in general (Paragraphs 8(g), 19.3 and 19.4), one of her colleagues particularly mentioned the call for a protocol (Paragraph 19.3). However, neither makes clear whether NGOs directly or indirectly influenced the language on biosafety, for example through the drafting of legal text or through introducing and debating the topic. A third respondent, though, refers to the latter. She acknowledges the fact that Malaysia was the initiator of the topic in the biodiversity arena, but adds that there was a lot of co-operation between this country and NGOs.

C Alter-perception

Three of the eleven government respondents indeed believe that NGOs played a role in elaborating the biosafety issue in the formation phase of the CBD, but their confirmations are rather weak. They assume some influence because NGOs were very active and noisy. One very carefully noted: "Maybe some of the Third World NGOs were influencing the language regarding the biosafety protocol, although there were also country delegates working to get that in." Another delegate, however, spontaneously rejected any NGO influence in this field:

> "For example the inclusion of biosafety was not a merit of NGOs. That was Malaysia. I was even involved in the brokerage of the Malaysia/US deal on the compromise formula about the protocol on biosafety, but the NGOs were definitely not involved there."

D *Causal analysis*

GOAL-ACHIEVEMENT — Early NGO documents do not reveal a great interest in biosafety (Docs II.65, II.78). It is therefore difficult to deduce initial policy goals with respect to this topic, although NGOs generally take a critical stand on genetic engineering and the testing and release of genetically modified organisms (GMOs). This also became obvious when Malaysia delivered its proposal: NGOs immediately agreed. According to one of their internal memos: "The Malaysian proposal on biosafety deserves a lot of support by NGOs" (Doc II.61, p. 3). Since this proposal fuelled the debate on biosafety in the INC, NGOs subsequently tried to achieve some additional objectives. They promoted a clause in Article 19 on strict liability concerning possible adverse impacts of imported GMOs on the environment (Doc II.68). In other words, providers of GMOs to other parties should pay compensation if their GMOs harm the environment in the recipient country. However, this provision was finally not included. Furthermore, NGOs wanted – just as several G77 countries did – industrialized countries to force their private sectors to keep to prior informed consent and full information supply on GMOs. These ideas, though, were opposed by the OECD, either on principle or on pragmatic grounds, and therefore rejected. Finally, some NGOs campaigned for a global moratorium on the release of GMOs, but such an objective was, of course, too idealistic given the interests of the biotech industry. In short, NGOs achieved only some of their policy goals, particularly as the language in the Biodiversity Convention was severely weakened. It does not contain strong obligations on biosafety.

NGO INTERVENTIONS — It seems that NGOs were quite reserved and, at best, reactive – rather than proactive – in the field of biosafety, at least in the first instance. Although the early UNEP drafts already included provisions on biosafety – for example an appeal for domestic codes of conduct on the release of GMOs – NGOs were silent on the subject (Doc II.8). Until G77 countries opened the debate, NGOs focused on other topics in the biodiversity arena. For as far as they were involved in biotechnology issues, organizations such as the Third World Network (TWN) were mainly committed to and engaged in the UNCED process. But as soon as Malaysia and others had introduced their views, NGOs supported and built on those ideas. The fact that NGOs only reacted was also vigorously stated by one Malaysian delegate:

> "The one thing the NGOs are now pursuing, which is on biosafety, was a Malaysian proposal, made by our government delegation. They made no contribution at all. So we were 'the' ones who brought it up."

Some other respondents agree with this observation, including some non-government ones. Therefore the question is whether NGOs indeed had an impact on the policy outcome. Probably not, because – firstly – some G77 countries were very proactive on their own and, secondly, because the final outcome was a compromise between these countries and the USA. Furthermore, the additional NGO proposals were not seriously considered. The following conclusion seems therefore justified: NGOs exerted no political influence on this topic, either directly or indirectly.

There may be two reasons to water down this conclusion. Firstly, several NGOs had raised their voices to prevent Southern countries being turned into testing grounds for GMOs. One government representative from Latin America touched upon this and admitted that domestic NGOs made the public and the government in his country more aware of the possible dangers of such test fields. But such messages were not generally heard or addressed in the biodiversity arena by NGOs or governments during the initial negotiation rounds. Apart from this one person, no government respondent referred to this role of NGOs. Secondly, the IUCN draft also includes an article on man-made organisms. It urges states to control the release of these and, moreover, *to pay attention to security guidelines* (Doc II.52, p. 7). In other words, the idea to have international biosafety regulations in the CBD was already there. Subsequently, this view may have been taken over by UNEP's experts meetings, but the UNEP proposals speak of national codes of conduct, which were already being discussed in other arenas as well (for example the OECD). Therefore the impact of the IUCN is questionable, and for as far this union may have been influential in this field, this impact has already been covered by case II on in-situ conservation (Article 8(g) on living modified organisms).

E *Extent of political influence*

Without NGO interventions, the policy outcome would have been exactly the same, as table 5.6 makes clear. NGOs did nonetheless attain some of their objectives, but this was due to the interventions of some Third World countries. Biosafety can definitely be considered a very relevant topic given the possible restrictions on biotechnology development and on genetic engineering. Therefore industrialized countries were reluctant to accept any binding regulations in that area within the CBD.

TABLE 5.6 *Scores on political influence in case 13 (Biosafety)*

Counterfactual (C)	similar (0)
Goal-achievement by NGOs (GA)	some (1)
GA due to NGOs' own interventions (OI)	no (0)
Political relevance of topic (PR)	high (3)
Directness of impact (D)	no impact (0)
Extent of political influence: PI = C [(GA x OI) + PR + D]	none (0)

F Final conclusion

Although NGOs were quite noisy on biosafety just before the CBD was agreed upon, their initial role was very limited, especially as NGOs mainly concentrated on the UNCED process as far as this topic was concerned. Not until Malaysia and some other countries introduced new proposals on the safe, transboundary transfer of genetically modified organisms (GMOs) at INC3, did NGOs wake up, supporting similar ideas and adding others. The new ideas were, however, not addressed by the Intergovernmental Negotiation Committee and the final policy outcome – Paragraphs 19.3 and 19.4 – was mainly determined by the clash between the G77 and the USA. True, NGOs co-operated with G77 delegates to strengthen and improve the proposals, but it is unlikely that without these NGO initiatives the outcome would have been different. Nevertheless, some NGO impact on Article 8(g) on biosafety – due to the IUCN draft – is accepted, but this political influence has already been covered by case 11.

— 5.4 NGOs in the implementation phase (1992-1995) —

Section 4.4 already underlined the fact that the treaty formation and implementation phases differ. For example, the decisions of INC5 (May 1992) and the first Conference of the Parties (COP 1, December 1994) are quite different in nature. To repeat, many of the latter are not *substantive* decisions in the literal sense. Such decisions either postpone actions or refer to *interim* products. Therefore the influence of NGOs cannot be assessed with regard to many *end* products, for example a protocol, as the scope of this study does not extend beyond COP 1. In addition, it should be taken into account that the CBD is a framework convention. This implies that several provisions are still rather broad and vague and that the COP needs to elaborate these. Such elaborations will often take the form of *agenda-setting* in the first place. Finally, it should be recognized that the number of NGO observers in the implementation phase of the CBD increased dramatically. For example, the four UNEP panels, the Trondheim Conference, the two sessions of the Intergovernmental Committee on the Convention on Biological Diversity (ICCBDs), the Mexico Meeting of experts, as well as the first Conference of the Parties (COP 1) witnessed in the period 1992-1995 an ever larger attendance by NGOs, with anything up to more than a hundred at any one meeting (see subsection 5.2.4).

5.4.1 Case 14: indigenous people

A Introduction

The convention topic of indigenous people and local communities was not one of the main items in the biodiversity arena in the period 1992-1995, unlike for example finance, technology transfer, bioprospecting and biosafety. As one government respondent noted:

> "In a way there has not been a debate on indigenous and local communities yet. There was an opportunity essentially to present views in the ICCBDs, but there has not been substantive work that has gone ahead on the issue yet."

For example, no UNEP panel explicitly went into this subject, although the topic was discussed under other headings, such as scientific and technical co-operation by Panel 1 (Doc II.5). The first session of the Intergovernmental Committee on the Convention on Biological Diversity – ICCBDI, Geneva, October 1993 – did not cover the role of indigenous and local communities either (Doc II.18). Some discussion nonetheless took place on the participation of indigenous groups in the conservation and sustainable use of biodiversity and NGOs also raised their voices there. For example the Indigenous People's Caucus made interventions at ICCBDI and the Congress of Guaymi People denounced the US for trying to patent a human cell-line virus from Indians in Panama (Doc II.49). Furthermore, IUCN and WRI had organized a Global Biodiversity Forum before ICCBDI. This forum also addressed community property rights and the need to recognize the value of traditional knowledge (Doc II.77). The outcome of this meeting was, moreover, presented at ICCBDI. But this first session of the Intergovernmental Committee was more or less a failure. Delegates could not even agree to adopt the reports of the two working groups, although some recommendations on financial and procedural matters could be given to the Global Environment Facility (GEF) and the interim secretariat of the CBD.

The second session of the ICCBD, which was held in Nairobi in July 1994, was more successful and the implementation of Articles 8(j) and 10(c) on indigenous knowledge, innovations, practices and benefit-sharing were indeed debated (Doc II.19). The meeting experienced a lot of state interventions in favour of indigenous people and local communities, including from countries in the EU and the G77, and from the USA (Doc II.83). Some delegates and NGOs suggested initiating a study on the implementation of Articles 8(j) and 10(c), considering the establishment of a protocol to the CBD on indigenous and local communities, and including the topic as a distinct item on the agenda of the first COP. Despite these suggestions, no concrete proposals were made and delivered to COP 1.

In the meantime, two supporting sessions of the ICCBD had also covered the topic: the Trondheim Conference (May 1993) and the Mexico Meeting (April 1994) (Doc II.5, Doc II.22). The former was a relatively open meeting to discuss all implementation aspects of the CBD and prepare for the first session of the Intergovernmental Committee (ICCBD1). Several workshops were held, among others on indigenous people. The Mexico Meeting resulted from the disagreement among delegates about establishing an interim Scientific and Technological Advisory Committee (ISTAC) to prepare for the first Conference of the Parties. Some countries favoured such an interim-STAC, others opposed it (Doc II.18). As a compromise, an Open-Ended Intergovernmental Meeting of Scientific Experts on Biological Diversity was held in Mexico City in April 1994. It had to identify programmes of international scientific and technological co-operation, prepare an agenda for research on biodiversity, and identify technologies for its conservation and sustainable use (Doc II.22). Indigenous and traditional knowledge and techniques were considered and research on the implementation of Articles 8(j) and 10(c) was included on the agenda. Furthermore, a specific annex to the meeting's report addressed the integration of modern and traditional knowledge, innovations and practices.

The first Conference of the Parties (COP 1) which convened in Nassau in November 1994 did, however, not make specific decisions on indigenous and local communities. The topic was nonetheless covered by decision I/2 on financial resources and decision I/9 on the medium-term programme of work of the COP (Doc II.32). According to the former decision, projects on the involvement of indigenous and local communities in the conservation and sustainable use of biodiversity should be eligible for funding by the GEF, whereas the implementation of Article 8(j) should, according to the programme of work, be considered by the third Conference of the Parties in 1996. However, references to indigenous people in the mandate of the Subsidiary Body on Scientific, Technical, and Technological Advice (SBSTTA) and in the COP's input in the work of the Commission on Sustainable Development (see case 16) were finally deleted. In addition, neither ICCBD2 nor COP 1 endorsed the research agenda, including the provisions on indigenous people, set by the Mexico Meeting (Doc II.19, II.32, II.95). It was left up to the SBSTTA to draw on the report of the scientific meeting in Mexico in the near future. In short, COP 1 postponed substantive decisions on this matter. Proposals to conduct a study and to discuss a protocol on indigenous people were not reiterated nor decided upon.

B Ego-perception

Two of the twelve NGO respondents claim influence on the elaboration of the indigenous people's provisions in the implementation phase. One reference, though, is very general and weak, whereas the other respondent's view quoted below is more specific:

> "The Latin American NGOs were very much focused on indigenous people and community rights. We had, I think, influence on the medium-term programme of work of the COP and on also on individual delegations."

The reference to the programme of work concerns Article 8(j) which is scheduled to be addressed by the COP in 1996 (Doc II.32).

C Alter-perception

Three of the eleven government respondents also believe that NGOs impacted the debate on indigenous people and local communities between UNCED and COP I. Two references are rather general in nature, whereas one is very specific:

> "The question of indigenous knowledge was also on the agenda of the Mexico Meeting. There was a group of scientists, however, not well experienced in such issues. The indigenous people's groups assisted us, which was very helpful, as they had already dealt with the Article 8(j). They produced a very good recommendation which was part of our message to ICCBD2. And it came up again at COP I, SBSTAI and COP 2. So you can see how certain ideas are developed and supported by NGOs with the consent and the interest of the COP."

One respondent also refers to the domestic level, where indigenous groups pushed for the elaboration of national laws and regulations in compliance with the convention. But this level of influence is not covered by this study.

D Causal analysis

GOAL-ACHIEVEMENT — In the run-up to COP I, NGOs and indigenous groups had launched several proposals regarding indigenous people, for example on the establishment of a prior informed consent procedure to be applied in advance of any access by third parties to indigenous resources and knowledge; on a protocol to establish the community rights of these groups; and on a greater involvement of these groups in national and international decision-making (Doc II.44, II.45, II.72, II.90, II.94). With regard to COP I itself, NGOs had asked for the immediate coverage of Articles 8(j) and 10(c) by the COP's programme of work, by the SBSTTA, by a specific working group, and by the interim secretariat of the CBD (Doc II.96, II.98). Furthermore, they wanted to directly link the rights of local people to the access to genetic resources in the medium-term programme of work of the COP

(Doc II.99). None of these objectives, however, were achieved. A protocol, a prior informed consent procedure and a special working group were not established. Moreover, the item of indigenous and local communities was scheduled separately for COP 3 in 1996, whereas NGOs would have preferred to already address this topic in relation to the access to genetic resources in 1995 (Doc II.26). NGOs were therefore very disappointed about the outcomes of the ICCBDs and COP 1 (Docs II.95, II.99) and it seemed that they had not achieved their policy goals. Nonetheless, a relative success may have been the upgrading of the topic in the programme of work. Whereas the interim secretariat in its proposal had scheduled indigenous and local communities on the agenda for 1997, COP 1 decided to deal with it one year earlier, in 1996 (compare Doc II.24 and II.26).

As far as the agenda of SBSTTA was concerned, the COP requested this body to prepare a proposal for their own programme based on the action priorities set by the COP 1. Therefore indigenous topics would not be immediately addressed at the first meeting of the SBSTTA (which was, by the way, held in September 1995). Again a preference of the NGOs was ignored by governments. In addition, it still remained to be seen at that time whether the SBSTTA would adopt the recommendations of the Mexico Meeting, including the one on indigenous knowledge. With regard to the *substance* of this recommendation, one may assume that NGOs achieved their goal, as they were the ones who wrote it. As far as the low *status* of this recommendation is concerned, they could hardly be satisfied. All in all, NGOs at best achieved *some* of their policy goals.

NGO INTERVENTIONS — One may wonder whether NGOs were able to impact the topic in the period 1992-1994, especially as substantive decisions were not yet being made. However, the aspects which NGOs may still have affected were: 1. the introduction of indigenous and local communities on the agenda of the ICCBDs; 2. the formulation of recommendations at the Mexico Meeting; 3. the transfer of the topic in the COP's programme of work from 1997 to 1996. As far as the first aspect is concerned, it should be emphasized that the topic had to be dealt with by governments sooner or later anyway, given the substance of the CBD and the need for an elaboration of Articles 8(j) and 10(c). Furthermore, several delegates and experts also continued to raise the topic at the UNEP panels, Trondheim Conference, ICCBDs and COP 1 (Doc II.5). Therefore it is unlikely that without the NGOs the topic of indigenous and local communities would not have been addressed in the implementation phase to date. But NGOs definitely contributed (process influence). Secondly, indigenous groups provided a text proposal at the Mexico Meeting, as the above mentioned key delegate admitted. Although the topic was already on the agenda, it is likely that the intervention of indigenous groups has led to a better articulation of their interests, given the lack of expertise on the delegates' side. Some political influence can therefore not be excluded.

However, the status of the Mexico recommendations remained unclear in the period which is covered by this study, as it was left to the SBSTTA to address these – or not – in the near future (Doc II.5, II.85, II.95). In short, NGOs impacted some text elements, but this language has been rather irrelevant up to the end of 1994.

Thirdly, NGOs may have caused governments to fix the topic in the COP's programme of work at an earlier date (1996 rather than 1997). Such an effect is indeed believed to have occurred, given the change in position of several countries, for example Brazil. Whereas Brazil was quite hostile to covering the rights of indigenous people in the CBD before UNCED, merely for domestic reasons, the country started to co-operate with indigenous groups both at national and global level after the Earth Summit. Since then Brazil has increasingly defended the indigenous cause in the biodiversity arena, partly because the country has been persuaded by arguments (e.g. the need to protect indigenous and local groups against patent piracy; indigenous knowledge as a means to preserve and use biodiversity) and partly because of government self-interest (e.g. green image building; better relationships with and more control over non-governmental groups). In Nassau the Brazilians even asked the COP to deal with the role and protection of indigenous and local communities in 1995 (the same demand as that of the NGOs; Doc II.85). As the EU (and others) favoured 1996 while UN officials had opted for 1997, the final compromise was 1996. Nonetheless, it is believed that indigenous groups and NGOs in general have impacted governments to give more priority to the topic, in particular Brazil, but European countries as well. This is also confirmed by the increasing presence of indigenous groups and local chiefs in the biodiversity arena in the period 1992-1995. These players were very visible and negotiated intensively with many country delegations (personal observations). Several government respondents also expressed praise for the extent and nature of the indigenous group's interventions (though these views may also be an expression of political correctness). Moreover, some NGOs were able to link local activism with global diplomacy in the biodiversity arena. As more cases of patent piracy became known and questioned by NGOs, this topic could no longer be ignored by governments, particularly when letters from victims reached the delegates (e.g. the Guaymi people; see above). In addition, the Indian activist Vandana Shiva visited several biodiversity sessions – among others the Trondheim Conference – and made delegates aware of what for example had happened in her country now several US and Japanese corporations had been granted patents on the medicinal and insecticidal properties of the *neem* tree (Doc II.5, II.44). Although these properties had already been discovered, used and innovated by local people for centuries, the free and local access to these applications was disputed by Western industry now patents had been granted. This led to farmers' protests. Some government respondents admit that delegations in the biodiversity arena turned out to be responsive to those local protests and they explicitly refer to Shiva as an intermediary. As a consequence, the topic has been given more intergovernmental priority.

To conclude, the interventions of NGOs, and in particular indigenous ones, has affected some policy outcomes a little, although the low status and non-substantive nature of these outcomes should not be overlooked (a recommendation without commitment and agenda-setting at some earlier point in time).

E *Extent of political influence*

Table 5.7 summarizes the above analysis. The counterfactual is considered different (probably a less pronounced recommendation and less priority for the topic). This outcome, however, addressed the NGOs' policy goals on indigenous and local communities to only a very limited extent. What is more, this outcome was only partly due to the NGOs themselves. After all, governments would have produced a recommendation anyway and some countries also preferred early coverage of the topic by the COP. In addition, the political relevance of the outcome is considered very low. It does not refer to any real substance yet and the political status of it is limited as well. Besides, countries generally did not disagree, so one may assume that there were no real interests at stake. In short, NGOs exerted *some* political influence.

TABLE 5.7 *Scores on political influence in case 14 (Indigenous people)*

Counterfactual (C)	different (1)
Goal-achievement by NGOs (GA)	some (1)
GA due to NGOs' own interventions (OI)	some (1)
Political relevance of topic (PR)	low (1)
Directness of impact (D)	direct impact (2)
Extent of political influence: PI = C [(GA x OI) + PR + D]	some (4)

F *Final conclusion*

The number of indigenous groups involved in the biodiversity arena grew substantially after UNCED. This increase turned out to be effective in terms of (some) political influence. Indigenous representatives produced a recommendation to the Mexico Meeting of experts on biodiversity, whereas NGOs pushed governments to prioritize the topic of indigenous and local communities a bit more. Substantive decision were, however, not taken on this topic. NGOs had also favoured to fuse concerns on indigenous people and access to genetic resources into one item in the COP's programme of work, but failed in convincing delegates to do so. They had also pressurized governments to immediately address the rights and knowledge of indigenous and local communities, for example through a protocol or a working group, but most governments so far have only paid lip service to this idea. In short, the NGOs' political influence was very limited in this specific case.

5.4.2 Case 15: marine biodiversity

A Introduction

As the successive drafts for the convention increasingly focused on national priorities and measures during the negotiations, the common concerns such as the oceans were slowly pushed into the background. Some NGOs such as the World Conservation Union (IUCN) and World Wide Fund for Nature (WWF), organizations such as the United Nations Environment Programme (UNEP), and some country delegations including, of course, the small island states, regretted that and tried to reintroduce these items onto the agenda in the biodiversity arena. Furthermore, the debate on the loss of biological diversity tended to focus mainly on land ecosystems and in particular tropical forests (see case 16). However, there are more 'hot spots' of biodiversity to distinguish, for example in coastal and marine ecosystems. Consequently, this topic came up at COP 1, where it was introduced as a new item in the medium-term programme of work of the Conference of the Parties. Whereas the initial draft of the interim secretariat did not mention marine biodiversity at all, the final decision was to schedule it for COP 2 (1995) (compare Doc II.24 and II.26).

There may have been further reasons for introducing this topic. First of all, COP 1 convened in Nassau, Bahamas, a member of the AOSIS. The COP physically took place near the beach and close to several coral reefs. All delegates could therefore experience the richness of marine biodiversity (at least if they wished). In addition, the marine topic seemed to be a 'nice alternative' to the forest debate. As the latter had become very politicized since UNCED, as will be outlined in the next case, the former could draw attention to 'safer' topics.

Although less sensitive, controversies were not absent in this area either. For example the G77 and China had still to agree to deal with the global nature of marine biodiversity. As outlined before, most developing countries did not want the convention to deal with global ecosystems. They preferred a treaty covering national problems and priorities. Therefore explicit references to global common concerns such as oceans, the atmosphere and Antarctica – present within brackets in earlier drafts of the CBD – were omitted during the course of negotiations (compare Doc II.4 and II.8). But now the marine issue had been reintroduced, the COP had to deal with a problem not only within national jurisdiction (coastal areas, territorial waters, exclusive economic zones), but also *outside* national jurisdiction (open seas, oceans).

It should be recognized, though, that the marine topic was only put on the agenda as something which the COP should elaborate on. No substantive decisions have as yet been made or measures agreed upon.

B Ego-perception

Three NGO respondents claim they were the ones who raised the issue. One of them noted: "Had it not been for NGOs, they might have forgotten this big ecosystem for another two years." An important step in getting marine biodiversity into the picture was, according to several respondents, the Global Forum in Nassau, which was organised by IUCN, World Resources Institute (WRI) and UNEP. It convened two days before COP I. This forum – a similar one was also organized before the first session of the Intergovernmental Committee (ICCBDI) – aimed at bringing together biodiversity experts from all sectors, government and non-government, to discuss the different aspects of the CBD. Although the forum in Nassau covered a lot of issues, marine biodiversity was the main theme. A recommendation to COP I was produced urging parties to consider the issue in their work programme (Doc II.32, II.54, II.97). As one NGO respondent said:

> "The way the marine issue was introduced into the medium-term work programme, I will bet came directly from our forum on Sunday. People in the room were rather impressed when they discovered that there was a real case for bringing marine affairs strongly into this convention. That slid into the programme. OK, it was pushed by Sweden. A Swedish delegate was at our meeting last Sunday. I cannot say he got the idea sitting in that room and took it to the COP. But you move a group dynamic, by having things like the forum."

Obviously, NGOs do only claim indirect political influence here, as they moved delegates to act. They did not draft texts for example.

C Alter-perception

Three government respondents also refer to the marine issue as one example of NGO influence. One key delegate on marine biodiversity even believes that the Global Forum played a decisive role in overcoming the objections of most parties:

> "We would not have been able to have marine biodiversity in the work programme of the COP if it had not been for the forum. There was such a resistance from the majority of the delegations. Most of them narrowly focused on the terrestrial side."

The other two respondents refer to the NGOs' non-controversial, co-operative and knowledge-driven approach on marine biodiversity, compared to for example the very politicized and confrontational NGO strategies on biosafety (case 17). They believe that as a result NGOs could have become influential with regard to this topic.

D *Causal analysis*

GOAL-ACHIEVEMENT — The most important objective of the Global Forum was not only to address several technical and scientific aspects of the Biodiversity Convention, but to raise particular interests in some topics as well (Doc II.54). Marine biodiversity was one. Furthermore, individual NGOs, such as Greenpeace, raised this topic as well (Doc II.97). Given its inclusion in the COP's programme of work afterwards, NGOs achieved their initial goal. As one NGO respondent put it: "Marine biodiversity is at the stage of simply bringing up the issue, and that is going well." True, NGOs would definitely have liked a more substantive debate on marine biodiversity at COP I, but that did not happen until one year later at the second session in Djakarta (Doc II.33).

NGO INTERVENTIONS — It is likely that NGOs played a role in the agenda-setting, although the intergovernmental contact group in which the medium-term programme of work was negotiated in Nassau closed its doors to NGOs (Doc II.85, II.99). Despite that, the causal links between the Global Forum, the Swedish delegation and the COP's decision on the medium-term programme of work could be verified through Swedish and Dutch government delegates who were involved. What is more, some other government respondents who were explicitly asked to either accept or reject the NGO impact on this topic were positive on their role as well. The conclusion therefore is that NGOs were able to exert indirect political influence with regard to marine biodiversity. It should nevertheless be recognized that several governments were very pleased with the inclusion of this topic in the programme of work. An example is AOSIS (Doc II.5, II.30), although this alliance mainly focused on inclusion of marine biodiversity in the programme priorities of the GEF. In consequence, the medium-term programme of work was for the most part covered by others (Sweden and, indirectly, NGOs).

E *Extent of political influence*

NGOs were jointly responsible – together with some countries (e.g. Sweden) – for raising the topic of marine biodiversity at COP I and, subsequently, its agenda-setting. Given the involvement of the countries, the extent to which the achievement of NGO goals was due to their own interventions (IO) is considered 'substantial' and not 'much'. In addition, the extent of goal-achievement (GA) does not amount to 'much' either, as NGOs had wanted a more substantive debate on the matter at COP I. The political relevance of the topic is low. After all, no binding, substantive decisions on marine biodiversity were made by governments at COP I, nor did the topic raise much conflict.

TABLE 5.8 *Scores on political influence in case 15 (Marine biodiversity)*

Counterfactual (C)	different (1)
Goal-achievement by NGOs (GA)	substantial (2)
GA due to NGOs' own interventions (OI)	substantial (2)
Political relevance of topic (PR)	low (1)
Directness of impact (D)	indirect (1)
Extent of political influence: PI = C [(GA x OI) + PR + D]	some (6)

F *Final conclusion*

Given the fact that the CBD mainly focuses on national programmes and priorities, the biodiversity-rich common resources such as the oceans were pushed into the background. NGOs, together with several countries, nonetheless succeeded in reintroducing this topic onto the agenda of the COP. An important step to achieve this goal was the Global Forum of IUCN, WRI and UNEP which was convened in Nassau, Bahamas, just before COP 1. This forum raised the issue and encouraged delegates to take some action. Consequently, marine biodiversity has been included in the COP's medium-term programme of work.

5.4.3 Case 16: forests

A *Introduction*

In the run-up to UNCED, three legally-binding international environmental treaties had to be negotiated and agreed before the Earth Summit: on biodiversity, the climate and forests (Johnson, 1993; Spector *et al.*, 1994). The last one, however, was not achieved. Industrialized and developing countries could not agree on a text due to different objectives and interests. Whereas the majority of the South favoured a *forestry* convention on the exploitation and management of forests all around the world, the majority of the North favoured a *conservation* treaty and emphasized the need to preserve the tropical forests in particular (Humphreys, 1994; Kolk, 1996). The G77 did not like this particular emphasis, as the exploitation of tropical forests is 'big business'. Besides, the developing countries accused the North of 'environmental colonialism'. After all, the rich countries had already exploited their primary forests for the sake of their economic development in the past and now they demanded that the poor should *conserve* these natural resources. In other words, the rich countries wanted to impose conditions for development on the poor which they had never applied themselves. Moreover, the question remained whether they were legally entitled to do so, as all nations are sovereign to exploit their resources as they wish (Article 3 of the CBD). The G77 also believed that the North favoured such a treaty for purely selfish reasons, as tropical forests are carbon sinks and 'hot spots' of genetic diversity. In short, the South – in particular Malaysia and India – opposed a binding forest convention at the end of the day, irrespective of the fact that the seriousness of the problem – a huge rate of deforestation around the globe – would certainly have legitimized such a legal instrument.

The failure to agree on this convention was to some extent masked by the adoption of a non-binding Authoritative Statement of Forest Principles at UNCED (Johnson, 1993). This statement includes quite general principles, and elements on for example sovereignty, the sustainable management of forests, energy, reforestation, scientific research, technology and trade. However, NGOs present in Rio condemned the statement, calling it a *Chain Saw* Charter (Johnson, 1993, p. 115). There is not much on conservation and sustainable management, nor does the statement deal adequately with the current trade in tropical timber, which accelerates deforestation so much. Furthermore, it does not include a provision which urges nations to start immediate negotiations on a binding legal instrument on forests, a provision which drafts of the statement *had* included, at least in brackets (Johnson, 1993).

The question after UNCED was how to proceed on forests, as most countries and NGOs still wanted to have an instrument as soon as possible. Several possibilities were heard:

1 a Forest Convention under the auspices of the FAO;
2 the elaboration of the UNCED Forest Principles in the Commission for Sustainable Development (CSD);
3 a Forest Convention negotiated outside the UN system (e.g. on the initiative of a group of countries);
4 the adoption of a forest protocol to the CBD.

The last possibility is, of course, the most interesting for this study.

The option to develop a forest protocol under the CBD was heard in the biodiversity arena immediately after UNCED. For example Sweden presented a proposal at the Trondheim Conference and several NGOs also promoted the adoption of such a protocol at the sessions of the Intergovernmental Committee (ICCBDS) and at the first session of the Conference of the Parties (Doc II.40, II.46, II.101). However: "They got nowhere", as one government respondent noted. The CBD option was opposed not only by many developing countries, but by several industrialized ones as well (Doc II.5). The latter preferred a different forum for negotiating a forest treaty. Finally, the Commission on Sustainable Development (CSD), established at UNCED, became the (provisional) body for this. In the period 1995-1997 the Intergovernmental Panel on Forests (IPF) met four times to try to achieve consensus on the forest issue among governments and to co-ordinate proposals for action to support the management, conservation and sustainable development of forests (Doc II.88). However, progress was slow and the IPF could not decide upon a new, legally-binding instrument. Additionally, UN members met in a special session of the UN General Assembly (UNGASS, June 1997) to consider the implementation of Agenda 21 and the Forest Statement five years after the Rio Summit (Doc II.89). This meeting decided to establish – again – an intergovernmental working group to continue debates on a forest treaty. So far a legally-binding instrument has not been adopted.

To contribute to – among other things – the debates on forests in the Commission on Sustainable Development (CSD) in 1995, the first session of the Conference of the Parties to the Convention on Biological Diversity (COP 1) had decided to send a message to this UN body. This message, adopted in Nassau, aimed at building on the linkages between the CSD and the COP/CBD. One such link was forests, dealt with in paragraph 15 of the message (Doc II.28). This section emphasizes the importance of forests for the conservation and sustainable use of biological diversity, encourages the CSD to implement the forest principles adopted in Rio and makes clear that the CBD is willing to co-operate with the CSD in this area.

B *Ego-perception*

Although they did not achieve their goal with regard to the forest protocol, three NGO respondents still believe that they impacted the debate on forests in the biodiversity arena:

> "The forest issue is one issue where the NGOs had an impact and most of our effort has been directed towards the CSD statement. There is some fairly strong language in paragraph 15 about the role of the convention in the forest issue. I think NGOs influenced that language through traditional diplomatic lobbying work and by oral statements in the conference."

C *Alter-perception*

Government respondents are less optimistic with regard to NGO influence. Two of them spontaneously commented upon the role of NGOs in the forest issue. To quote one: "The very, very strong message that came from the NGOs for the need for a forest protocol was totally ignored by the COP." Another government respondent also refers to the 'failure' of NGOs regarding the protocol, but nonetheless does not exclude the possibility that they influenced the text in the CSD statement.

D *Causal analysis*

GOAL-ACHIEVEMENT — It is rather difficult to distinguish 'the' NGOs' policy goal on forests, as not *all* organizations favoured an emphasis on this topic in the biodiversity arena, let alone on the adoption of a protocol (personal observation). In particular Latin American NGOs, which were able to co-operate quite well with their respective governments on a lot of other topics, such as indigenous people, did not want to put this co-operation at risk by emphasizing the forest item too much. These NGOs also believed that conservation priorities should not overrule development goals to too large an extent. In addition, some organizations like IUCN were indeed in favour of a binding instrument, but not necessarily in favour of a CBD *protocol* (Doc II.5). Consequently, there was a split in the NGO community on forests, with among others Greenpeace and WWF as the most prominent spokesmen in favour of a forest protocol. It is more than obvious however that they did not achieve their goal. As far as the CSD input was concerned, most NGOs agreed that references to forests should be included in that message (Doc II.59). Given the fact the COP did so, they more or less achieved their goal with respect to that, although they had preferred a much stronger and more elaborate message. In addition, NGOs were displeased that forests were not separately mentioned in the medium-term programme of work, but only covered by the item 'terrestrial ecosystems', to be dealt with not before 1996, eighteen months *after* the special session of the Commission on Sustainable Development (CSD) on forests. NGOs considered this a lost opportunity to really link the CSD and the COP (Doc II.50, II.98, II.99). Furthermore, they believed that there was a danger that the CSD would already have elaborated on and determined the international agenda on forests before the COP could even have developed its own approach.

NGO INTERVENTIONS — It is more than obvious that NGOs were not influential with regard to the forest protocol. But in as far as NGOs achieved their goals on the CSD input, one may wonder whether they themselves caused the inclusion of this subject. It would seem not, as the draft proposal from the interim secretariat for the CSD input already contained a paragraph on forests (Doc II.23). Of course, NGOs could have influenced this document, but they did not claim this themselves, even after additional consultation. Furthermore, the final outcome changed substantially compared to the note from the secretariat (compare Doc. II.23 and II.28). The forests paragraph was extended, but a (hidden) reference to a possible protocol under the CBD in the secretariat's note was deleted by the COP. As the extension of the paragraph appeared to be an improvement from an ecological point of view, it might have seemed possible that this change had been co-determined by NGOs, had they themselves not condemned the final outcome (Doc II.99). Furthermore, several delegations – perhaps even the majority – were convinced that the CBD had a role to play in the conservation and sustainable use of forests, although nearly all opposed a protocol (Docs II.32). Therefore it is no surprise that the COP would address the forest topic in the CSD statement and that many wished to improve the forest paragraph in the secretariat's note. What is more, several countries had wanted to send a much stronger message, Denmark for example, and were as disappointed as the NGOs at the end of the day (Doc II.5).

Finally, it should be mentioned that as some NGOs were very disappointed with the COP decisions on forests, they therefore themselves did not believe that they had any impact. For example WWF stated that the COP had turned its back on forests (personal observation). To conclude, NGOs seemed not to have been influential with regard to the forest paragraph in the CSD statement. True, NGOs may have affected the establishment and mandate of the Intergovernmental Panel on Forest under the CSD and the more substantive decisions on forests at COP 2 in Djakarta, but the former is a different political arena and the latter is outside the period covered by this study.

E *Extent of political influence*

Table 5.9 makes it perfectly clear that NGOs did not impact policy outcomes on forests, although they welcomed the fact that the COP had addressed the topic in the CSD statement. Forests are considered of high relevance in global politics, irrespective of the fact that the legal status of the CSD statement is low (it is a voluntary message from one UN body to another and the CSD may do with it as it likes). But given the fact that North and South have continued to clash on the use and conservation of forests during the last decade and given all the political sensitivities and economic interests at stake, any decision of any UN body – however weak – leads to severe confrontations between developing and developed countries. It is therefore one of the (unresolved) core issues in international environmental policies.

TABLE 5.9 *Scores on political influence in case 16 (Forests)*

Counterfactual (C)	similar (0)
Goal-achievement by NGOs (GA)	some (1)
GA due to NGOs' own interventions (OI)	no (0)
Political relevance of topic (PR)	high (3)
Directness of impact (D)	no impact (0)
Extent of political influence: PI = C [(GA x OI) + PR + D]	none (0)

F *Final conclusion*

Some NGOs – notably Greenpeace and WWF – were quite noisy on forests in the CBD arena, but they were unable to influence the COP decisions on forests in Nassau. The option of a protocol under the CBD was ignored and the substance of the forest paragraph in the message to the Commission on Sustainable Development (CSD) was written by governments and UN officials. In short, NGOs did not exert any political influence in this case.

5.4.4 Case 17: biosafety

A Introduction

Genetically modified organisms (GMOs), their possible effects on biodiversity and human health, and the establishment of international standards to regulate the transboundary movement of GMOs, determined the post-Rio biodiversity debate to a very large extent. After all, Article 19.3 of the CBD urges parties to consider the need for and modalities of a biosafety protocol. As was outlined in case 13, Malaysia introduced the idea and experienced fierce opposition from the USA and some other countries (Japan, Switzerland, the Russian Federation). Nonetheless, Paragraph 3 of Article 19 was finally accepted due to pressure from the G77 and a few industrialized (mainly the Nordic) countries. As this paragraph states that parties should first consider the *need* for a protocol – and not immediately *negotiate* it – it is no surprise that biosafety was on the agenda of the two sessions of the Intergovernmental Committee (ICCBDs). In addition the United Nations Environment Programme (UNEP) had dedicated one of its panels to this topic (Panel 4). This panel produced a report which made clear that a majority of the participants were in favour of a legally-binding protocol to the CBD on biosafety. However, a minority was strongly opposed (Doc II.5). The status of this panel report, though, remained unclear, especially as the panels were open to all kinds of experts, including NGOs. That may have been the reason why the reports were not officially tabled at the ICCBD, except for a summary by the secretariat. In the end however they *were* presented, due to G77 and, probably, NGO pressure (Doc II.18, II.90).

Despite the preparations done by this UNEP panel, biosafety was briefly shortly discussed at the first session of the Intergovernmental Committee (ICCBD1). Countries reiterated their positions and NGOs also had an opportunity to make oral statements on the topic (Doc II.18). They reminded the delegates of the conclusions of Panel 4 and urged the parties to start negotiations on a protocol as soon as possible. ICCBD2 experienced a more extensive debate, but the positional divide among countries – immediate protocol negotiations versus a step-by-step approach – remained. Furthermore, the (voluntary) international technical guidelines on biosafety proposed by the UK and the Netherlands as a contribution to Agenda 21 were also taken into consideration (Doc II.6). NGOs – again – made statements on the immediate need for a protocol (Doc II.93), in addition they opposed the UK/Dutch initiative on voluntary guidelines. They feared that these would prejudice the negotiation of a binding instrument (Doc II.25, II.75, II.81).

In Nassau it was decided that biosafety should be dealt with by COP 2. A special open-ended ad hoc expert group was established to prepare for this (Doc II.27). The group should reconsider the need for and modalities of a protocol in accordance with Article 19.3. This outcome did not please the G77 and NGOs at all, because to them the need had already been established (Doc II.97). Furthermore, there was some dismay about money, as the COP had not allocated a budget to the expert group on biosafety. In addition, NGOs were, together with some G77 countries, also 'noisy' – as one government respondent said – with regard to a moratorium on the transboundary movements of GMOs until the new biosafety protocol could enter into force (Doc II.64). This plea was more or less ignored, just as NGO proposals on liability if genetically modified organisms (GMOs) exported by one party harmed other parties' biodiversity turned out to be premature (Doc II.97). Nonetheless, the expert group met after COP I and – again – found that a majority of the delegations favoured the development of a protocol. Ultimately, COP 2 decided that negotiations on such a binding instrument would start and be concluded in 1998 (Doc II.33). A specific open-ended ad hoc working group, which had a mandate to negotiate a legally-binding instrument, was therefore established.

B *Ego-perception*

Five of the twelve NGO respondents believe that NGOs impacted the debate on biosafety in the period between June 1992 (UNCED) and December 1994 (COP I). One of them refers to the establishment of the expert group at COP I: "NGOs clearly played a role in pushing the governments to take a decision to establish this working group." The others, though, claim influence, but are unable to point to specific aspects of biosafety which were impacted by them, for example the wording of the documents or the nature of the mandate of the working group. They note that NGOs pushed the topic very hard and had a substantial input. However, another NGO respondent also acknowledges the fact that they did not get what they wanted, at least not at COP I:

> "On the issue of biosafety, I think, we took a step back, although the fact that biosafety is on the international agenda is a success. It is a step back, because we are very worried about this expert working group considering the *need* for a protocol. Everybody says there *is* a need but many do not want it for obvious reasons. We are very worried that the opening of the discussion on 'need or no need' will end up with very bad, voluntary guidelines."

We know that this worry, expressed by one of the NGO representatives in Nassau, was unfounded. After all, the parties to the CBD are now negotiating a biosafety protocol.

C *Alter-perception*

Four of the eleven government respondents also perceive NGOs as influential on the biosafety topic, which, by the way, does not mean that they are all equally pleased to admit it, on the contrary. As one notes:

> "Certainly NGOs are influential when they make very strong statements on very political issues and biosafety is a good example. They certainly played

> into the fears on biotechnology. In a case where people have very little information and very little experience, one is responsive to that type of NGO interventions. These have largely been scare stories, you know. Their interventions are not based on substance but on fears. Well, I should not denigrate that because there are a lot of issues where I would be on their side. And of course they should fan the flames."

All four respondents, though, speak of NGO influence in quite general terms. They note that NGOs were able to play a substantial role in the UNEP panel on biosafety, pushed – together with the G77 and China – the ICCBD to table the report of this Panel 4, and made strong interventions in favour of a protocol at several meetings. The specific impact of these activities in terms of substance remains, however, unclear. In the words of one of the respondents: "It is hard to say how they precisely influenced the debate, but they certainly set the tone in a lot of cases."

D *Causal analysis*

GOAL-ACHIEVEMENT — NGOs had campaigned for an immediate start to negotiations on a biosafety protocol after UNCED and – at a later stage – after COP 1, but the working group established only had a mandate to first consider the need for such an instrument (Doc II.32). Hence, NGOs did not get what they wanted. Furthermore, other NGO ideas on biosafety were ignored, for example a moratorium on the release of GMOs before a protocol was in place and proposals on liability and redress in case harm was done to other parties (Doc II.64). Therefore it seems as if they did not achieve any of their goals. Still, NGOs could have – from their perspective – prevented a worse situation, in which biosafety might have been suspended as a topic with which the COP should deal. This possibility will be explored below.

It should be recognized that not all NGOs were equally pleased with the nature of the debate on biosafety, although all agree it is an important topic. As one representative said:

> "The comments on biosafety were so general and so broad, and generally so ideological, that they were very hard to translate into specific changes or pieces of proposals. NGOs have not done a good job at all in putting the case for biosafety. I mean there is very little factual basis for what statements have been made about biosafety."

In general, there was a split among NGOs such as Greenpeace, Third World Network (TWN) and Rafi on the one hand and IUCN, WWF and the World Resources Institute (WRI) on the other on how to proceed and how to intervene: loud and confrontational or low-profile and expertise-oriented.

NGO INTERVENTIONS — The basic question is this: did NGO pressure, among other things, cause governments to establish the working group on biosafety at COP 1? Although NGOs definitely applied pressure on this, it seems likely that such a working group would have been established anyway, given the identical – if not stronger – policy position of and political pressure from most developing countries on biosafety. As one delegate noted:

> "I think on particular issues some governments have been even more activist than a lot of the NGOs. This is an observation that needs to be made: some of the most vocal and most radical positions on biosafety have been expressed by developing countries."

One can also see the decision to establish the working group as a victory for the opponents of a protocol: the USA, the Russian Federation, the United Kingdom and others. Ever since Malaysia had proposed the prior informed consent procedure on the introduction of GMOs abroad at INC3, the 'threat' of an international and binding safety regulation on GMOs had hung over the biodiversity arena like a Sword of Damocles. Despite the fact that the proponents reaffirmed the need for the protocol at the ICCBDs and despite their increasing pressure, the opponents succeeded in postponing the start of negotiations, at least at COP 1. Given this alternative perspective, it would be foolish to consider the establishment of the working group as an NGO success.

Three additional government respondents were also explicitly asked to comment on the NGOs' biosafety claims. All three agree that NGOs had initiated new ideas and pressure, but add that the option of a protocol was already in the treaty. As one said:

> "I think for the momentum a number of NGOs like Greenpeace have played an important role in keeping the idea alive. There is no doubt about that. But I think the claim may be a little bit exaggerated. The idea of a protocol was already in the document."

COP 1 did just what could have been expected from the provisions in Article 19.3: establish a working group to consider the need for and modalities of a protocol. Therefore the policy outcome was a quite natural and predictable result from the perspective of the entire implementation process. In as far as NGOs mobilized additional information and pressure, they did so in support of those countries – mainly developing and Nordic ones – which were already in favour of immediately negotiating a protocol. To conclude, NGOs definitely contributed to the debate on biosafety, but it is very unlikely that the policy outcome at COP 1 would have been different without them.

E Extent of political influence

Again, NGOs failed to exert political influence on a highly relevant topic (see table 5.10). They did not even achieve any of their policy goals, although they definitely preferred having at least a working group on biosafety to having nothing. But others were responsible for the establishment of this group. The question of political relevance, incidentally, was already highlighted in case 13 (see subsection 5.3.6).

TABLE 5.10 *Scores on political influence in case 17 (Biosafety)*

Counterfactual (C)	similar (0)
Goal-achievement by NGOs (GA)	no (0)
GA due to NGOs' own interventions (OI)	no (0)
Political relevance of topic (PR)	high (3)
Directness of impact (D)	no impact (0)
Extent of political influence: PI = C [(GA x OI) + PR + D]	none (0)

F Final conclusion

Several NGOs played a very activist role on biosafety, but so did several developing countries. Despite pressure from these two sides, the first session of the Conference of the Parties (COP 1) decided not to start immediate negotiations on a protocol, due to opposition from countries which either rejected any instrument (USA) or favoured voluntary and technical guidelines under the UNEP or the Commission for Sustainable Development (United Kingdom, Netherlands). Instead, COP 1 established a working group to consider the need for a protocol. The NGOs were disappointed about that. In addition, some of their other proposals were not taken up either (moratorium on GMOs). Given this outcome, NGOs did not exert any political influence on the biosafety topic in the period 1992-1994, although their interventions on the dangers of GMOs and the need for strong international biosafety regulations definitely fuelled the debate (process influence).

5.4.5 CASE 18: FINANCIAL MECHANISM

A Introduction

As section 3 in chapter 1 on international environmental policies revealed, developing countries formulated their own agenda and priorities in the course of the 1970s and 1980s (see page 23). One priority was additional financial assistance from the industrialized countries. Such assistance would enable developing countries – which generally have major concerns other than environmental problems – to implement measures required under international regulations. Most of these countries even go as far as to make implementation dependent on the extent to which the rich countries effectively raise and hand over such money. The industrialized countries, though, are either reluctant to supply new resources in addition to ODA flows (which stands for Official Development Assistance) or demand strict conditions under which such resources can be transferred. The latter is, in turn, generally opposed by the South, as it fears environmental conditionality with regard to projects. According to developing countries, they should be able to set their own (national) priorities in spending environmental assistance, instead of being subject to the wishes of the donor community. It is therefore no surprise that North and South clash on the financial topic every time a new environmental agreement is negotiated. The Biodiversity Convention was no exception (and neither was the FCCC).

The financial debates in the biodiversity arena paralleled those in the climate arena (see subsection 4.2.3). Again, two topics were in the forefront: the wording to be accepted in the treaty text on additional funding by the industrialized countries and the financial mechanism to be established. Despite opposition from the USA, governments finally accepted the clause '*new* and *additional* financial resources' for inclusion in the CBD. And despite opposition from the G77, which was in favour of an independent biodiversity fund, the INC adopted the Global Environment Facility (GEF) – which is linked to the World Bank – on an interim basis. But the developing countries demanded that the GEF should be democratized soon.

Finance remained controversial after Rio. UNEP Panel 3 covered this topic and, again, faced all the North-South contradictions (Doc II.5). The G77 reiterated their demands on specific CBD institutions (which means: *no* GEF), on a broad definition of so-called incremental costs which arise out of the implementation of the CBD (which implies: *all* biodiversity projects in the South should be financed by the industrialized countries), and on national priorities as criteria to design projects under the CBD (which means: eradication of poverty should be paid for by the financial mechanism as well). The OECD countries counterbalanced these demands, supporting a permanent status for the Global Environment Facility (GEF) to the CBD, a limited interpretation of incremental costs, and project criteria linked to 'narrow' biodiversity objectives. These debates continued at both meetings of the Intergovernmental Committee (ICCBDs). In addition, other issues were addressed as well: for example the relationship between the financial mechanism and the Conference of the Parties (COP) and the possibility of using sources of funding other than the GEF under the convention. In particular the last topic was also promoted by the NGO community (Doc II.18, II.19, II.90, II.97). They stressed the need to broaden the funding debate and they presented other options, such as NGO sources, national trust funds and international taxation of the use of natural resources (including genetic ones). ICCBD2 came up with the idea of doing a survey of other financial institutions – such as the World Bank, Regional Development Banks, Montreal Protocol Fund and others – regarding their possible contribution to both the funding and objectives of the CBD. Therefore COP 1 decided to instruct the secretariat to conduct such a survey, a decision which was reconfirmed by COP 2 (Doc II.32, II.33).

But the most hotly debated issue remained the GEF, the 'Interim Institutional Structure to Operate the Financial Mechanism of the CBD' (ISOFM), as it was called in UN jargon. According to Article 39, the GEF should operate as the mechanism between the entry-into-force of the CBD and COP 1. COP 1 should then decide on what to do thereafter and the aim was to design a permanent structure in Nassau. Consequently, the ICCBDs had to prepare for such a decision. However, this turned out to be extremely difficult. Countries had not really changed their positions after UNCED, irrespective of the fact that the GEF had been restructured in the meantime. A GEF Assembly and a GEF Council of both donor and recipient countries were established and the council was handed some policy control by the World Bank (Doc II.100). However, neither the G77 nor many NGOs were satisfied with the restructuring process, as no real democratization was achieved (Doc II.19, II.92). They were also dissatisfied with the record to date of the GEF on biodiversity projects. Consequently, the 'old' North-South divide on the GEF remained. A compromise was nevertheless found, as the option to keep the GEF as the interim mechanism after Nassau was presented by a group of developing countries at ICCBD2 (Doc II.5). This option was finally accepted by the parties and even extended to COP 3 (1996). This implies that other options had failed, for example pledges to give permanent status to the GEF or proposals that favoured the establishment of an alternative and independent fund.

B *Ego-perception*

Four NGO respondents believe that they impacted the financial debate at the ICCBDs and COP 1. Three aspects are mentioned: 1. the survey which the secretariat was instructed to conduct on financial institutions other than the GEF after COP 1; 2. the legal incompatibility of the CBD and the GEF; 3. the restructuring of the GEF. As far as the first aspect is concerned, one respondent makes the following observation:

> "In Nairobi there was a decision by the ICCBD that a survey of financial institutions should be conducted with more or less a two-card agenda: 1. to look at additional financial resources out there, apart from the GEF, and 2. to look at the policies and projects of international financial agencies to see if they are consistent with the objectives of the CBD. And in Nassau there was a decision to instruct the secretariat to carry out this study. I think NGOs played a very valuable role in getting this language into the decision paper."

In particular the second part of the agenda – the question of consistency – was stressed by NGOs (Doc II.36, II.90, II.99). After all, it would be foolish to have a small fund on biodiversity, while the big ones on development – e.g. those of the World Bank – would continue to finance large projects with adverse effects on nature.

At COP I, Greenpeace made a strong intervention on the legal incompatibility of the CBD and the GEF, although others had referred to this incompatibility before, both NGOs and government representatives (Doc II.48). The text of the treaty says that the financial mechanism should function *under the authority and guidance* of the Conference of the Parties (Article 21.1). However, the GEF – even after it was restructured in 1994 – cannot function under the authority of the COP, as it is a governed by the World Bank, the United Nations Development Programme (UNDP) and the United Nations Environment Programme (UNEP). Hence, according to Greenpeace, either the CBD text or the GEF constitution should be amended or a different fund should be established under the authority of the COP. This intervention is now believed to have impacted the financial debate at COP I. As the respondent concerned said: "We forced the COP to reconsider its relationship with the GEF." Another respondent claims that NGOs contributed to the restructuring of the GEF. This restructuring was done to please the G77 and China. However, the developing countries demanded such institutional change partly due to NGO pressure, according to this respondent. As the restructuring of the GEF took place *outside* the biodiversity arena, this claim could not be verified. Therefore such NGO influence may have occurred, but is not assessed, and only the secretariat's survey and the COP-GEF relationship are covered below.

C *Alter-perception*

No government respondent spontaneously confirmed any of the above claims.

D *Causal analysis*

GOAL-ACHIEVEMENT — If one considers the policy outcomes with regard to the above topics, the picture is mixed. The survey on other financial sources, which the NGOs had pushed for, was indeed positively decided upon, although the substance of the COP decision is less clear, for example on the consistency aspect, than the ICCBD recommendation (compare Doc II.19 AND II.32). The relationship with the GEF, though, was not really reconsidered: the interim status remained. Hence, NGOs which had favoured the establishment of a biodiversity fund under the COP did not achieve their goal. However, it should be recognized that a number of NGOs, including IUCN, World Resources Institute (WRI), Environmental Defense Fund (EDF) and WWF, did not agree with Greenpeace's anti-GEF intervention (Doc II.56, II.58). They believed that the whole GEF debate was no longer useful. It would be better to accept the reality of the GEF as the (interim) financial mechanism and look for additional alternatives to meet the financial needs of the South. However, most NGOs and in particular the Southern ones remained hostile towards the GEF.

NGO INTERVENTIONS — It is accepted that NGOs had an impact on the COP decision to order a survey of multiple sources, as a causal chain between NGO initiatives on the one hand and the policy outcome on the other could be reconstructed. First of all, it appeared that NGOs were the first to raise the topic of financial diversity and multiple sources of funding after UNCED. UNEP Panel I did for example not address it, but the NGOs' Global Forum before ICCBD1 brought the need for financial diversity to the first session of the Intergovernmental Committee (Doc II.77). Moreover, an NGO intervention on multiple sources was followed by a recommendation from the ICCBD on this topic (Doc II.18). ICCBD2 witnessed a similar pattern: NGOs made several interventions on the need for multiple sources of funding and, subsequently, a survey was proposed by one of the delegates (Doc II.19). Finally, ICCBD2 produced a recommendation for COP I to conduct a survey on other financial institutions and additional sources of funding, a recommendation which was taken up in Nassau (Doc II.32). Besides, NGOs had organized a Global Forum on environmental funds in Santa Cruz, Bolivia, in June 1994 (Doc II.53), the output of which was fed into ICCBD2. Given this sequence of events, NGOs seem to have impacted the COP decision concerned.

One may wonder, though, whether the COP would have instituted such a survey anyway, as Articles 20.3 and 21.4 of the CBD create the legal possibility for the parties to provide financial resources through channels other than the treaty mechanism and to stimulate other financial institutions to provide resources for biodiversity as well. No wonder that the delegate who proposed the survey at ICCBD2 considered it as a first step to the implementation of these provisions of the CBD (Doc II.19). However, the CBD does not indicate *when* and *how* the COP should address these provisions. It is therefore assumed that NGOs contributed to giving substance to this 'when and how', on the one hand by agenda-setting at ICCBD1, on the other by enriching the agenda on multiple sources (for example through proposals on consistency, on NGO sources of funding, etc.; Doc II.36, II.77, II.90, II.95).

This political influence is a good example of how NGOs might contribute to a (small) breakthrough in a heavily politicized debate by presenting some innovative ideas. It seems that the proposal to have multiple sources of funding opened a window of opportunity in the North-South deadlock on the GEF. Whereas proponents and opponents of the GEF had stuck to their position in the past, the idea to have several funds – *including* the GEF – instead of one – *only* the GEF – may have caused (some of) them to join forces and, subsequently, commission a survey.

The legal incompatibility of the mandates of the Conference of the Parties and the Global Environment Facility, next, was not so much introduced by the NGOs at COP 1 as well as 'mobilized' by them in the dispute over the GEF. After all, the Chair of the ICCBD had already brought this topic to the attention of the GEF Council as well as ICCBD2 earlier (Doc II.19). It was also an issue which was emphasized by the G77 and China. Nonetheless, the Greenpeace intervention in Nassau *did* have an impact, in the sense that industrialized countries believed that it played, according to them, the wrong tune (personal observation). As the developing countries became more and more divided on the GEF, the North had hoped that it would be finally accepted as the institutional structure in Nassau. After all, ICCBD2 had already witnessed a group of developing countries announcing agreement to retain the GEF on an interim basis after COP 1. However, the G77 came up with an united and rather strong anti-GEF statement the day after the Greenpeace intervention in Nassau and some industrialized countries blamed the NGOs. But the final policy outcome was what the OECD countries had hoped for and the Greenpeace intervention and G77 statement were stifled by the follow-up negotiation process at COP 1.

E *Extent of political influence*

NGOs were able to exert some political influence in this specific case for the following reasons. Firstly, it is believed that the counterfactual would have been different without them (probably no survey on multiple sources of funding of this nature and at this time). Secondly, NGOs achieved some of their policy goals, although the GEF was retained and the COP decision on the survey was diluted compared to NGO proposals. Thirdly, the fact that this survey was ordered at COP 1 was substantially due to the NGOs' specific interventions in the ICCBDs, although it owed to governments as well. Fourthly, the topic of the financial resources and the funding mechanism is of the utmost importance in the biodiversity arena. Whether developing countries feel committed to the treaty is completely dependent on how satisfactorily this topic is resolved by the industrialized countries. Hence, the success or failure of the CBD is directly linked to it. Although a survey is no guarantee that policy changes by the parties in favour of a diversification of funds will follow, it is still considered a very relevant proposal, given the huge interests at stake. Finally, it is obvious that NGOs only had some *indirect* impact on the COP decision.

TABLE 5.11 *Scores on political influence in case 18 (Financial mechanism)*

Counterfactual (C)	different (1)
Goal-achievement by NGOs (GA)	some (1)
GA due to NGOs' own interventions (OI)	substantial (2)
Political relevance of topic (PR)	high (3)
Directness of impact (D)	indirect impact (1)
Extent of political influence: PI = C [(GA x OI) + PR + D]	some (6)

F *Final conclusion*

The financial mechanism to assist developing countries in the implementation of the CBD at home turned out to be 'the' divisive topic in the biodiversity arena. North and South had quite opposite views. The former preferred the World Bank-linked Global Environment Facility (GEF) to be the mechanism, whereas the G77 and China as well as most NGOs preferred the establishment of a separate biodiversity fund governed by the Conference of the Parties to the CBD. Although a compromise was reached just before Rio – the acceptance of the GEF on an interim basis – parties remained divided. So far no alternative has been found or established. Nonetheless, COP I decided to conduct a survey on other sources of funding and NGOs were jointly responsible for this idea. Therefore it is believed that their political influence amounts to 'some' in this specific case.

TABLE 5.12 *Overview of the ten CBD cases*

CBD	*Ego-perception (n respondents)*	*Alter-perception (n respondents)*	*Causal analysis (confirmation?)*	*Extent of political influence*
9 Preamble	n = 2	n = 1	yes	substantial
10 Indigenous and local communities	n = 3	n = 7	no	none
11 In-situ conservation	n = 4	n = 0	yes	some
12 Access to genetic resources	n = 3	n = 2	no	none
13 Biosafety	n = 3	n = 3	no	none
14 Indigenous and local communities	n = 2	n = 3	yes	some
15 Marine biodiversity	n = 3	n = 3	yes	some
16 Forests	n = 3	n = 0	no	none
17 Biosafety	n = 5	n = 4	no	none
18 Financial mechanism	n = 4	n = 0	yes	some

— 5.5 An overview of the ten cases —

Table 5.12 summarizes the outcome of the ten cases in terms of scores and conclusions. The implications of the small number of respondents, of the different numbers of (NGO) respondents in the formation and implementation phase, as well as of the phenomenon of temporal distortion, have already been discussed in chapter 4 (section 4.4). All the remarks made there are also valid here. However, one aspect of the table may put the impact of temporal distortion into perspective. It was argued that this phenomenon might explain the trend of generally higher scores in the implementation phase (1992-1995) than in the formation phase (1990-1992). However, if one compares the alter-perceptions of cases 10 and 14 on indigenous and local communities, it will become obvious that respondents gave more weight to NGO impact on a specific topic in the formation phase than in the implementation phase. It is remarkable, however, that the most notable perceptions of NGO influence on the part of both government and non-government respondents could *not* be confirmed by causal analyses (see cases 10 and 17 and shaded cells in table 5.12). This phenomenon was not found with regard to the FCCC (see table 4.11, page 154). There the most prominent perceptions of the respondents were indeed confirmed by subsequent analyses.

Again half of the NGOs' claims of political influence could be confirmed. Obviously, NGOs overestimated their own impact. The same applies, by the way, to government respondents. NGOs nonetheless exerted influence on:

– the inclusion of the notion of *intrinsic value* and of the paragraph on women in the Preamble of the CBD (case 9);

– some aspects of in-situ conservation measures in Article 8 (e.g. on the compatibility of present use of biodiversity and conservation objectives and on living modified organisms; case 11);

– the scientific recommendations on indigenous and local communities of the Mexico Meeting, as well as the decision to reschedule this topic in the medium-term programme of work of the COP from 1997 to 1996 (case 14);

– the inclusion of the marine biodiversity topic in the medium-term programme of work of the COP (case 15);

– the survey on multiple sources of funding and the compatibility of existing funds with the objectives of the CBD (case 18).

The first two topics refer to the formation phase, the other three to the implementation phase. The topics were generally influenced to *some* extent only, except for case 9 which was *substantially* influenced by NGOs (Preamble). All NGO impact was *indirect* in nature.

Half of the NGOs claims could, however, not be confirmed. With regard to the provisions on indigenous and local communities (Preamble, Articles 8(j) and

10(c)), on access to genetic resources (Article 15) and on biosafety (Article 19.3 and 19.4), NGO interventions were not effective in terms of policy outcomes, although they were generally able to contribute to the debates (process influence). In one case, an NGO was even able to directly impact some text, but without creating a more positive policy outcome from the NGOs' perspective (legal advice; case 12). The following topics in the implementation phase were not influenced either: forests and biosafety (cases 16 and 17). Again, NGOs were generally able to contribute to the debates on these topics, but not in such a way that their views made a difference in intergovernmental decision-making.

To conclude this section, a reference to the political influence of other policy players should be made (as was done in chapter 4). It is obvious that this political influence is not assessed here, but implied in the limited impact of NGOs. Other players were obviously more able to determine the policy outcomes (and this is, of course, no surprise). Examples are the governments themselves, the business lobby (although less prominent in the biodiversity than in the climate arena), scientists and international organizations such as the UNEP, FAO, UNDP and World Bank.

— 5.6 Conclusions —

The assessment of the NGOs' political influence on the Convention on Biological Diversity, which was based on the so-called EAC Method, produced the following conclusions:

1. NGOs exerted *indirect* influence on some provisions in the CBD: the Preamble (intrinsic value of biodiversity; the role of women) and Article 8 (in-situ conservation measures). Such impact was also claimed with regard to articles or paragraphs on indigenous and local communities, access to genetic resources, and biosafety. But these NGO claims proved to be invalid.
2. With regard to the period after the adoption of the treaty text (1992-1995), NGOs influenced decisions on indigenous and local communities, marine biodiversity, and financial institutions. On the other hand they failed to have an impact on the following topics in this period: forests and biosafety.
3. In general, NGOs exerted *some* political influence. An exception is case 9 on the Preamble of the Biodiversity Convention (*substantial* influence).
4. Again, NGO representatives definitely overestimated their own political influence. Only 50% of their claims could be confirmed (5 out of 10 cases).
5. Nonetheless, NGOs contributed to the debates in the biodiversity arena on all the above topics (process influence).

CHAPTER 6

The Explanation of Political Influence

— 6.1 Introduction —

Chapters 4 and 5 yielded several examples of on the one hand the NGOs' political influence and on the other examples of a lack of such NGO impact on policy outcomes. Such differences raise the question of which factors may explain the political influence of global NGOs (see research question 2 in subsection 1.5.2). To produce an answer, a comparative case-analysis will be employed below. This analysis consists of three steps, as was outlined in chapter 3 (subsection 3.3.2). First of all, the 18 cases will be positioned on the basis of the extent of political influence. The aim is to have a hierarchy of cases which ranges from a high to a low level of NGO influence. Secondly, the cases will be linked to a number of factors which may explain political influence. These factors are deduced from the theoretical model which was designed in chapter 2 (see figure 2.2, page 71). The possible linkages will be examined in a data-matrix, checking whether correlations can be discovered between the level of political influence on the one hand and specific factor outcomes on the other (so-called *vertical* comparison of cases in the data-matrix). Thirdly, the factors which provide a positive outcome for the possible explanation of NGO influence are linked to and compared with each other in the data-matrix (so-called *horizontal* comparison of factors). An examination is then made of the extent to which the enabling and constraining factors are linked or even overlap. Moreover, it will be assessed which of these factors are best able to explain NGO influence.

Apart from the comparative case-analysis, the views of the government and non-government respondents with regard to the explanation of NGO influence are taken into account as well. Several respondents were asked why they think NGOs are influential in some cases and not in others. In as far as these views yield knowledge additional to the comparative case-analysis, they will be included in the final conclusions of this chapter. The views of government respondents in particular are considered relevant, as government delegates are the ones to be influenced by NGOs.

The format of this chapter is based on table 3.3 in chapter 3 (page 90). This table builds on the theoretical framework designed in chapter 2 and distinguishes three levels of factors which may possibly explain political influence: factors at the level of the political arena, the policy outcome, and the context (international system). These levels will be chronologically elaborated on in separate sections. First, however, the hierarchy of cases based on the extent of political influence will be argued for in section 6.2. After sections 6.2 to 6.5 have yielded a number of conclusions on the explanation of NGO influence in the climate and biodiversity arenas, section 6.6 will discuss the arguments of the respondents and produce some additional conclusions. Next, section 6.7 will identify the most important enabling and constraining factors for NGO influence and assess their links. To visualize the possible impact of these (related) factors, a causal model will be drawn. Finally, section 6.8 will summarize the chapter.

It should be emphasized that the analysis in this chapter is related to the climate and biodiversity arenas in the period 1990-1995. Conclusions on the explanation of the NGOs' political influence therefore do not apply to other arenas and other periods, as one should be careful in generalizing empirical findings of case studies (see subsection 3.3.2). However, one may consider the conclusions of this chapter as general hypotheses on NGO influence in international environmental policies, because theoretical generalizations *can* be made on the basis of case studies. Therefore the causal model which will be drawn at the end of this chapter may be generalized in theoretical terms (yet not in empirical terms). As a theory, this model may become the input of new research.

— 6.2 A hierarchy of cases —

The first column of table 6.1 shows the 18 cases which were analysed in chapters 4 and 5 in a specific sequence. This sequence is based on the extent of political influence which was assessed on a case-by-case basis. In table 6.1, all scores derived from chapters 4 and 5 are grouped together (see tables 4.3 to 4.10 and 5.2 to 5.11). Given this new hierarchy of cases, the previous, chronological enumeration of FCCC and CBD cases is intermixed in table 6.1. Note that individual cells may be shaded grey to point out outstanding scores in a column. The specific meaning of these shaded cells, though, will be highlighted below. The cases above the double horizontal line in the table are characterized by a positive score on the NGOs' political influence, those below this line by an absence of such influence.

As well as a comparison of the *final* scores on political influence, a comparison of the cases' scores on the different variables – such as goal-achievement, political relevance and directness of influence – can be illuminating too. This might already tell us something about the background of political influence. For example, a comparison of columns 2 and 3 in table 6.1 gives a deeper understanding of what has been pointed out before, namely that goal-achievement by NGOs (see colum 3) and political influence (which implies a 'positive' counterfactual in column 2) do not necessarily coincide, although the former is still a prerequisite for the latter (see subsection 2.3.1). The shaded cells in column 3 make obvious that NGOs may achieve (some of) their goals without having impacted the policy outcomes themselves (see cases 1, 2, 7, 10, 13, and 16). In all these cases, states or UN bodies promoted more or less the same policy goals as NGOs, and they were responsible for the fact that some of the NGOs' goals were reflected in the outcomes. Examples of such players are the Intergovernmental Panel on Climate Change (IPCC) and the European Union (EU) with regard to the objective of the FCCC (case 1), the G77 and some Western nations with regard to the principles of the FCCC (case 2), the interim secretariat of the FCCC with regard to the first review of the implementation of the stabilization target by the OECD (case 7), and Malaysia with regard to the biosafety topic (cases 13 and 17). These so-called *like-minded* – as they shared some of the NGOs' policy goals – states and UN bodies however dominated these topics, so that NGOs did not get a chance to intervene. Therefore one may conclude that the presence of such like-minded but dominant (inter)governmental players prevent NGOs from becoming influential themselves. True, the final policy outcomes may be (partly) preferred not only by states but by NGOs as well, but from the perspective of political influence such like-minded players constrain NGOs. Because 6 of the 9 cases with a negative score on political influence refer to such 'goal-achievement without political influence', the following conclusion seems valid:

> *Conclusion 1 — The presence of like-minded, dominant states – or other players – constrain NGOs in exerting political influence.*

It should, however, immediately be noted that NGOs at the same time need like-minded states, at least those which are responsive to NGO demands. After all, NGO influence is always realized due to 'friendly' government delegations which carry (aspects of) the NGOs' expertise, views or proposals through the intergovernmental negotiation and decision-making process. This is a prerequisite for any NGO influence. The fact that NGOs need additional interventions by governments is also expressed by the variable 'own intervention' (OI) in column 4 of table 6.1, which lacks a score of 3.

Only three cases are characterized by a total lack of goal-achievement by NGOs (see the topics implementation, genetic resources and biosafety 2; respectively cases 4, 12 and 17). Obviously, the goals which were formulated by NGOs in these specific cases were not taken over or shared by the majority of policy- and decision-makers and, hence, none of the NGOs' goals were in the end achieved. This may be explained by the fact that NGOs defended far too radical or unrealistic a policy goal

TABLE 6.1 *Hierarchy of cases based on the extent of the NGOs' political influence*

Columns: 1	2	3	4	5	6	7
Cases	*Counter-factual*	*Goal-achieve-ment*	*Own inter-vention*	*Political relevance*	*Directness*	*Extent*
5 AOSIS Protocol	1	2	2	2	2	8
9 Preamble	1	3	2	1	1	8
6 Joint implementation	1	1	2	3	1	6
15 Marine biodiversity	1	2	2	1	1	6
18 Financial mechanism	1	1	2	3	1	6
3 Target	1	1	1	3	1	5
8 Berlin Mandate	1	1	1	3	1	5
11 In-situ conservation	1	1	1	2	1	4
14 Indigenous /locals 2	1	1	1	1	2	4
1 Objective	0	3	0	2	0	0
2 Principles	0	1	0	2	0	0
4 Implementation	0	0	0	3	0	0
7 First review	0	1	0	3	0	0
10 Indigenous / locals 1	0	1	0	2	0	0
12 Genetic resources	0	0	0	3	2	0
13 Biosafety 1	0	1	0	3	0	0
16 Forests	0	1	0	3	0	0
17 Biosafety 2	0	0	0	3	0	0

Legend

counterfactual (C): 0 = similar; 1 = different
goal-achievement (GA): 0 = no; 1 = some; 2 = substantial; 3 = much
own intervention (OI): 0 = no; 1 = some; 2 = substantial; 3 = much
political relevance (PR): 0 = no; 1 = low; 2 = moderate; 3 = high
directness (D): 0 = no; 1 = indirect; 2 = direct
extent political infl. (PI): 0 = none; 3-6 = some; 7-10 = substantial; 11-14 = much
(PI = C [(GA x OI) + PR + D])

Cases number 1 to 8 belong to the FCCC
Cases number 9 to 18 belong to the CBD

from the perspective of (most) states. A good example is the wish to adopt new binding regulations on the community rights of indigenous people in relation to the access to biological resources (case 12). Therefore the following conclusion also seems valid:

> *Conclusion 2 — As long as NGOs defend too radical or unrealistic a policy goal, they will fail to exert political influence on the topics concerned.*

Although this conclusion is more than obvious, it should immediately be put into perspective. After all, the formulation of far-going demands does not necessarily *exclude* political influence. An example is case 1 on the target of the FCCC. Here NGOs defended quite radical demands on emission reduction targets, but their pressure did still move the negotiations and decision-making a bit in their direction. Obviously, there are cases in which radical demands by NGOs are ignored by governments, and other cases in which such demands are not – or cannot be – ignored. It is likely, however, that other variables also play a role in this.

Column 5 of table 6.1 addresses the political relevance of the various policy topics. A number of aspects are striking here as well. It appears that all cases of low relevance yield positive scores on political influence (see shaded cells of cases 9, 15 and 14 in column 4). This might imply that topics of lower relevance are more easily impacted by NGOs. However, cases of high relevance produced positive outcomes on political influence as well (see cases 6, 18, 3 and 8). In consequence, a high relevancy of topics does not necessarily *exclude* NGO influence. On the other hand, 6 of the 9 cases *not* politically influenced by NGOs were of high relevance (see cases 4, 7, 12, 13, 16, 17), whereas the same applies to only 4 of the 9 topics showing a positive score on political influence (see cases 3, 6, 8 and 18). In other words, the factor 'political relevance' may explain NGO influence, although the correlation is definitely weak:

> *Conclusion 3 — The more politically relevant a policy topic is, the more difficult it seems to be for NGOs to impact it.*

This conclusion is obvious. For example, sensitive topics which may have huge implications for governments and raise diplomatic conflicts are generally negotiated in small contact groups behind closed doors. NGOs and other observers are then excluded and it is difficult for them to impact the compromises which may result. In other cases governments may simply stick to their national interests when sensitive topics are dealt with, and again, it is hard for NGOs to influence the outcomes.

Column 6 deals with the directness of NGO influence. Obviously, most cases were only indirectly influenced. This means that NGO proposals or views were not taken over literally by governments, but did nonetheless impact the process of intergovernmental negotiations and decision-making in such a way that the policy outcome in the end reflected (a bit) more of the NGOs' goals than if they had not intervened. However, there are three cases of direct impact, in which NGO representatives were engaged in drafting text proposals that even survived decision-making (see the shaded cells: case 4 on the AOSIS Protocol, case 14 on indigenous people and case 12 on genetic resources). One of these cases, however, does not reflect the political influence of NGOs because: 1. the input of the NGO representative concerned was limited to details and to legal advice, and 2. NGOs did not achieve their policy goals (case 12 on genetic resources). Therefore, participation of NGOs in the drafting process of formal texts does not automatically lead to political influence. Or in other words:

> *Conclusion 4 — Direct impact of NGOs on policy outcomes, for example through the drafting of legal text, is not a necessary or a sufficient precondition for political influence.*

This may be a surprise to some, as it is said in NGO circles that the co-writing of text elements is 'the' way of exerting political influence. It is, however, not!

Finally, column 7 of table 6.1 contains the final data on the extent of political influence. Three groups can be distinguished: a group of two cases characterized by *substantial* political influence (PI = 8); a group of 7 cases characterized by *some* political influence (PI ranges from 4 to 6); a group of 9 cases characterized by an *absence* of political influence (PI = 0). It will be examined in the following sections whether explanatory factors deduced from the theoretical model may explain this specific hierarchy of cases, in addition to the factors which have been elaborated on in this section (like-minded states, radical/unrealistic demands and political relevance; see conclusions 1 to 3).

— 6.3 Explanations at the level of the political arena —

6.3.1 NGOs and their interventions

As was argued in chapter 2 (sections 2.1 and 2.6) and elaborated on in chapter 3 (subsection 3.3.1), the theoretical model of this study functions as a lens through which the empirical domain is both observed and *explained* (see figure 2.2, page 71). Given this function of the model, its elements are considered potential explanatory factors of the NGOs' political influence. As a consequence, a set of factors was deduced from the theoretical model in chapter 3 (see table 3.3, page 90). This list, in turn, functions as the format for the analysis in this chapter. To repeat, three levels are distinguished (arena, outcome, context). Within these levels, separate factors are recognizable which will be dealt with one by one in the sections 6.3-6.5. To examine their explanatory capability, the substance of the factors will be related to the hierarchy of cases – based on political influence – in the data-matrix, a procedure which was highlighted in subsection 3.3.2. If a pattern – or correlation – between the substance of a factor and the extent of political influence is discerned in the data-matrix, it will be concluded that this factor is capable of explaining political influence. In this section, the level of the political arena is examined. The following factors will be dealt with: the players involved (NGOs, states and others), the players' interventions (in particular those of the NGOs), the policy issues involved (climate change or biodiversity loss), and the nature of the negotiation and decision-making processes (see subsections 6.3.1-6.3.4 and tables 6.2-6.4).

The extent to which NGOs are successful may be dependent on the nature of the organizations involved as well as on their interventions. Some NGOs are probably more experienced and have a better reputation than others, so that their access to and persuasion of policy-makers is more likely. Also some interventions are definitely more appropriate than others in specific circumstances, so that they are seldom equally effective. To check whether these links are valid, the organizations, the types of NGOs and the interventions concerned are brought together in the table below (table 6.2). Column 1 gives the hierarchy of cases constructed in the previous section, column 2 the NGOs which intervened in those cases, column 3 the types of organization, and column 4 the nature of the interventions. It should be mentioned, though, that the information in this table is far from complete. Not all NGOs and interventions are included, as this was impossible, but only the dominant ones in each case. Note that individual cells may be shaded grey in table 6.2 – and in further tables as well – to point out (possible) patterns in the data-matrix. The specific meaning of each pattern will be illuminated in the text.

In searching for correlations in the columns, one has to conclude that these cannot be found. Similar NGOs and, hence, types of organization are successful in one case, but fail in another. Examples are Climate Action Network (CAN), World Conservation Union (IUCN), Rural Advancement Foundation International (Rafi), Genetic Resources Action International (GRAIN), World Wide Fund for Nature (WWF), Greenpeace, Bionet and Third World Network (TWN). The number of NGOs covering a topic seems not important either, as one NGO (e.g. case 11) or two NGOs (e.g. case 5) can be as successful as a group of NGOs (e.g. cases 6, 14, 18). Besides, the fact that the *quantity* of NGOs is not decisive already became obvious in discussing the biodiversity and climate arenas in the period 1990-1992. To repeat, a small number of NGOs in the former arena were able to impact more cases than their more numerous colleagues in the latter. In addition, the nature of interventions cannot explain success or failure either, especially as it was quite a difficult

TABLE 6.2 *Factors at the arena level I: NGOs and their interventions*

	Cases	*NGO*	*Type organization*	*Prominent interventions*
5	AOSIS Protocol	FIELD Greenpeace	legal advice + environmental organization	legal advice, oral statements
9	Preamble	IUCN, WRI, WWF	conservation organizations	concrete proposal, draft text
6	Joint implementation	CAN	environmental action network	position papers, oral statements
15	Marine biodiversity	IUCN, WRI, Greenpeace	conservation + environmental organizations	parallel conference, concrete proposal
18	Financial mechanism	WWF, IUCN, Bionet, EDF	conservation organizations	oral statements, concrete proposal
3	Target	CAN	environmental action network	protest acts, dissemination of science
8	Berlin mandate	CAN	environmental action network	protest acts, contribution to compromise
11	In-situ conservation	IUCN	conservation organization	IUCN draft
14	Indigenous / locals 2	GRAIN, Rafi, Indigenous groups, TWN	agricultural + environmental action groups	draft text, concrete proposal, protest acts
1	Objective	CAN, WWF, Greenpeace	conservation + environmental organizations	oral statements, dissemination of science
2	Principles	CSE, BCAS, Greenpeace	environmental (research) groups	position papers, oral statements
4	Implementation	SNM, NRDC	environmental organizations	concrete proposals
7	First review	CAN	environmental action network	reports
10	Indigenous / locals 1	Grain	agricultural action groups	position papers
12	Genetic resources	Grain, IUCN	agricultural + conservation organizations	legal advice, position papers
13	Biosafety 1	NGO-task group	conservation + environmental organizations	position papers, draft text
16	Forests	WWF, IUCN, Greenpeace, Bionet	conservation + environmental organizations	oral statements, concrete proposal
17	Biosafety 2	Greenpeace, TWN, Rafi	environmental organizations	oral statements, dissemination of science

task distinguishing different NGO interventions in the various cases. *Any* case was characterized by a combination of many possible interventions, as set out in chapters 4 and 5 (see table 4.1). For example *Eco* published material on all subjects, several topics were covered by oral statements in plenary sessions as well as by many position papers and publications, and delegates were approached with specific proposals in nearly all cases. Nonetheless, some prominent interventions could – after consulting respondents, documents and personal notes – be distinguished in each case, as NGOs did specialize to a certain extent (see column 4, table 6.2). Again, no pattern that points to a correlation between specific NGO interventions on the one hand and the extent of political influence on the other can be distinguished. This was already confirmed by conclusion 4, but with regard to one activity only (drafting of legal text).

> *Conclusion 5 — Political influence is neither guaranteed nor excluded by a specific (type of) NGO, a specific number of NGOs, or by a specific (type of) NGO intervention.*

However, there is one type of intervention which only appears in the table in relation to a positive score on political influence and which is mentioned more than once: *protest acts* (see shaded cells). One can think here of marches, sit-ins, public tribunals, the disturbance of political meetings etc., in defence of a cause or to show discontent with current policies. The shaded cells might point at a supportive contribution from such acts outside formal political arenas to the impact of advocacy and lobbying activities inside. It is nonetheless obvious that protest is not a necessary precondition for political influence, given the fact that it lacks in other 'successful' cases. Therefore the linkage is weak (3 out of the 9 cases).

> *Conclusion 6 — Protest activities of NGOs outside formal political arenas may be supportive to advocacy and lobby activities of NGOs inside.*

It should be added that protest in itself is seldom enough. For example media coverage of actions and public support are also important prerequisites (Van Noort *et al.* 1987). The preconditions under which protest may be effective in terms of political influence will be dealt with elsewhere (chapter 7).

6.3.2 States and other players

Column 2 of table 6.3 lists the states – or groups of states – which were most prominent in individual cases. It seems likely that the key representatives of these (groups of) states were the most obvious players at whom NGOs had to direct their efforts to gain influence on the topics concerned. One may assume that some states are more accessible in this respect than others and, hence, that the dominance of for example country A in case X might enable NGOs, whereas that of country B in case Y might constrain them, irrespective of which NGOs are present and what they actually do in these cases. However, it is not possible to distinguish a pattern of countries among the different categories of political influence, especially as nearly all delegations were involved in all topics one way or the other.

TABLE 6.3 *Factors at the arena level II: states, other players, policy issues*

	Cases	*(Groups of) States*	*Other players*	*Issue*
5	AOSIS Protocol	AOSIS	-	climate
9	Preamble	a.o. Brazil, Germany, Netherlands	UNEP	biodiversity
6	Joint implementation	CEE, G77, OECD	-	climate
15	Marine biodiversity	AOSIS, Sweden	UNEP	biodiversity
18	Financial mechanism	G77, OECD	World Bank, UNEP, UNDP, GEF	biodiversity
3	Target	CEE, EU, Japan, USA	IPCC, WMO, UNEP, BINGOS	climate
8	Berlin Mandate	CEE, G77, EU, JUSCANZ	IPCC, WMO, UNEP, BINGOS	climate
11	In-situ conservation	EU (France), G77, Nordics	UNEP, UNESCO	biodiversity
14	Indigenous / locals 2	Brazil, EU	FAO	biodiversity
1	Objective	EU, USA	IPCC, UNEP, WMO	climate
2	Principles	G77, OECD	UNEP, UNDP	climate
4	Implementation	G77, OECD	-	climate
7	First review	G77, OECD	Secretariat FCCC	climate
10	Indigenous / locals 1	Asia (India, Indonesia), Canada, Latin America (Colombia), USA	FAO	biodiversity
12	Genetic resources	G77, OECD	FAO	biodiversity
13	Biosafety 1	Malaysia (G77), USA (OECD)	UNEP	biodiversity
16	Forests	G77, OECD	FAO	biodiversity
17	Biosafety 2	G77, OECD	UNEP, CSD, BINGOS	biodiversity

Besides, states tend to operate as regional groups or under the banner of economic integration organizations (e.g. AOSIS – Alliance of Small Island States; CEE – Central and Eastern Europe; EU – European Union; OECD – Organization for Economic Cooperation and Development; G77 – the Group of 77 and China; Nordics – Denmark, Finland, Norway, Sweden; JUSCANZ – Japan, USA, Canada, Australia, New Zealand).

However, one aspect in column 2 of table 6.3 is striking, namely the fact that in cases where the OECD countries on one side and the G77 on the other each acted as the more or less united and dominant group, NGOs generally exerted no influence (see shaded cells). This may be due to the fact that most of these cases were characterized by North-South confrontations or – in other words – by *traditional block politics.* The notion 'traditional' is used here because these confrontations are as old as the history of international environmental policies (see chapter 1). Such block politics seem to diminish the chances for NGOs to successfully intervene. Examples are cases 2 and 4 on principles and implementation mechanisms of the FCCC, in which the rich and the poor countries defended quite opposite interests. Therefore the negotiations and decision-making concerned were quite difficult. Power games were played and the topics concerned became heavily politicized. This seems a bad environment for NGOs to operate successfully.

Conclusion 7 — Traditional block politics among the rich and the poor countries diminishes the chances for NGOs to become influential.

Again, there are exceptions. NGOs were for example able to impact decisions on joint implementation and the financial mechanism (cases 6 and 18), two topics which were also very much dominated by conflicts among the rich and the poor. These two cases, though, are exceptional in that NGOs were able to come up with new ideas which were subsequently adopted by one or both blocks (respectively a survey on financial mechanisms and criticism on joint implementation). In addition, NGOs were also influential in one case where the poor and the rich did more or less agree (e.g. case 14 on indigenous people). In the other cases in which the North and South were dominant as united blocks, however, NGOs either defended unattainable goals so that they were overruled by the interests of the rich and the poor (e.g. implementation mechanisms), or they simply supported the views of one of the blocks, which was usually the South (e.g. biosafety).

Conclusion 8 — NGOs seem able to impact traditional block politics if they can come up with innovative proposals that may either strengthen the position of one block or offer an option for compromise.

If conclusion 7 is indeed valid, the opposite may be as well: a split in the blocks or a dominance of individual countries in specific cases enables NGOs to influence outcomes. For example, on the FCCC target the North was heavily divided (EU versus USA), as was the South on the Berlin Mandate (AOSIS versus OPEC versus the rest). In both cases NGOs exerted political influence. So non-traditional, new, or ad hoc coalitions of states create new opportunities for NGOs. This may be explained by the fact that such coalitions are less subject to traditional political rituals and sensitivities. In consequence, the power games played may be fewer and the politicization of the topic concerned may be less (although this is not necessarily the case, given the EU-USA confrontation on the FCCC target; see case 3). Furthermore, individual countries or smaller groups of states are more accessible to NGOs than large blocks, one may assume. Individual countries may even need the support of NGOs to strengthen their position vis-à-vis others.

NGOs might be enabled or constrained in the political arena by players other than states as well, e.g. scientific bodies, intergovernmental organizations (IGOs), or the business lobby (BINGOs). Through coalitions with supportive players, through competition with like-minded groups, and through actions of opposing groups, the impact of NGOs might be increased or limited. To examine such possible linkages, the most prominent players other than states and NGOs are – with regard to each topic – listed in column 3 of table 6.3. Again, no obvious pattern is visible at first sight. The position and role of other players seem not to have impacted the NGOs' chances of success, as similar ones go hand in hand with positive and negative scores on NGO influence (e.g. IPCC, FAO, UNEP). Individual cases, though, may be explained by the fact that such groups can act as like-minded and dominant players (see conclusion 1). Examples are the IPCC in case 1 and the FCCC secretariat in case 7, but these examples are already covered by that conclusion.

Remarkably enough, NGOs were perfectly well able to exercise influence irrespective of whether the business community was active on the same policy topics (see for example cases 3 and 8). This is remarkable, as it is generally assumed that the BINGOs overrule the non-profit NGOs due to their power, interests and policy goals. However, this did not lead to an 'elimination' of NGOs, although the BINGOs were – according to several government respondents – able to impact decision-making in both arenas.

Conclusion 9 — The presence of business NGOs in environmental issue arenas at global level does not prejudice the chances for non-profit NGOs to become influential.

It is striking that NGOs were influential despite the fact that the coal and oil lobby was quite active in the climate arena (cases 3 and 8), whereas they were not in one case where the biotech lobby was present in the biodiversity arena (case 17). This might point at more effective lobbying by the latter industry, as their arguments may have been more persuasive to governments than those of the non-profit NGOs. In the climate arena, though, the situation was exactly the reverse. But it remains to be seen whether the data may be interpreted in this way. After all, the biosafety topic was resolved mainly among Northern and Southern governments, with few possibilities for both NGOs or BINGOs to intervene. At the same time, it would not be correct to deny any (indirect) impact of the coal and oil industry on the climate negotiations. In short, although business interests do always play a role in international environmental negotiations, this does not exclude NGO influence.

6.3.3 Policy issues

Political arenas as understood in this thesis are organized around policy issues (see section 2.2.3). Each issue has its own characteristics, which may lead to distinctive conflicts among political players and to different policy outcomes (see subsections 2.4.2 and 4.2.3). This given fact, though, may also create different opportunities and constraints for NGOs to intervene in political arenas and become successful in terms of political influence. One could for example assume that biodiversity loss is politically less controversial than climate change and, therefore, more easily influenced by NGOs. After all, the issue of biodiversity is related to traditional policies on nature conservation, which are quite accepted nowadays and considered to be economically feasible, whereas the issue of climate change – apart from being scientifically disputed – implies policies which might affect the heart of economic growth (energy and transport sectors). Besides, NGOs such as IUCN and WWF have played important roles in the nature conservation area for decades, whereas the involvement of NGOs in the climate arena is rather new. Therefore one might assume that countries are more responsive to intergovernmental agreements on biodiversity than on climate change, and also to the claims of NGOs in the former arena. However, this assumption seems wrong, given the data in column 4 of table 6.3. As the climate and biodiversity cases are equally divided between the categories of political influence, the following conclusion needs to be formulated:

> *Conclusion 10 — The different nature of the climate change and biodiversity loss issues cannot explain the extent of political influence of NGOs.*

It is incidentally no surprise that the above assumption on the less controversial nature of the biodiversity issue turned out to be wrong, as the content of the CBD in the end exceeded traditional conservation policies and provoked debates as economically sensitive as those with regard to many topics in the climate arena. To repeat, the CBD also covers items such as the rights over and access to genetic resources, the transfer of modern (bio)technology to poor countries, and the transboundary trade in and handling of biotechnology products (biosafety). These topics impinge on vested national and/or industrial interests, just as greenhouse gas emission reductions do in the energy and transport sector. Therefore many topics under the CBD are politically as important as the ones under the FCCC, a fact which was also illustrated by Bush's refusal to sign the CBD in Rio de Janeiro in 1992.

6.3.4 Negotiations and decision-making

Given the model on treaty formation and implementation as designed in chapter 2, the global political arenas consist of policy players who negotiate and decide on the formation and implementation of conventions. Such negotiation and decision-making processes differ depending on the policy phase and, moreover, may be characterized by either integration or confrontation among the parties (Young, 1975; 1980; 1982). To start with the policy phase, a difference was already diagnosed in chapters 4 and 5. After all, there is a distinction between the establishment of binding rules in a treaty and the initial interpretation and implementation of it. The former results in binding law, whereas the latter does generally not exceed post-treaty 'agenda-setting'. This was also referred to as the making of *end products* as against the making of *interim products.* It may be assumed that such interim products are more easily influenced by NGOs than end products and the data seem to confirm this belief. Column 2 of table 6.4 below reveals that only 3 of the 9 topics which refer to the formation phase were impacted by NGOs, whereas they influenced 6 of the 9 topics which refer to the implementation phase. In other words:

Conclusion 11 — Policy outcomes are more easily influenced by NGOs during (initial) treaty implementation than during treaty formation.

This conclusion is valid for both issues. As far as the FCCC is concerned, 1 out of the 4 cases in the formation phase was affected by NGOs, whereas 3 out of the 4 cases were so in the implementation phase. With regard to the CBD, these figures are: 2 out of 5 in the formation phase and 3 out of 5 in the implementation phase. True, numbers and differences are small, but the trend is at least similar in both arenas.

A next factor is the integrative or confrontational nature of the negotiation and decision-making process with regard to a specific topic. Although the distinction between these two categories is not always totally obvious in a specific case, integ-

TABLE 6.4 *Factors at the arena level III (policy phases, negotiations) and at the outcome level (regime elements, domestic effects)*

	Cases	*Policy phase*	*Negotiations*	*Regime element*	*Short-term, domestic effects*
5	AOSIS Protocol	implementation	integrative	regulation	none
9	Preamble	formation	integrative	principles	none
6	Joint implementation	implementation	confrontational	procedure	some
15	Marine biodiversity	implementation	integrative	principle	none
18	Financial mechanism	implementation	confrontational	procedure	none
3	Target	formation	confrontational	regulation	substantial
8	Berlin Mandate	implementation	confrontational	procedure	none
11	In-situ conservation	formation	confrontational	regulations	some
14	Indigenous / locals 2	implementation	integrative	norms	none
1	Objective	formation	integrative	norm	none
2	Principles	formation	confrontational	principles	none
4	Implementation	formation	confrontational	procedures	substantial
7	First review	implementation	confrontational	regulation	some
10	Indigenous / locals 1	formation	confrontational	norms	some
12	Genetic resources	formation	confrontational	regulations	substantial
13	Biosafety 1	formation	confrontational	regulation	some
16	Forests	implementation	confrontational	principles	none
17	Biosafety 2	implementation	confrontational	procedure	none

rative negotiations and decision-making generally result when parties are more or less committed to a cause, but still need to compromise on substantive and legal details. Confrontational negotiations and decision-making are, on the other hand, characterized by parties which do not agree on the nature of a topic or on the policy outcome. The former negotiations are therefore generally more technical in nature, the latter more political. One might assume that these different circumstances provide different opportunities and constraints for NGOs to intervene. Integrative negotiations will probably enable NGOs to impact outcomes, confrontational negotiations will constrain NGOs from doing so. Given the results in column 3 of table 6.4, this assumption seems invalid. Of the 9 cases which were positively impacted by NGOs, the majority is characterized by confrontational negotiations. Still, there *is* a difference, because of the 5 cases characterized by integrative negotiations 4 were indeed affected by NGOs and only 1 was not.

> *Conclusion 12 — Integrative negotiations and decision-making may enable NGOs in exerting political influence, but this does not mean that confrontational negotiations and decision-making exclude such NGO influence.*

It should be mentioned, however, that this conclusion has a lot in common with conclusion 7 on traditional block politics among North and South. After all, such block politics are characterized by confrontations. Therefore conclusions 8 and 13 are to some extent mirror images, although it should be realized that confrontations may also occur *within* the Northern and Southern blocks.

— 6.4 Explanations at the level of the policy outcome —

6.4.1 Regime elements

This section explores explanations of political influence at the level of the policy outcome. In conformity with the theoretical model (figure 2.2, page 71) and the list of possible explanatory factors (table 3.3, page 90), the following aspects will be addressed: the nature of the policy outcome and its possible policy effects at domestic and global levels.

As was already outlined in section 3.3.1, the specific (legal) nature and expected policy effects of the outcome which is aimed for in a political arena probably co-determine the possibilities for NGOs to intervene. For example, if a specific policy topic will probably lead to the prescription of certain measures which are binding upon all states and, moreover, are very costly, governments may think twice about letting NGOs – which always want more stringent policies – successfully intervene. Therefore one may assume that NGOs can more easily influence soft rather than hard law, and – in terms of the regime theory – more easily influence general principles and norms than specific regulations and procedures which generally have consequences in terms of state behaviour. As both the FCCC and CBD are binding law, the possible difference made by soft and hard law with regard to NGO influence cannot be deduced on the basis of the cases studied here. Consequently, only the potential explanatory ability of the different regime elements are examined in this section.

To repeat, regimes consist of principles, norms, regulations and procedures operative in an issue-area, many of which are laid down in a wide range of documents, including conventions (see subsection 2.4.2). Therefore examples of the four different regime elements may also be distinguished in an individual treaty. Principles refer to the policy beliefs of the parties involved and set the basic premises of a convention and its elaboration (e.g. indigenous people and women are important players to address); norms refer to the more general rights and obligations of parties as included in the treaty concerned (e.g. parties have sovereign rights over their resources); regulations define more specific pro- and prescriptions for action (parties should establish protected areas); and procedures are rules for making and implementing collective choices (e.g. targets may be implemented jointly). Column 4 in table 6.4 examines under which regime element the policy topics concerned may be classified. This classification, though, was not without problems, as the nature of a policy topic as well as the distinction between the different regime elements is not always that obvious. Some cases cover more than one element – e.g. the Berlin Mandate was both on procedure and substance – and for example norms and regulations are not always easily distinguishable. But as long as one keeps in mind that procedures – unlike the other elements – are *mainly not* on substance, that principles and norms – as opposed to regulations – do not refer to measures parties should undertake *immediately* or *soon*, and that principles are more *vaguely* formulated than norms, one can indeed categorize.

From the content of column 4 in table 6.4, the nature of a policy topic in terms of the regime theory, however, does not seem an explanatory factor, given the lack of a clear pattern in the categories of political influence. Consequently, the assumption that principles and norms are more easily influenced by NGOs than regulations and procedures seems not verified here. To give an example: whereas NGOs were able to impact Article 4.2 of the FCCC (regulation), they were unable to do so with regard to Article 3 (principles). This, incidentally, can be explained by the fact that principles may prejudice on state actions. The principle that the CBD is 'the' instrument to deal with forests may – if accepted and elaborated – in the long term lead to sustainability measures binding upon tropical timber producers, a possibility which is not very much to the liking of these players (case 16). So principles may stand for future regulations and they therefore lead to negotiations as tough as on regulations which should be implemented immediately. In other words:

Conclusion 13 — Principles, norms, regulations and procedures are all equally easy – or difficult – to influence for NGOs.

6.4.2 Domestic policy effects

Governments probably anticipate the potential policy effects of the treaty element they negotiate and decide upon, as was shown above. These effects may be individual, felt at the 'individual' level of a government which needs to adapt its national policy to international rules, or they may be felt at global level in the sense that a convention might change aspects of the international system structure or produce new trends and events within different sectors of that system (see theoretical framework, section 2.6). In addition, the individual level generally refers to short-term effects, whereas the global level generally covers effects in the (very) long term. To start with the former, it is assessed in column 5 of table 6.4 to what extent the different policy topics are expected to have short-term effects on individual government policies. True, such effects are difficult to assess, especially as neither treaty contains very strict and binding provisions. Nonetheless, some differences are traceable. For example, provisions in a preamble are not binding, and these do not have any direct and short-term policy consequences. The same applies to proposals not yet adopted (case 5 on the AOSIS Protocol) and to studies which are commissioned (case 18 on the financial mechanism). Other provisions, though, have short-term effects at domestic level, for example Article 8 of the CBD (in-situ conservation) and Article 4.2 of the FCCC (target). These provisions urge parties to undertake measures and design or change national policies and laws, *although these national effects should not be exaggerated.* Therefore the scores on domestic effects in table 6.4, ranging from 'none' to 'substantial', should be interpreted as relative. Furthermore, the expected effects are not necessarily valid for all parties. For example, only so-called Annex I Parties to the FCCC – the industrialized countries – should, as a moral obligation, implement the stabilization target (case 3).

One might assume that the more governments expect new international rules to impose policy changes upon them, the less they want others – including NGOs – to co-determine the outcome. This assumption, though, has a lot in common with previous ones, for example on 'political relevance'. It was already concluded on the basis of the comparative case-analysis that the less politically relevant a policy outcome is, the more easily NGOs can impact such an outcome (see conclusion 4, page 234). As political relevance refers to the importance of *key* topics in an issue arena and as topics which urge parties to immediately undertake policy initiatives at domestic level are generally *key* aspects of negotiation processes, there is probably some overlap. Such relationships between explanatory factors, however, will be elaborated on below (section 6.7).

Given the information in the fifth column of table 6.4, NGOs *can* impact topics which will probably have some or substantial short-term policy effects and *cannot* always impact those *without* such effects. The above assumption seems therefore invalid. Still, the division of scores seems to point to a (very weak) relationship. Only 3 out of the 9 cases which were politically influenced by NGOs refer to some or substantial effects on state policies, whereas effects occur in 5 out of the 9 cases without such political influence (see shaded cells in column 5 of table 6.4). Therefore it seems that:

Conclusion 14 — The more effects new international rules will (probably) have on domestic policies in the short term, the less easily they can be impacted by NGOs.

6.4.3 GLOBAL EFFECTS

It is obvious that it is far too early – given the fact that both treaties have only recently entered into force – to try to assess the global and long-term effects of the FCCC and CBD on the international system (as expressed by arrow (f) in the theoretical model, figure 2.2, page 71). One may even wonder whether such effects will ever occur. For example, it is quite unlikely that these two environmental treaties will transform the division of resources and the rules of the game at global level, although the substantial participation of NGOs in the climate and biodiversity arenas might raise precedents as far as other arenas are concerned and change rules on the formal position of NGOs in world politics in the future. In addition, long-term effects on the international system are generally unintentional, as it is for example not always obvious whether rules of treaties will transform an existing regime. One may consider Article 15 of the CBD on the access to genetic resources in this respect (case 12). Legally, this provision reaffirms traditional international law based on national sovereignty. In practice, this rule may – if indeed implemented – change the customary free access regime to the genetic resources of (tropical) regions (Witmeyer, 1994). Legally, negotiators seemed not to have intended to change anything, only reaffirming existing law. In practice, a regime change may be an unintended consequence. Given the unpredictability and – sometimes – unintended nature of the global and long-term effects of both treaties, it is rather unlikely that negotiators anticipate such effects during day-to-day policy-making on the ground. It is even more unlikely that they try to 'control' NGO input and NGO success on the basis of such possible long-term effects. Therefore no additional conclusions on the explanation of political influence have been formulated at this level.

— 6.5 Explanations at the contextual level —

6.5.1 UNCED AND OTHER CONTEXTUAL FACTORS

Section 6.5 covers the factors at the contextual level. In conformity with the theoretical model (figure 2.2, page 71) and the list of possible explanatory factors (table 3.3, page 90), the following aspects will be addressed in this section: distribution of resources, regimes, rules of the game, and events and trends in the international system (economy, politics, environment).

In the previous section the potential future effects on the international system of new treaties was briefly dealt with. However, there is a reciprocal linkage, as the given context of policy-making does co-determine the outcome as well (see arrow (a) in the theoretical framework). This second relationship is examined here. It is assumed that the extent to which NGOs are (un)able to influence policy outcomes is also dependent on the context in which these policies are negotiated and decided upon. With regard to treaty formation and implementation, this context consists of the international system. As contextual factors such as distribution of resources, rules, events and trends may differ in various arenas and eras, NGOs are not similarly enabled or constrained by these factors in all cases. Therefore political influence might be explained by such contextual factors.

However, this study only covers one short timespan (1990-1995), a fact which implies that many factors are more or less constant phenomena in both issue arenas. A comparative case-analysis can therefore not yield explanations based on these constant factors. For example, both the biodiversity and climate treaty were negotiated parallel to the UNCED process. This conference can rightly be considered a very important event in the history of global environmental policies. It definitely influenced the climate change and biodiversity negotiations, as was highlighted in several case descriptions in chapters 4 and 5. In addition, the UNCED process probably enabled the NGOs to intervene in global environmental politics, as – due to this Earth Summit – the environment stood high on the international agenda at the beginning of the 1990s. But the plausibility of this assumption on UNCED cannot be checked on the basis of this study's data, as all cases are situated in the same context – including the Earth Summit – and in the same period.

The same probably applies to the general distribution of resources in the international system. This factor does not change that easily, and in as far as the system has changed in the period studied (1990-1995) – for example the collapse of the Eastern bloc after the fall of the Berlin Wall in 1989 – these developments impacted world politics, including environmental policy-making, as a whole. In general, NGOs are constrained by the existing power relations in world politics, given their relative lack of resources compared to other political players, such as states and BINGOs. To conclude, it is very likely that several contextual factors which either remained constant or changes of which were felt in both arenas co-determined NGO influence, but this could not be checked on the basis of this study's data. It was therefore impossible to formulate individual and detailed conclusions on such factors. But given the likely assumption that they may have intervened in both arenas, one should be careful not to overlook these contextual factors.

> *Conclusion 15 — It is very likely that several contextual factors such as the UNCED process and the distribution of resources in the international system co-determined the extent of NGO influence, but this could not be verified by this study.*

Alongside these general contextual factors, more issue-specific ones may exist. Examples are regimes which are related to specific issue-areas, different rules of the game which characterize different political arenas, as well as very specific trends and events relating to climate change and biodiversity loss. Given these possible contextual variations in the studied cases, linkages between such issue-specific contextual factors and NGO influence might be discovered. These linkages are therefore examined more closely below.

6.5.2 Resources and regimes

Although the general distribution of resources in the international system is more or less given in any era, the distribution of issue-specific resources may vary in arenas. Theoretically, it is possible that NGOs have access to for example expertise which business organizations and country delegations lack (for an overview of 'NGO resources' see table 2.2, page 52). In that case, one might speak of *resource dominance* of NGOs. Column 2 of table 6.5 addresses this aspect. It is obvious that such dominance is rare, given the fact that resources such as expertise, contacts or legitimacy are seldom the exclusive domain of NGOs as many political players may share similar kinds of knowledge, contacts and political legitimacy. Moreover, it is hard to decide whether a certain player has more resources at his or her disposal than another. Therefore the substance of column 2, table 6.5, is mainly based on rough assessments, and doubts are included. It is nonetheless believed that in three cases NGOs might have been dominant vis-à-vis other players with regard to a specific resource. Firstly, NGOs definitely became involved in the AOSIS Protocol due to their substantive and legal expertise, which the Small Island States lacked (see case 5). That is not to say that the latter were entirely ignorant, on the contrary, but NGOs brought in additional expertise, as one AOSIS delegate also admitted. Secondly, NGOs generally have more contacts with indigenous and local grassroots

TABLE 6.5 *Factors at the contextual level: resource dominance and regimes*

	Cases	*Resource dominance*	*Regimes*
5	AOSIS Protocol	expertise	Atmosphere
9	Preamble	-	Environment and Development, Nature conservation
6	Joint implementation	-	Atmosphere
15	Marine biodiversity	-	Nature conservation
18	Financial mechanism	-	Environment and Development (GEF); Ozone Fund
3	Target	public opinion (?)	Atmosphere
8	Berlin Mandate	supporters (?)	Atmosphere
11	In-situ conservation	expertise (?)	Nature conservation
14	Indigenous / locals 2	contacts	Indigenous people
1	Objective	-	Atmosphere
2	Principles	-	Environment and Development (Rio Declaration)
4	Implementation	-	Environmental law (implementation mechanisms in general)
7	First review	-	Environmental Law (review mechanisms in general)
10	Indigenous / locals 1	contacts	Indigenous people
12	Genetic resources	expertise (?)	FAO Global System
13	Biosafety 1	-	Environment and Development (Agenda 21)
16	Forests	-	Forest principles (ITTO, Agenda 21)
17	Biosafety 2	-	Environment and Development (Agenda 21)

organizations than governments do. This may have been to the NGOs' advantage (see cases 10 and 14). Furthermore, there were doubts on resource dominance in other cases. It may have been the case that public opinion just before UNCED was more in favour of NGOs' views on strong climate regulations than of other players' preferences (case 3); it may also have been the case that NGOs mobilized more supporters in Berlin than other players did (case 8); and it may have been the case that NGOs which had a lot of experience on nature conservation and agricultural issues – notably IUCN and GRAIN – had more expertise to offer than many delegates did (cases 9, 11 and 12). All in all, resource dominance is no prerequisite for political influence, but it may be helpful (see shaded cells in column 2 of table 6.5). This is no surprise, as the phenomenon of resource dominance implies that NGOs have something to offer to policy- and decision-makers or have something to threaten states with.

> *Conclusion 16 — In cases where NGOs are dominant vis-à-vis other players with regard to a specific resource, this enables them to become influential.*

The next contextual factor to be examined is the impact of specific regimes on NGO influence. It is obvious that the climate and biodiversity treaties were negotiated within the context of different regimes (although some very general rules on the environment may have impacted both). Whereas the FCCC was co-shaped by the atmosphere regime, within which several older treaties can be distinguished such as the Convention on Long Range Transboundary Air Pollution (LRTAP) and the Vienna Convention and Montreal Protocol on the ozone layer, the CBD was co-shaped by given sets of rules on nature conservation, agriculture, indigenous people and trade (e.g. rules from the World Heritage Convention, FAO Global System on Plant Genetic Resources, ILO Convention 169 on Indigenous People, General Agreement on Traffic and Trade, and others). These supposed linkages, incidentally, are not mere assumptions, but recognizable in the preparatory documents of both treaties. As negotiators generally look for precedents, for ideas as well as for consistency in international law, they seek inspiration from related legal instruments. For example, there was a lot of discussion on how to apply the 'ozone model', including the financial mechanism, to climate change (Goldberg, 1993; Doc I.68). There were also discussions on how to copy the 'global list instrument' of other nature conservation treaties into the CBD, although this inclusion failed (Doc II.7). In other words, existing sets of rules inspire policy- and decision-makers to copy things or, to the contrary, to do things differently. Hence, existing regimes function as frames of reference.

NGOs may also use existing rules. One way to do this is to marshal arguments from existing treaties. For example, they tried to legitimize the need for and possibility of strong regulations within the FCCC with the help of the Montreal Protocol, just as they tried to legitimize global lists by referring to CITES and the World Heritage Convention (Glowka *et al.*, 1994; Docs II.52, II.78). It should be realized, however, that although players may intentionally use and legitimize arguments from existing regimes, given rules may impact negotiations and decision-making without the deliberate intervention of players as well. For example, negotiators may routinely reproduce principles on environmental protection and management – such as the precautionary principle – without any explicit inspiration from existing instruments.

The above arguments already disclose some differences between the FCCC and CBD. Apart from the substance itself, the FCCC seemed to be another step in expanding one regime (atmosphere), whereas the CBD appeared the basis for integrating several (nature conservation, agriculture, trade). As one respondent noted, the CBD may be considered the first sustainable development treaty. This implies that, on the one hand, various and sometimes contradictory regimes were frames of reference for the CBD, and, on the other, that no precedents for such regime integration were available. Therefore one can assume that the processes of governance by existing regimes were quite different within the climate and the biodiversity arenas. In short, the atmosphere regime was probably a rather obvious frame of reference for the FCCC, whereas the CBD lacked a clear frame of reference.

Apart from the different regimes which governed the formation and implementation of the FCCC and CBD as a whole, the various policy topics may have been affected by different sets of rules as well. Column 3 in table 6.5 summarizes these. Again, the biodiversity cases are – unlike those on climate change – characterized by a broad set of contextual rules (nature conservation, indigenous people, FAO Global System, etc.). Given the fact that no real differences were found with regard to NGO influence on the two issues, these different contexts do not at first sight seem to be explanatory factors of NGO influence. A closer look at the data in column 3, though, reveals that NGOs seemed to be enabled by existing rules on atmospheric pollution, on nature conservation and on indigenous people, while they were constrained by others, e.g. on environment and development, forests and genetic resources. This may be explained by the similarity of NGO demands on the one hand and given rules on the other. As was outlined at the theoretical level in chapter 2, the more these correspond, the more legitimate and influential an NGO demand becomes. The fact that the Montreal Protocol – as well as the LRTAP instrument – already contained strong regulations on emission reduction of substances reinforced the arguments of NGOs – and others – to adopt similar rules in the FCCC. The fact that ever more bodies started to recognize the important role of indigenous people in agriculture, forestry and nature conservation helped those who propagated similar provisions in the biodiversity arena. The same applies to nature conservation measures, but an exception does exist there. Although the instrument of global lists was accepted in other treaties, and therefore a strong and legitimate demand, it was not in the end accepted. This could be explained by the fact that most negotiators did not want to have a purely 'green' treaty and therefore did not accept the nature conservation regime as the one and only source of inspiration. Therefore whether a similarity between NGO demands and existing rules becomes effective also depends on the negotiators' views on which existing rules the new instrument should be built on. Delegates in the climate arena did accept the Montreal Protocol as a legitimate source of inspiration, although many did not agree on specific emission reductions, whereas the majority of delegates in the biodiversity arena did not simply want to produce a treaty in the tradition of nature conservation.

> *Conclusion 17 — If there is (some) similarity between NGO demands on the one hand and the rules of an existing regime on the other, NGOs are enabled to impact policy outcomes by these given rules, at least where they are accepted as legitimate frames of reference by (most) political players.*

The validity of this conclusion also depends on its ability to explain the lack of NGO influence. To return to table 6.5, NGOs seemed to be constrained by the environment and development regime, in which for example the Rio Declaration and Agenda 21 are nested, by environmental law in general, as well as by the FAO regime on genetic resources. Coincidentally, or not, many of these rules are non-binding – or soft law – and also quite recent, particularly the regime on environment and development. These two characteristics might suggest to those involved in treaty formation and implementation that such rules are less valid and legitimate than binding and longer-established ones. Furthermore, NGOs tended to demand far more progressive rules on environment and development, agriculture, and implementation and review mechanisms than those which had already been adopted in international law and agreements. In other words, the similarity between NGO demands and given rules was less in these cases, and the topics of access to genetic resources and biosafety are good examples in this respect (see cases 12, 13 and 17 in chapter 5). To sum up, the substance of the above hypothesis seems to be able to explain the *lack* of NGO influence as well. As a consequence, 'regime similarity' is considered a relevant explanatory factor of political influence.

6.5.3 Rules of the game, trends, events

The formal rules of the game describe – among other things – the legal position of NGOs and their role in intergovernmental negotiations and decision-making (which is generally quite limited in nature). To repeat, the UN General Assembly (UNGA) functioned as the organizational context for the formation of the FCCC, and the United Nations Environment Programme (UNEP) for the CBD. Traditionally these two bodies, which are so different in extent and scope, operate under different rules on NGO participation. The UNGA, the supreme and political body of the UN, more or less excludes NGO participation in its sessions, whereas the UNEP – as a small, sectional and technical body – accepts it, and perhaps increases and strengthens it (Willets, 1982; 1996). Therefore formally the opportunities for NGOs to intervene in UNEP sessions are better than in UNGA. Nonetheless, the Intergovernmental Negotiation Committee (INC) of the FCCC as established by the UNGA introduced its own rules of procedure, whereby those on NGO participation were based on the rules of procedure of the UNCED Preparatory Committee (PrepCom; Doherty, 1994). These latter rules recognize the observer status of NGOs, but exclude any official negotiating role for them. Furthermore, the first session of the INC – which was held in the USA with its liberal attitude towards pressure groups – opened its doors to NGOs completely. This set a precedent for subsequent sessions. In fact, the rules on NGO participation in the climate and the biodiversity arenas became quite similar after all, despite different organizational contexts.

One should, however, also distinguish between the rules of the formation and the implementation phases respectively. The previous paragraph obviously deals with the first phase, in which treaties are negotiated and adopted within the framework of UN bodies. In the implementation phase, though, the provisions of the treaty itself and – at a later stage – the rules of procedure of the Conference of the Parties, replace the formation rules. With regard to the period after 1992, the formal provisions on NGO participation are completely the same in the climate and the biodiversity arenas and, moreover, not very different from those of the formation phase (see Article 7.6 in the FCCC and Article 23.5 in the CBD). In short, the rules on NGO participation coincided in both arenas and in both policy phases, at least in substance. Therefore this factor cannot be linked to different outcomes of NGO influence. True, UN rules may have (negatively) affected the chances for NGOs to intervene in the formation and implementation of the FCCC and CBD – as they have no formal negotiating role – but this could not be assessed here, given the fact that these rules have remained the same in the political arenas, policy phases and period studied.

> *Conclusion 18 — It is very likely that the rules of the game in the UN system co-determined the extent of NGO influence, but this could not be verified by this study.*

It is obvious that this conclusion has a lot in common with conclusion 15 (page 249). Again, the impact of a contextual factor on the NGOs' political influence could not be assessed.

The last set of contextual factors which should be elaborated on in this section covers issue-specific trends and events in international environmental, economic and political affairs, all factors which may have enabled or constrained NGOs during the period of this study. Table 6.6 presents an overview of the factors which might have played a role, including the general trends/events already mentioned above (UNCED, Berlin Wall). It should, however, immediately be noted that the substance of table 6.6 is far from complete and somewhat arbitrary. Only those relevant events and trends which were documented in the literature and in government and non-government files were taken into consideration. It is perfectly possible that additional analyses – for example of the media – would have yielded new information.

TABLE 6.6 *Trends and events in the international system in the period 1990-1995*

	General	*Climate*	*Biodiversity*
Environment	- UNCED process	- extreme weather conditions - more scientific certainty	- loss of biodiversity - patent piracy
Economy	- recovery of the world economy	- economic costs	- InBio – Merck deal
Politics	- End of the Cold War	- (FCCC)	- (CBD)

To repeat, the comparative case-analysis cannot produce conclusions on the impact of general events and long-term trends on NGO influence, since these are by their very nature contextual factors in *all* cases. Nonetheless, events or processes such as the UNCED, the recovery of the world economy in the 1980s, and the end of the Cold War, definitely impacted environmental policy-making at global level in the period 1990-1995. For example, all three factors may have contributed to the rise of the environmental issue on the global agenda in the early 1990s, and this may as a result have favoured the environmental movement in that period. However, this assumption, which seems so obvious at first sight, cannot be linked to NGO influence on the basis of this study. But linkages with NGO influence might be discovered if more specific and short-term trends and events are examined.

To do this, it was among other things checked whether so-called *environmental shock events* – in the category of Cjernobyl, Bhopal or the Exxon Valdez – with a bearing on climate change and biodiversity occurred in the period 1990-1995. As far as the climate issue is concerned, various (minor) shock events did indeed happen around the globe – droughts, floods, typhoons, extremely hot summers – but there was no single outstanding event after which NGO influence tended to increase. Furthermore, such events had also occurred before the negotiations started in 1990. NGOs, of course, utilized these shock events as arguments to adopt a strong treaty and that may have helped them, but this is – again – difficult to decipher from the data available. A similar story applies to biodiversity. Bad news on the dramatic decrease of biological diversity was heard during the entire 1980s and the early 1990s, so we deal with a more or less constant trend rather than distinctive shock events here. But there may be one exceptions in the field of biodiversity: the news on patent piracy which reached the world press in 1992 (see case 14). The fact that local species, land varieties and even human cell lines from indigenous people were freely appropriated and subsequently patented by transnational corporations raised indignation and anxiety among most delegates in the biodiversity arena. It is not unlikely that these events strengthened the NGOs which campaigned for regulated access to genetic resources as well as for recognizing the rights of indigenous people and local communities. The fact that NGOs influenced policy outcomes on this topic after 1992 (case 14), whereas they had failed to do so in the period before (case 10) might point to an impact of the 1992 shock event 'patent piracy' on NGO influence. However, this difference in political influence can be explained by other factors as well, namely the presence and absence of indigenous groups and chiefs in the biodiversity arena at the respective times. Besides, NGOs were not able to achieve their goals with regard to regulated access, despite this shock event (see case 12).

This last may, incidentally, also have been caused by another event, but in the economic field: the 1991 Costa Rican InBio-Merck deal on the access to genetic resources (see case 12). This deal became a model for the delegates in the biodiversity arena and is reflected in Article 15 of the CBD. Access to genetic resources is based on bilateral contracts, whereas NGOs favoured a multilateral mechanism. So this event may have constrained the NGOs in achieving their policy goals. To sum up, the impact of (shock) events on NGO influence is not obvious, especially as the above examples only relate to a few cases in this study.

Conclusion 19 — NGO influence seems neither positively nor negatively related to environmental (shock) events.

Apart from shock events, *scientific discoveries* which may have impacted negotiations and NGO influence on the FCCC and CBD were looked for by the researcher, but these are absent in the period studied. Again, there may be one exception, namely the increase in scientific certainty with regard to global warming. It seems that any new report from the IPCC and any new intervention by its chairman Bert Bolin in the INCs and COP eliminated a part of the uncertainty, although some has remained until today (Doc 1.28). This trend of growing certainty seems to run alongside with a trend of growing NGO influence in the climate arena in the period 1990-1995 (compare the formation and implementation phases of the FCCC in this regard). True, this does not necessarily imply that there is a causal link, especially as delegates may have been directly influenced by the IPCC. Besides, the increase of NGO influence in the biodiversity arena since 1992 cannot be linked to an increase of scientific certainty. But a (weak) linkage of science with political influence cannot be fully excluded, given the fact that NGOs need good, science-based arguments to become persuasive in the eyes of policy-makers. Therefore the following conclusion is (cautiously) formulated:

> *Conclusion 20 — It seems that the more scientific support NGOs can mobilize, the more influential they will become.*

Finally, issue-specific trends and events in the economic and political sectors were examined. As far as biodiversity is concerned, the InBio-Merck deal has already been mentioned, but other relevant economic events seemed to be absent. With regard to climate change, several economic scenarios of different policies to prevent or mitigate climate change were published (Hayes and Smith, 1993). Most of these revealed that strong policies would imply huge economic costs. As a consequence, the fear of economic costs circulated throughout the climate arena in the period studied and may have constrained NGOs, but in general, not with regard to specific cases. In the political sector, next, very obvious trends or events were not traceable with regard to either issue. In fact, the CBD and FCCC were themselves 'the' political events on these two policy issues at global level in the period 1990-1995 (apart from the UNCED). But these treaties are the object of study here, not contextual factors, and therefore are bracketed in table 6.6. To sum up, outstanding, relevant and issue-specific trends and events in the economy and in politics were more or less absent.

— 6.6 The respondents' views —

The conclusions above were based on a comparison of the 18 cases presented in chapters 4 and 5. But insights of experts may yield some additional hypotheses on the background of NGO influence. Therefore the non-government as well as government respondents were asked to sum up those factors on which they believe the political influence of NGOs is based. Respondents were also asked to answer a similar question on the *lack* of such influence. Their arguments are classified in tables 6.7 and 6.8. Before elaborating on these data, it should be stressed that not all respondents were able to deal with these two questions, so that the number of respondents varies in the columns of the tables. In particular the number of NGO respondents is small. The background to these omissions has already been reviewed in subsection 3.3.3. In addition, it is not always easy to classify the arguments of respondents, as each uses his or her 'own' jargon. Nonetheless, more or less similar factors to explain NGO influence brought forward by different respondents were grouped under headings such as 'knowledge' or 'contacts'. To check whether this procedure of classification was correctly used by the researcher, it was repeated once. As the outcome remained more or less the same, it was believed to be appropriate.

TABLE 6.7 *Factors mentioned by respondents to explain the political influence of NGOs*

Respondents	*Government*		*NGO*		
Arguments	*FCCC*	*CBD*	*FCCC*	*CBD*	*Total*
knowledge/info	5	7	1	5	18
NGO-NGO contacts/ co-operation	-	1	4	6	11
skills	1	4	1	2	8
legitimacy	1	2	2	3	8
watchdog	1	2	-	2	5
NGO-state interactions	-	-	-	3	3
media attention	-	-	1	1	2
public opinion	1	-	1	-	2
other factors (only mentioned once)	2	2	3	1	8
Number of respondents	8	8	4	6	26

Table 6.7 summarizes the factors which may, according to the respondents, explain the political influence of NGOs. Remarkably enough, the respondents refer mainly to factors at the player level, in particular to characteristics of the NGOs. Arena or contextual factors are barely included, with the exception of media attention and the support of public opinion, but these were only rarely mentioned. More important are: knowledge and information on the policy topic concerned; co-operation and contacts between NGOs; skills in lobbying and negotiating; legitimacy of the NGO(s) concerned; their watchdog function. One factor is obvious: knowledge and information (18 references). If NGOs bring solid facts to the negotiations, they will become influential. This argument was particularly put forward by government respondents. To quote one:

"They are influential as long as they are providing facts instead of wild speculation or accusation. Anybody with authoritative facts and a good discussion is welcome. I think that is the role of NGOs: to provide information apart from the official channel. As long as they are factual, they can play a role and influence the process."

Substantive knowledge on the topic is important, but so is procedural knowledge to be able to bring it forward at the right place and time. This latter is referred to as 'skills' of NGOs in for example lobbying, negotiating, international diplomacy etcetera. The two categories of knowledge can be jointly classified as *expertise.* If this is done, the factor 'expertise' is considered the most important explanatory variable by far for NGO influence (26 references).

Conclusion 21 — The more expertise NGOs bring to the negotiating table, the more influential they will be.

The second most important factor is contacts and co-operation among NGOs (11 references). However, this factor is *not* considered relevant by government respondents. Obviously, delegates do not believe that NGO effectiveness is dependent on their mutual contacts and co-operation. This may indicate that they believe that one NGO can be as influential as a whole NGO coalition. What is more, it may even indicate that delegates are as responsive to one NGO (with good arguments) as to a whole group of NGOs. Therefore the widespread believe among NGOs that co-operative arrangements are definitely more effective may be a misconception, as government delegates – the ones who have to be influenced after all – believe otherwise.

Conclusion 22 — Political influence does not seem more likely when NGOs co-operate more or maintain more mutual contacts.

This conclusion has some common ground with the fifth one, which denies the linkage between – among other things – the number of NGOs and their influence (see page 237).

Other factors are considered relevant by the respondents as well, but to a lesser extent: for example legitimacy and the watchdog function. The former refers to the issue of representation. It is believed that NGOs are influential because they represent in global political arenas legitimate interests of both the environment and large sections of society. Remarkably enough, fewer government respondents believe this argument to be valid than do the NGOs themselves. As one of the delegates noted:

"I am not impressed by the claims of some NGOs to be representatives of anybody in particular. (...) NGOs are more a source of ideas and opinions."

In addition, some respondents believe that NGOs are influential because they function as watchdogs. NGOs critically monitor the statements and actions of states and – if necessary – publicly condemn their lack of consistency. They also monitor the implementation of international agreements and criticize any failings. As states are generally concerned about their public image, they are rather easily responsive to the interventions of such watchdog NGOs. Nonetheless, the number of references to this 'source of influence' is quite low (5 out of 26).

Conclusion 23 — Legitimacy and the ability to function as watchdogs are additional sources of influence for NGOs.

Additionally, interactions between NGOs and states, coverage of NGO views and actions by the press, and the support of public opinion are mentioned by some respondents. Surprisingly enough, only a few (NGO) respondents refer to the importance of interactions between NGO and state representatives. Despite the fact that such contacts are a prerequisite for political influence, this factor is hardly mentioned by the respondents. The last category, 'other factors', in table 6.7 refers to 8 arguments which were only mentioned once and were therefore considered of little relevance.

TABLE 6.8 *Factors mentioned by respondents to explain the lack of political influence of NGOs*

Respondents	*Government*		*NGO*		
Argument	*FCCC*	*CBD*	*FCCC*	*CBD*	*Total*
'Wrong approach'	7	6	1	-	14
Lack of legitimacy	1	1	1	1	4
Legal position	1	1	-	-	2
Other factors (only mentioned once)	-	-	-	2	2
Number of respondents	8	8	2	2	18

Table 6.8 summarizes the respondents' arguments on the NGOs' *lack* of political influence. It should be immediately noted that the number of respondents, in particular of NGO respondents, is smaller than in the previous table. Consequently, the NGO arguments are barely representative. Still, the table highlights one aspect in particular, namely that government representatives dislike what they among other things call emotional, speculative, idealistic, confrontational, accusing or radical arguments and approaches of NGOs (what is grouped and labelled as 'wrong approach' in table 6.8). As long as NGOs behave like this, delegates are unlikely to be responsive and accessible and, hence, political influence seems out of the question.

Conclusion 24 — If NGOs employ a 'wrong' approach – an approach that is for example based on emotion, speculation, ideals or confrontation – their chances of exerting political influence decrease.

This conclusion has a lot in common with the third one on the link between radical or unrealistic policy goals of NGOs and less political influence (see page 233). Moreover, it can be considered a mirror image of the governments' positive view on expertise highlighted in the above. After all, expertise is generally based on rational, non-speculative and pragmatic arguments and people. A second mirror image of one aspect of table 6.7 is the lack of legitimacy. Doubts of NGO legitimacy in the eyes of policy- and decision-makers might, of course, harm the NGOs' credibility and their chances of success, but this aspect is already covered by conclusion 22. Finally, two government respondents refer to the legal position of NGOs, which formally excludes them from intergovernmental negotiations and decision-making. This makes it difficult for NGOs to intervene and impact policy outcomes. However, the small number of references to this factor does not justify the formulation of an additional conclusion.

— 6.7 Enabling, constraining and neutral factors —

Table 6.9 summarizes all those factors which were dealt with in the previous sections and which may – or may not – explain NGO influence. Distinguished are enabling, constraining and neutral factors deduced from both the comparative case-analysis and respondents' views. The first category is hypothesized as enabling NGOs to exert political influence, the second category is assumed to constrain them from doing so, and the third is believed not to have any positive or negative impact in this regard. This section will use a specific procedure, as was outlined in section 3.3.2 and in the introduction to this chapter, namely the *horizontal comparison* of factors in the data-matrix. The aim is to compare the explanatory capability of the various factors and assess their linkages.

In the previous sections, 24 conclusions were formulated which relate specific factors to NGO influence. As has already been, these relationships are not always obvious. Some seem to be stronger than others. In other words, the correlation between NGO influence and these factors varies. To assess this difference, two formulae were employed, one with regard to the comparative case-analysis, another regarding the respondents views. The first formula checks to what extent the factor concerned seemed to be able to explain the outcome of the different cases. It is examined how many cases within the group confirm the correlation of factor X and NGO influence as well as how many negate it. For example, traditional block politics may explain the lack of NGO influence in 7 out of 9 cases (see table 6.3, page 238). However, 2 of the 9 cases which yielded positive outcomes of NGO influence are also characterized by such North-South block politics. Therefore, 7 out of 9 cases confirm the potential impact of this factor, whereas 2 out of 9 cases, though, negate it. A score to express the explanatory capability of this factor is: (7 : 9) - (2 : 9) = 0.56. In mathematical terms:

(1) $$EC = (Nc : Nt) - (Nn : Nt)$$

EC: explanatory capability of factor X
Nc: number of cases which confirm the explanatory capability of X
Nn: number of cases which negate the explanatory capability of X
Nt: total number of cases within the group concerned

Again, this formula is just a heuristic tool to position the various factors on the basis of their explanatory capability and it does not aim at an objective quantification or a statistical test. The figures are thus relative in nature and aim at an ordinal hierarchy of factors.

The second formula highlights the number of government respondents who referred to factor X compared to the size of the group of respondents as a whole. Here the basic assumption is that the more respondents refer to a specific factor X, the more its explanatory capability is confirmed. Note that only the references of government delegates are taken into account. Given the fact that these are the ones to be influenced by NGOs, they are considered the ultimate referees on the explanation of NGO influence. To give an example, the factor 'knowledge' is mentioned by 12 out of 16 government respondents. This gives a score of 0.75. In mathematical terms:

(2) $$EC = Nr : Nt$$

EC: explanatory capability of factor X
Nr: number of government respondents who mention factor X
Nt: total number of government respondents within the group concerned

It should immediately be noted that the two formulae – one related to the cases and one to the respondents – are very different in nature and therefore difficult to compare. Nonetheless, they are combined in table 6.9 just to get an *impression* of the hierarchy of factors based on their explanatory capability.

Given the fact that the maximum outcome of both formulae would be a score of 1.00 – if all cases/respondents confirm the impact of factor X – the scores in table 6.9 are generally quite low. Only a few exceed 0.50. In consequence, the correlation between most factors and NGO influence is rather limited. This should be kept in mind while one is reading the rest of this book. It is therefore no coincidence that the factors are labelled 'enabling' and 'constraining', since this implies that the factors do not *guarantee* or *exclude* NGO influence, but that their occurrence may make it easier or harder to achieve.

TABLE 6.9 *Enabling, constraining and neutral factors on the explanation of the NGOs' political influence*

	Enabling factors	*Constraining factors*	*Neutral factors*
Comparative case-analysis	-regime similarity (0.56) - integrative negotia tions (0.33) - protest 'outside' (0.33) - implementation phase (0.33) - innovative proposal (0.22) - scientific certainty (0.22) - resource dominance (0.11) - UNCED process (?)	- like-minded and dominant states or other players (0.67) - traditional block politics (0.56) - unrealistic/radical policy goals (0.33) - formation phase (0.33) political relevance (0.22) - domestic policy effects (0.22) - distribution of resources in international system (?) - rules of the game (?)	- direct impact/drafting - specific (type of) NGO - specific (type of) NGO intervention - number of NGOs - presence of BINGOs - nature of the environmental issue - nature of the regime element - environmental shock event (?)
Respondents' views	- knowledge/info (0.75) - skills (0.31) - legitimacy (0.19) - watchdog (0.19)	- 'wrong approach' (0.81)	- NGO-NGO cooperation and contacts

Legend
(0.22): score based on formula (1) (comparative case-analysis) or (2) (government respondents' views)
(?) : status of factor uncertain but nevertheless likely

It is obvious that many factors in table 6.9 are not mutually exclusive or independent. There are many linkages to distinguish. Furthermore, some seem synonymous, whereas others seem to contradict each other. In order to picture the various linkages and contradictions, the factors in table 6.9 were examined in pairs through and across the subsequent columns. The following could be deduced from this comparative factor-analysis:

1 There is a linkage between 'integrative negotiations' and 'traditional block politics'. As the latter excludes the former, both are each other's mirror image, although confrontational negotiations may also occur among (blocks of) states other than North and South, for example between the EU and US (see case 3 on the target of the FCCC).

2 Protest activities outside issue arenas may be supportive to NGOs which try to exert political influence. This seems to contradict the conclusion that a specific type of NGO intervention is a neutral factor (see column 4, table 6.9). The same applies to the enabling factors 'innovative proposals', 'knowledge' and 'skills', which also cover specific NGO activities. Obviously, the type of an NGO intervention does matter. However, the neutral factor mentioned above refers to *types* of activity *within* issue arenas – drafting, advice, oral statements, publications, drafting etc. – whereas the enabling factors refer to either the *quality* of these activities (innovative, knowledgeable, skilful) or to a type of activity *outside* the formal political arena (protest).

3 The factors 'knowledge/info' and 'scientific certainty' are probably linked as well. After all, the more certainty in this regard, the more solid knowledge or information NGOs can use.

4 A similar relationship can be distinguished between 'skills' and 'innovative proposals'. Whether NGOs are able to come up with the latter is probably dependent on their skills to correctly intervene in negotiation processes.

5 The fact that the legitimacy of an NGO may matter seems to deny the observation that a 'specific (type of) NGO' is a neutral factor. But, again, the former refers to a *quality*, the latter to *classes* of NGOs (environmental group, nature conservation organization, agricultural group etc.).

6 'Traditional block politics' and 'political relevance' are not mutually exclusive. Given: a. the definition of the latter which includes the conflictual nature of a policy topic and b. the fact that traditional block politics always implies conflict, these two factors overlap (see subsection 3.2.5).

7 This also applies to 'political relevance' and 'domestic policy effects'. Both are (partly) interpreted in terms of *binding* provisions: the more binding a provision, the more it is considered politically relevant, and the more domestic policy effects are expected. Hence, these factors overlap as well.

8 The factors 'radical/unrealistic policy goals' and 'wrong approach' are to some extent synonymous, as was already noted in one of the earlier sections. If NGOs formulate such goals, they are definitely on the wrong track in the eyes of (most) policy- and decision-makers.

To sum up, it is not individual factors which may explain (the lack of) NGO influence, but groups of these. This conclusion is visualized in figure 6.1. Six groups of more or less interrelated factors are distinguished which co-determine the political influence of NGOs. The composition of the groups is based on the above analysis of linkages among the factors, as well as on the notions of the theoretical model in chapter 2. The enabling factors are characterized by the sign (+), the constraining factors by (-). Moreover, those factors which have the greatest explanatory capability are underlined (only those which yield a score of 0.50 or higher in table 6.9). There are, next, linkages between the six groups of factors to distinguish. Thus, the impact of factors on NGO influence may be individual in nature or co-productive with another factor. Apart from the many links between the factors *within* the six groups, the following two groups in figure 6.1 are (partly) interrelated as well: 1. the groups 'negotiations' and 'policy topic' (through traditional block politics and political relevance) and 2. the groups 'specific context' and 'NGOs' (through scientific certainty and the NGOs' expertise). There is, in addition, one separate group of factors: the 'general context', the impact of which was only speculated about (see dotted arrow in figure 6.1).

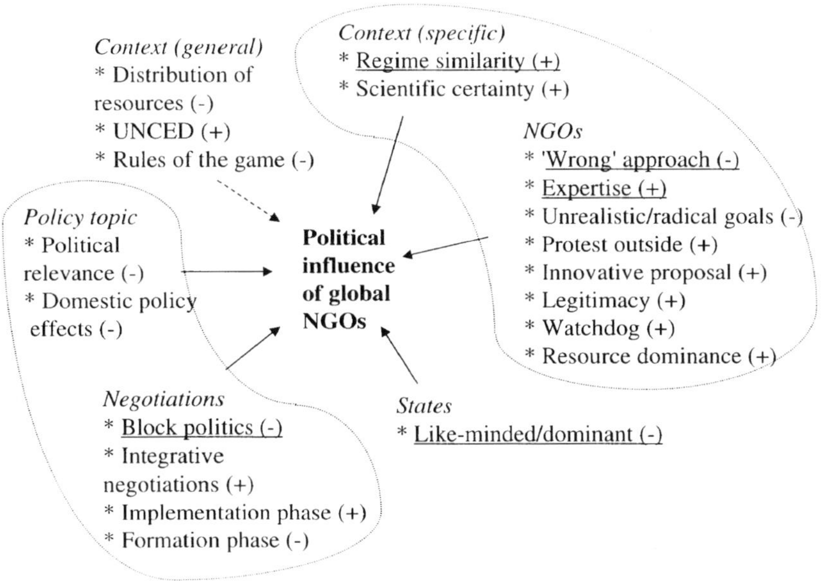

FIGURE 6.1 *Causal model on political influence*

One may consider figure 6.1 as the final conclusion of this chapter. It is a causal model on the political influence of global NGOs which follows from the comparative case-analysis and the respondents' views. On the basis of these methodologies, several factors which seemed relevant in advance – such as the number of NGOs and the extent of their co-operation – were eliminated and a hierarchy among those which indeed seemed to have at least some explanatory capability was constructed. This causal model will function as a guideline to formulate recommendations in the next chapter.

As was already outlined in the introductionary section to this chapter, one should be careful in generalizing figure 6.1. In empirical terms, the model only covers the climate and biodiversity arenas in the period 1990-1995. In theoretical terms, though, it may be seen as a general model for the explanation of NGO influence. As such it may become the input of new research, by which its causal links can either be verified or negated on the basis of empirical domains other than the FCCC and CBD.

— 6.8 Conclusions —

The analyses in this chapter, which included the positioning and comparison of the FCCC and CBD cases, the comparison of possible explanatory factors, as well as the interpretation of respondents' views, produced the following conclusions:

1 The two most important factors which enable NGOs to exert political influence are the quality of their interventions (in particular based on their knowledge and skills) and the similarity between NGO demands and existing regimes (which increases the legitimacy of these demands). Less important but still enabling factors are: the integrative nature of negotiations, protest activities of (other) NGOs outside the political arena, the implementation phase of a treaty, resource dominance of NGOs vis-à-vis (some) states, scientific certainty on a policy issue, the legitimacy of NGOs, their watchdog role, and – probably – a political process such as the UNCED.

2 The main three important factors which constrain NGOs in exerting political influence are: the use of a 'wrong' approach, the presence of like-minded but dominant states (which means that they dominate a topic and are not dependent on any NGO input), and negotiations based on traditional block politics of North and South. Less important constraining factors are: radical and unrealistic policy goals of NGOs, the formation phase of a treaty, the political relevance of a policy topic, the (potential) domestic policy effects of a decision, as well as – probably – the uneven distribution of resources in the international system.

3 A number of factors which seemed relevant in explaining NGO influence at first sight proved not so when examined. Examples are: a specific (type of) NGO, a specific (type of) NGO intervention (including the drafting of legal texts), the number of NGOs which intervene, the extent of NGO co-operation and of their mutual contacts, the presence of business NGOs which may neutralize the input of NGOs, the nature of an environmental issue, the nature of the regime element which NGOs try to affect, as well as – probably – the occurrence of an environmental shock event.

4 Several of the above factors are interdependent or overlap. Besides, one should not generalize these findings in the empirical sense – beyond the climate and biodiversity arenas in 1990-1995 – although the conclusions may still offer some general, theoretical insights into the explanation of the political influence of global NGOs.

CHAPTER 7

Recommendations

— 7.1 Introduction —

This chapter elaborates on the second objective of this study and on research question 3 (see subsection 1.5.2). It also builds on the conclusions of chapter 6 (see table 6.9 and figure 6.1). Its aim is to formulate recommendations for NGOs on how to increase their political influence. To do this, the explanatory factors identified in the previous chapter will be considered in the light of two judgement criteria: *anticipation* and *extrapolation.* The first covers the judgement of the extent to which NGOs might anticipate on the enabling, constraining and neutral factors of political influence and, consequently, increase their impact on treaty formation and implementation. In as far as NGOs are indeed able to anticipate, recommendations can be made. The criterion of extrapolation, secondly, implies a judgement on the external validity of these recommendations. In other words, to what extent can they be extrapolated in time and to other issues, respectively to the biodiversity and climate arenas of the near future and to other arenas? As some recommendations are probably more time-sensitive and issue-specific than others, some conclusions can be extrapolated and others not, although one should be very careful with generalizations (see section 1.5.3). Another source of information, which might yield additional recommendations for NGOs, is the views of the 36 respondents who reflected on the question of how NGOs might increase their political influence. Again, the views of government representatives will be given special attention, as they are the ones to be influenced by NGOs.

Before discussing the recommendations themselves, something should be said about their nature and level. Firstly, their nature is in line with the qualitative methodology of this thesis. This implies that the recommendations which follow should be put into the right perspective. They will be far from predictions along the lines of: '*If* intervention X, *then* political influence Y'. Therefore there is no pretence of presenting a kind of recipe for NGOs to become influential. On the contrary, the recommendations are at best considered an (unrequested) input into the strategic debate of global NGOs. Ultimately, NGOs are the ones to judge the value of this chapter's conclusions. Secondly, the level of the recommendations is limited as well. The study is not primarily aimed at the improvement of NGO participation, but at the assessment of political influence. This objective implied a meso-level of analysis, namely that of organizations and negotiation processes. The individual level was therefore not thoroughly studied (although individuals were nonetheless units of observation; see chapter 3), and the evaluation will yield quite general recommendations at meso-level. In other words, this chapter will not provide suggestions on how for example an individual lobbyist can perform better. Recommendations will, instead, refer to NGO strategies in general.

The format of this chapter is as follows. Section 7.2 will provide a number of recommendations based on the outcome of the previous chapter. In subsection 7.2.1 the nature and use of the judgement criterion of anticipation will be highlighted. Then this criterion will be applied in subsections 7.2.2, 7.2.3 and 7.2.4 with regard to the enabling, constraining and neutral factors respectively. This will be done through taking into consideration the 24 conclusions identified in chapter 6, the explanatory capability of each factor (table 6.9, page 263), and the format of the causal model on political influence (figure 6.1, page 265). Section 7.3 will deal with the respondents' views, from which some additional recommendations will be derived. Section 7.4 will fuse all recommendations from both the comparative case-analysis and the views of the respondents into one strategic model, while the question to what extent this model might be generalized over time and policy issues will be reflected upon in section 7.5 (extrapolation). Finally, a number of conclusions will be formulated in section 7.6.

Please note that the 24 conclusions in the previous chapter and the 19 Recommendations in this chapter do not directly correspond, either in number or in sequence. This is due to the fact that chapter 6 is built on the theoretical model and the list of potential explanatory factors (table 3.4), whereas this chapter will elaborate on the list of enabling, constraining and neutral factors as well as on the causal model (table 6.9 and figure 6.1). There was also some overlap in conclusions in chapter 6, which is neutralized below, and specific insights of the respondents have produced some additional recommendations. Nonetheless, the substance of all the conclusions in the previous chapter is covered by the recommendations in the following sections.

— 7.2 From explanatory factors to recommendations —

7.2.1 Anticipation

Chapter 6 provided an overview of factors which explain the political influence of global NGOs. In as far as NGOs are able to anticipate on – or even manipulate –

these factors, they might increase their impact on treaty formation and implementation. For example, the fact that expertise is an enabling factor for NGOs implies that their impact will probably grow the more expertise they bring to the negotiating table. Consequently, one might recommend that NGOs strengthen their abilities in this respect. Therefore explanatory factors can be translated into recommendations. However, it is not immediately obvious to what extent NGOs can indeed anticipate on all those variables identified in chapter 6, as some are definitely within but others are beyond their control. For example, it is obvious that it is easier for NGOs to change their own strategies than to change the rules of the game in a political arena. Hence, not all factors – for example at the level of agents, political arenas, policy outcomes or contexts – are equally easily anticipated on by NGOs. Furthermore, the nature of the anticipation which is required varies with the nature of the explanatory factor concerned. After all, an enabling factor should – in as far as possible – be strengthened, a constraining factor be weakened, and a neutral factor be ignored, if one aims at more political influence. Therefore the factors are dealt with below in accordance with these three categories. It will be judged whether NGOs are potentially able to anticipate on them – or not – and whether recommendations can be made.

7.2.2 Enabling factors

Table 6.9 in the previous chapter identifies a number of enabling factors, which differ in explanatory capability and in the level at which they are operative. The conclusion has been drawn that the more these factors are present or function in the practice of treaty formation and implementation, the more influential NGOs may become. It is obvious that NGOs are not passively subjected to such factors, but that they may intervene and try to enhance them. In as far as this is possible or likely, strategies to purposely increase political influence are to be deduced. The sequence of the enabling factors below is based on the explanatory capabilities of each factor and the format of the causal model on political influence. We will start with factors at the agent level and end with those at the contextual one (dealing with factors relating to NGOs, states, negotiations, policy topics and the context; see figure 6.1, page 265). In addition, the factors characterized by the greatest explanatory capability will come first in each text segment below (for the explanatory capabilities of the enabling factors see table 6.9, page 263).

A NGOs

Expertise, as the sum of knowledge and skills on the part of the NGOs, is the most important explanatory enabling factor. Conclusion 21 in chapter 6 revealed that the more expertise NGOs bring to the negotiating table, the more influential they will be (see page 259). This conclusion almost resembles a recommendation in nature. It implicitly says: "Increase your substantive knowledge and political skills, and you will increase your political influence." This factor is a good example of one over which NGOs potentially have a great deal of control, as it concerns one of their *own* qualities. This is not to say that more expertise necessarily guarantees better outcomes or that it is an easy job for NGOs to improve their expertise, as that which is required in political arenas is generally very specialized and complex, both in substantive and political terms. But nonetheless NGOs can decide to invest more resources in it.

> *Recommendation 1 — Increase your expertise, both in substantive knowledge on policy topics and in political skills in international diplomacy.*

At the same time, it should be recognized that NGOs already have acquired and developed a lot of expertise. The number of studies which are conducted or financed by NGOs and disseminated among governments and delegations is impressive, for example on climate change data (Greenpeace), energy efficiency (WWF), biodiversity strategies (WRI), nature conservation (IUCN, Birdlife International), and international environmental law (FIELD). Moreover, many NGO observers are as skilled in international policy-making and diplomacy as many delegates. Nonetheless, things can always improve. The strategic question then is whether more resources should be mobilized to improve expertise among NGOs, even at the cost of other activities. This question, though, will be dealt with at a later stage, in the light of other recommendations to be formulated below.

However, there is another side to the coin here, as NGOs are not established as scientific organizations. To repeat, they are promotional pressure groups which seek to influence global policies in accordance with their societal and environmental objectives. Knowledge which might persuade or surprise decision-makers is one of the *means* to achieve this, but not an *end* in itself. However, in order to have such an effective means one should ensure that the knowledge concerned is backed by scientific arguments and data, otherwise it will not convince others. Or even worse, it may become counterproductive, as the case of Greenpeace and the wrong data on the polluted oil reserves in the Brent Spar has shown. One way to guarantee credibility is the involvement of more scientists in the work of NGOs. For example, NGOs organized a workshop together with scientists in Nassau (COP 1) on biosafety. The latter did not agree entirely with all the NGO views, but nonetheless presented a critical assessment of the pros and cons of genetic engineering and test-fields of genetically modified organisms. This seemed to have inspired a number of less knowledgeable delegates (personal observation). In short, one NGO strategy should be to bring more supportive science into politics.

Apart from expertise, the following (less) relevant enabling factors on part of the NGOs were identified: protest activities, innovative proposals, legitimacy, watchdog function, and resource dominance (see figure 6.1, page 265). According to conclusion 6 in the previous chapter, parallel protest activities outside formal political arenas may put additional pressure on policy- and decision-makers and strengthen the efforts of NGO advocates and lobbyists to impact policy outcomes inside arenas (see page 237). Therefore NGOs should try to combine protest, advocacy and lobbying. This is, incidentally, not an easy job for several reasons.

First of all, protest in itself is seldom enough. Media coverage of actions and public support, for example, are also important prerequisites (Bramble and Porter, 1992; Lowe, 1983; Van Noort *et al.*, 1987; Potter, 1996). The fact that protest activities seem to have contributed to the NGOs' impact on the FCCC's target as well as on the Berlin Mandate (cases 3 and 8) is certainly linked to the fact that both the conferences in New York (INC5b of the FCCC) and in Berlin (COP 1 of the FCCC) were landmarks in the climate debate, were very well attended meetings covered by a large media presence, and took place while the environment was relatively high on the international agenda. Therefore, the effectiveness of protest is very much dependent on specific circumstances over which NGOs do not have full control.

Secondly, it is also possible that protest activities by NGOs may annoy policy- and decision-makers and, consequently, limit the possibilities for NGOs to impact intergovernmental policies. For example, the Climate Summit in Berlin saw some activities *inside* the conference building – among other things a disturbance of the final plenary session by displeased activists – which put the relationships between the participating NGOs and government delegations under pressure (personal observation). Therefore it seems best to spatially separate protest activities on the one hand (outside conference buildings) and advocacy and lobbying activities on the other (inside). Furthermore, protest activities should never be too radical or lack broad public support, as many diplomats and delegates are on the conservative side of the political spectrum.

Thirdly, protest, advocacy and lobbying activities are generally conducted by different groups with – more often than not – different ideologies, although there are exceptions. An example is Greenpeace International which, for example, combined the occupation of a chimney of a German energy plant to condemn the emission of greenhouse gases with active lobbying to adopt strong targets in Berlin (COP1, April 1985). Nonetheless, many protest groups share a quite radical ideology and do not see much purpose in NGO participation in international policy-making. According to them, mainstream policies are generally meant to further the interests of the strong – e.g. USA, transnational corporations – and do in fact postpone real political action and social change (Chatterjee and Finger, 1994; McCoy and McCully, 1993). Many of these activists are consequently quite hostile towards lobbyists (and vice versa). Coordination of activities is therefore a rather difficult task and it remains to be seen whether activists would indeed agree to limit their protests to 'the outside'. In short, for protest to become effective in terms of political influence a number of preconditions should be met: 1. protest is spatially separated from advocacy and lobbying; 2. the protest is not too radical; 3. protest and lobbying groups succeed in co-ordinating their activities; 4. media coverage and public support are present or mobilized. It is obvious that these preconditions are probably difficult to arrange by NGOs at the same time. But if the circumstances seem right, the following recommendation is valid:

Recommendation 2 — Organize protest activities outside formal political arenas, at least when the relevant preconditions are met.

A further enabling factor at the agent-level is 'innovative proposal', but this is very much related to recommendation 1. After all, the NGOs' ability to introduce innovative proposals to reframe a policy topic (as was the case with regard to joint implementation) or to overcome a deadlock in the negotiations (as was the case with regard to the survey on financial mechanisms) is very much dependent on substantive as well as procedural knowledge regarding the specific topic. One should not only be able to discern a policy window, but be capable of entering it with the right substance in the right place and at the right time. Therefore the chance that NGOs will indeed introduce innovative proposals will increase if they have enhanced their expertise. But this by itself is certainly not enough, as NGOs still have to decide to use their expertise in favour of such proposals. Again, a critical note should be sounded, as NGOs are not solely scientists or technical advisors. They have not primarily been established to solve substantive or technical problems for governments, but nonetheless may do so if it serves their own cause. This last should, of course, be kept in mind.

Recommendation 3 — Use your expertise to introduce innovative proposals which may, apart from furthering the NGOs' policy objectives, offer new substantive or procedural insights to policy- and decision-makers.

One should keep in mind too that an element of 'luck' is also required for an innovative proposal to become effective. The 'right' policy window, substance, time and place are never entirely known in advance, let alone fully manageable by NGOs.

Being legitimate representatives of environmental interests and 'strong' watchdogs are additional enabling factors for NGOs (conclusion 23). Both the NGOs' legitimacy and their watchdog function should therefore be increased if political influence needs to be increased, although these factors are less important by far than expertise (see table 6.9, page 263). Legitimacy may be improved, first of all, by an extension of the NGOs' constituencies. But the recruitment of members and donors is already a permanent area of attention and concern to NGOs and, moreover, it mainly refers to an activity executed at national and local levels. Therefore it does not seem a primary task for pressurizers and lobbyists in global arenas. Another way to increase the NGOs' legitimacy lies in bringing more 'victim-interests' to the negotiating table, either in terms of people (e.g. indigenous people) or in terms of environmental data (e.g. the extinction of species). The latter coincides with offering more environmental knowledge which supports the claims of NGOs (see recommendations 1 and 3). The former covers a strategy which in the meantime has already become important. Ever more victims of environmental destruction – in particular people from the South such as small farmers, women and indigenous groups – are asked and enabled by NGOs to engage in international environmental diplomacy, in order to make government delegations aware of their situation (personal observation). A good example in this respect is case 14 on indigenous people, as CBD/COP1 experienced a relatively large participation of local groups and chiefs (see subsection 5.4.1). This tactic seems to have strengthened the claims of NGOs, although the effectiveness of it is difficult to assess on the basis of one case. The following recommendation should therefore be considered as 'additional' and 'tentative', especially as it became obvious that delegates consider knowledge and ideas more important NGO resources than the legitimacy of their claims.

Recommendation 4 — Mobilize more local 'victim-interest' in global political arenas.

The watchdog function, next, may be improved by monitoring the fulfilment of promises and the compliance with policies by governments more intensively. It should immediately be recognized, though, that many NGOs already consider this function as one of their main tasks. What is more, it is widely recognized that this role is one of the foundation stones of their existence (Princen and Finger, 1994; Jamison, 1996). Apart from influencing policy plans or instruments, pressure groups monitor the consistency of what governments promise on paper and what they in fact do. The danger of public embarrassment makes governments potentially vulnerable as a result of such monitoring and the watchdog function is therefore a possible source of influence for NGOs. Given the fact that NGOs already perform this role quite extensively, a recommendation to increase it is a bit superfluous. Nonetheless, one has been formulated for the sake of completeness:

Recommendation 5 — Continue the critical monitoring of the policy performance of governments.

However, case 7 revealed that the watchdog role is not necessarily a success (see chapter 4). At the time of the first formal review of the Annex I countries' performance on the stabilization target, NGOs published their own comments on those countries' performance and condemned the fact that they will not attain the target in the year 2000 at all. Here NGOs acted as watchdogs, but without becoming effective in terms of political influence. Therefore, a correlation between the watchdog function and political influence could not be found in this case, irrespective of the more positive views on this correlation of some respondents (see table 6.7, page 258).

A further – and equally weak – enabling factor is 'resource dominance'. In cases where states need resources which are at the disposal of NGOs – for example public support, contacts or knowledge – the latter are in a good position to exert political influence. Although this seems obvious, a position of resource dominance still turned out to be ineffective (see cases 10 and 12 in chapter 5). Besides, political influence is not guaranteed, especially as states may use NGO resources in their own way. In such cases, NGOs are 'used' as suppliers of expertise, contacts or public support without being paid in return in political terms. Furthermore, it is very difficult for NGOs to become dominant with regard to a specific resource, as states – and other policy players – are generally more capable of doing so, and the access to a specific resource is seldom the privilege of one player. Therefore one may wonder whether NGOs should indeed try to focus on dominating a resource. Instead, it seems better to use resources properly in order to have well-formulated, scientifically-backed, broadly-supported and innovative proposals, rather than try to control resources vis-à-vis other players. Again, this last is quite difficult and success is not guaranteed.

B Negotiations

At the level of states, no enabling factors have been distinguished (see causal model on political influence in figure 6.1, page 265). At the level of negotiations and decision-making, though, two factors have been found, namely 'integrative negotiations' and 'implementation phase'. These two factors are examples of variables over which NGOs generally have less control, given the fact that other players determine their nature. After all, states – and not NGOs – in particular determine how intergovernmental negotiations proceed and how the policy cycle is organized.

As far as the first factor is concerned, it is difficult for NGOs to increase integration between the parties in a negotiation process. Of course they may try to mediate and build bridges between opposing groups of countries, as they did in the case of the Berlin Mandate (case 8), but it is quite unlikely that NGOs alone can change the confrontational nature of a policy process, particularly as it seems inherent in several topics to provoke confrontations. However, NGOs could anticipate on this factor in a different way. For example, they might avoid an involvement in confrontational negotiations and concentrate on policy topics which are, conversely, characterized by integration. Particularly when NGOs 'need' a success to legitimize their presence in global arenas or to satisfy their constituency, such a focus seems a good strategy. But a 'cheap' success may turn into an empty one, as mainly less relevant policy topics are characterized by integrative negotiations. A focus on these might therefore imply a neglect of the 'real stuff' by NGOs. Assuming that NGOs had followed that strategy in the climate arena, they would have avoided any involvement in the negotiations on targets and timetables, which were after all confrontational. However, NGOs *did* become involved and, what is more, impacted the outcome. To recommend that NGOs should particularly engage in integrative negotiations would therefore be bad advice. But they should be aware of the fact that the chance to really impact confrontational negotiations is small. In short, NGOs should not ignore such topics, but adapt their expectations. Furthermore, they might look for the optimum mix of a focus on really important but confrontational substance on the one hand and some integrative policy topics which at least might guarantee some success on the other.

> *Recommendation 6 — Focus efforts to exert political influence on those policy topics which are characterized by integrative negotiations, but do not neglect those confrontational ones which are, in your opinion, the most important.*

A second enabling factor relating to the negotiation process is the 'policy phase'. Conclusion 11 in chapter 6 (page 242) revealed that (initial) treaty implementation offers more opportunities for NGOs to influence policy outcomes than treaty formation does, although the latter can indeed be affected (see cases 3, 9 and 11 in chapters 4 and 5). However, it is difficult to derive any recommendation from this conclusion, as a policy phase is a given phenomenon in a certain moment of time. Therefore it is not a factor NGOs can simply anticipate on, except by either ignoring the phase concerned or participating in it. But it would be bad advice to recommend to NGOs to ignore treaty formation, and concentrate instead on treaty implementation. After all, a treaty text becomes 'fixed' as soon as it is adopted by states at the end of the formation phase. It is then too late for NGOs to try to change provisions in it (although the Conference of the Parties is generally allowed to amend the wording of its treaty during the implementation phase, but this is a difficult procedure and not employed very often in international law). Consequently, it is important for NGOs to be around when the language of a treaty is formulated by governments.

Recommendation 7 — Focus on both treaty formation and implementation, although policy outcomes of the latter are more easily impacted. But you cannot afford to ignore binding law-making by governments, especially as there is a (small) chance you may impact outcomes during this formation phase as well.

C Context

The contextual level yielded three enabling factors: regime similarity, scientific certainty, and the UNCED (see figure 6.1, page 265). The impact of the last, though, remained a mere assumption, as it could not be correlated to various outcomes on NGO influence. Besides, it is not a factor NGOs can easily anticipate on, as UN members are the ones to initiate such a conference. Nonetheless, NGOs might still launch a call for it and lobby governments as well as UN bodies and UN officials to have one. But even then, a positive impact of such an event on (NGO) efforts to have stronger environmental policies is no longer assured, since there appears to be an overkill of UN conferences on various issues recently. The 'magic' of such meetings – also referred to as the 'Rio Spirit' in the case of UNCED – seems to have been diminished. An affirmation of this proposition was the special session of the UN General Assembly to assess the implementation of Agenda 21 five years after Rio in June 1997 (Doc II.89). This meeting turned out to be a failure and received a bad press. It is very likely that governments are tired of UN meetings and environmental issues. Maybe one has to conclude that climate and biodiversity NGOs were simply lucky to have the UNCED process out there while the FCCC and CBD were negotiated in the period 1990-1992.

In addition, NGOs may have been lucky for another reason, namely the presence of some 'similar' rules in international environmental policies. 'Similar' rules are those which strengthen the NGOs' demands on a specific policy topic. An example is the Montreal Protocol which foresees the complete abolition of CFC production. Again and again, climate NGOs referred to this example of strong, global, environmental regulation and it is plausible that their demands on climate change policies became more convincing due to the mere existence of these CFC regulations (Doc I.68). Without the ozone precedent, the struggle in the climate arena would probably have been even more difficult. But whether such similar rules already exist or not seems just a question of luck. However, NGOs might still anticipate on this factor, as was outlined before. They may for example try to utilize related rules already adopted in international environmental policies, in order to pressurize governments to adopt similar provisions in other arenas. As governments already agreed upon such rules at an earlier stage, it will be difficult for them to reject similar rules now.

Recommendation 8 — Utilize 'similar' rules of existing environmental regimes as arguments to strengthen and defend your demands. If the legitimacy of such rules is disputed by governments, organize a debate and try to convince your opponents.

The second sentence has been added to recommendation 8 because chapter 6 also revealed that success is not guaranteed in cases where NGOs use existing rules as arguments (see conclusion 17, page 253). The example of the global lists has made this perfectly clear. As was outlined there, a regime should be accepted as a legitimate frame of reference by a majority of the delegations, otherwise the above strategy will fail. Therefore NGOs should try to convince delegates when they do not accept certain rules as premises for treaty formation or implementation.

The third contextual enabling factor concerns 'scientific certainty'. It has been concluded that the greater this certainty is, the more successfully NGOs can press their demands (at least if science tends to support these demands). Given the fact, though, that NGOs are not in the first place scientific organizations, one cannot expect them to improve scientific research themselves. That is mainly up to the scientists. Therefore no additional recommendation on science has been formulated here (e.g. in terms of: '... NGOs should increase scientific certainty ...'). Nonetheless, NGOs can contribute to collecting and interpreting new environmental data as well as utilize (more) supporting science for political use, but this function has already been covered by recommendation 1 on the strengthening of expertise. Furthermore, NGOs can emphasize the precautionary principle where scientific uncertainty on an environmental issue exists. This principle states that a lack of complete scientific certainty should not be used by governments as a reason for postponing measures. It has been laid down in several international instruments, including the Rio Declaration, the CBD and the FCCC. As this precautionary principle has already been institutionalized in the existing international regime on the environment, NGOs could utilize it as an argument to compensate for lack of scientific certainty during future negotiation sessions. But this call for the use of existing rules has already been covered by the previous recommendation 8.

7.2.3 Constraining factors

The relevant constraining variables with regard to political influence were identified in table 6.9 in chapter 6 (page 263). Whereas NGOs should anticipate on enabling factors in terms of using or improving them, they should react differently to constraining factors. These should be weakened or avoided in order to increase political influence. Below the various factors are, in as far as possible, converted into recommendations. The format parallels the one of the previous subsection and therefore factors at the agent-level and with the greatest explanatory capability will come first.

A NGOs

Two constraining and, moreover, related factors were identified on the NGOs' side: the so-called 'wrong approach' (relatively strong factor) and 'unrealistic/radical policy goals' (relatively weak factor). The first refers to strategies and attitudes which policy- and decision-makers generally dislike: for example emotional, accusatory, confrontational, radical, idealistic, speculative or one-sided approaches. Consequently, it seems obvious that NGOs should avoid these and focus on a pragmatic approach based on expertise (see recommendation 1 in this chapter). The question, though, remains to what extent NGOs should adapt their approaches to all the preferences of delegates. On the one hand, they should do so to a large extent, at least if they wish to become influential policy players. On the other hand, they need to go further and be more idealistic or radical, for example, than the average policy-maker, otherwise things will never change. In other words, there is some in-built tension between adaptation and confrontation for NGOs in political arenas. But here again, some specialization can be applied (which is, incidentally, already being done). The more closely NGO representatives operate with policy- and decision-makers, the more they can use pragmatic approaches. But the more distance there is between NGO representatives and delegates, the more radical the demands they can formulate. Furthermore, a distinction between short-term and long-term policy goals should be made as well: activists deal with long-term concerns and ecological risks outside formal arenas, whereas lobbyists focus on short-term political gains within. This is not, however, to say that these activities or different groups should be *unconnected*, although it is believed that spatial separation is an important prerequisite for success (see previous subsection). On the contrary, different groups and their individuals should be linked even *closer*, so as to harmonize operations inside and outside political arenas.

> *Recommendation 9 — Avoid approaches which policy- and decision-makers dislike (speculation, accusation, confrontation etc.), but remain linked to the more radical and long-term demands of 'outside' groups and individuals.*

This recommendation, though, seems unworkable, because the history of the environmental movement teaches us that pragmatic and radical groups tend to follow their own and opposite paths (Cramer, 1989; Jamison, 1996). Therefore an example might be illuminating. On greenhouse gas (GHG) emission reductions, NGOs formulated very radical demands at the end of the 1980s (although still based on interpretations of atmospheric sciences by the Intergovernmental Panel on Climate Change), but adapted their policy goals in 1994, given the lack of progress in the climate change negotiations. They did so as a result of national targets set by the most progressive countries, which were variants of the Toronto Target. This target seemed both economically and politically feasible, and therefore the NGOs' position in the climate arena became stronger from the moment they started to promote it. 'Outside' activists nonetheless continued to emphasize their original and more radical demands, thus putting additional pressure on the process, while 'inside' lobbyists were prepared to do business on lower targets. The lobbyists however remained linked to the activists through the Climate Action Network (CAN) and were therefore responsive to the claims of more radical groups. It is such a positional dynamic that should be developed by NGOs on other topics in pursuance of the above recommendation (and maybe in other political arenas as well).

The second constraining factor 'unrealistic/radical policy goals' has much in common with the previous one. After all, NGOs which follow a 'wrong' approach in the eyes of policy- and decision-makers generally defend extreme goals at the same time. Therefore no new recommendation is formulated here, but a small addition is made to recommendation 9: a 'wrong approach' as referred to in that recommendation should also be taken to mean the defence of unrealistic and too radical policy goals by NGOs.

B *States*

Conclusion 1 formulated in chapter 6 (page 231) revealed that the presence of like-minded but dominant states diminishes the chances for NGOs to become influential. This was explained by the fact that governments, which further more or less similar interests to NGOs and moreover dominate a policy topic, make the input of the latter superfluous. What is more, in a few cases some countries turned out to be more activist than NGOs, an example being Malaysia with regard to the biosafety topic. Political influence is out of the question in such circumstances. Therefore it seems advisable for NGOs to leave topics which are already covered by like-minded but more powerful players alone and divert their attention to other topics which are not. In other words, NGOs should try to find new policy windows and create their own niches in an issue arena. An example is the agenda-setting on marine biodiversity just before COP1 of the CBD (see case 15, subsection 5.4.2).

> *Recommendation 10 — Leave alone policy topics which are well covered by other like-minded players and concentrate on ones which are not.*

It should be remembered that this recommendation refers to political influence only. After all, public support of like-minded but dominant states may still serve other objectives. For example, the NGO community in both arenas tended to support the South on a number of policy topics, e.g. on the climate change principles (Article 3 of the FCCC). Such support was not very effective in terms of political influence, but it probably strengthened friendly relationships between developing countries and NGOs, which was needed given the fact that (Western) environmental organizations were distrusted by several poor countries. The latter considered the former's ideology to be anti-development, but broad NGO support for some of the G77's goals proved otherwise.

C Negotiations

At the level of the negotiation process, two constraining factors have been identified: 'block politics' (with a relatively high explanatory capability) and 'formation phase' (with a relatively low explanatory capability). To start with the first, confrontations between in particular North and South diminish the possibilities for NGOs to intervene. This was one of the conclusions of the previous chapter, and is no surprise. After all, the North-South development debate predates the environmental one and has its own historical dynamics, old rituals and predictable clashes, which generally have little to do with the environmental issue itself. Where such North-South clashes become dominant, other policy players will not have many opportunities to intervene. True, NGOs may still play a (minor) role in the definition of policy positions by individual states and in the creation of like-minded coalitions, but whether these can survive the internal discipline which is generally urged by the traditional blocks of states is another question. OECD countries, the European Union (EU) and the G77 group each tend to unite as much as possible, irrespective of whether internal views differ. There is tremendous pressure to obey group discipline, certainly when issues which are politically sensitive are dealt with.

Although it is more than obvious that NGOs cannot impact the behaviour of state blocks, they nonetheless have room for manoeuvre. They may decide to waste no more time and energy on topics which are most dominated by traditional North-South clashes. Instead, they can concentrate on those topics which are less affected by this divide and, consequently, offer more opportunities to impact decision-making. For example, NGOs in both the biodiversity and the climate arena invested a lot of resources in getting an independent financial mechanism for the treaties, but this topic soon became caught up in traditional North-South games. At that time it would have been a better option for NGOs to shift all their efforts to topics which were less caught up in such games, for example the commitments in the FCCC or the indigenous peoples' issue in the CBD. True, these topics were also characterized by differences between developing and developed countries, but to a lesser extent.

Recommendation 11 — Ignore policy topics which are most characterized by traditional North-South confrontations and divert your attention to ones which are less affected by this.

There may be two exceptions, in which NGOs should not ignore block politics but try to intervene. In cases where there are *strong* internal divides dormant in blocks over a *long* period, NGOs might make these known to possible allies in different groups and try to create a situation in which new coalitions can emerge. This is what they did in Berlin, where a break-up of the G77 had become a real possibility. NGOs pressurized the most progressive developing countries to go their own way and mediated between the latter and some Western countries in order to find common positions. This NGO initiative helped make the final compromise on the Berlin Mandate possible (see case 8, subsection 4.4.4).

The constraining factor 'formation phase' has already been – briefly – discussed above, in relation to the enabling factor 'implementation phase'. Recommendation 7 says that NGOs should focus on both treaty formation and implementation, although the outcomes of the latter policy phase are more easily influenced by NGOs than those of the former. But NGOs cannot afford to ignore the formulation of binding law by states.

D *Policy topic*

The analysis in chapter 6 yielded two correlations between aspects of policy topics and political influence. One aspect refers to 'political relevance', the other to 'domestic policy effects'. Both turned out to be constraining factors in the sense that NGOs will experience greater difficulty in influencing global rules the more these are politically relevant and the more they have domestic effects. Note, though, that these correlations are weak compared to some of those dealt with above (see table 6.9, page 263). The recommendations which follow below are therefore less relevant than previous ones.

Conclusion 3 in the previous chapter states that the more politically relevant a topic is, the more difficult it is for NGOs to impact the policy outcome concerned (page 233). It is therefore tempting to advise NGOs to avoid such topics and shift their attention to others. But that would probably be bad advise, given the fact that the most relevant topics from a political point of view may at the same time be the most relevant ones from an environmental point of view. For example, the target of the FCCC, the Berlin Mandate, the GEF, the implementation mechanisms, the first review of implementation and biosafety were all hotly debated topics in the negotiations, but equally they were all relevant from the perspective of environmental policy performances. To concentrate on less relevant topics would thus imply that the real environmental issues might be neglected by NGOs, which does not accord well with their main intentions. Therefore NGOs simply have to cope with politically relevant topics in some way or other. But they should not expect easy successes, although some impact is definitely not impossible (e.g. see case 3, subsection 4.3.4).

> *Recommendation 12 — Do not ignore the most politically relevant policy topics, but do not expect any 'easy' success in impacting these either.*

The next constraining factor at the level of policy topics is the 'domestic effects' of international agreements. The more these are expected to be (immediately) binding upon individual states, the less they can be influenced by NGOs. A similar dilemma in formulating a recommendation to that over the previous factor is evident here. One might advise NGOs to primarily look at topics with no or only some domestic effects, but then they will only focus on symbolic politics with no effects on the environment whatsoever. Besides, the value of international agreements is dependent on what will be done with the rules at national and local levels, since it is there that implementation will mainly take place (apart from some international initiatives). Hence, domestic effects are a prerequisite for effective international rules and, consequently, NGOs should focus on these topics. Again, easy success cannot be expected given the constraining nature of this factor.

> *Recommendation 13 — Do not ignore policy topics which are meant to have the greatest domestic policy effects, but – again – do not expect any 'easy' success either.*

E *Context*

The distribution of resources in world politics and the formal UN rules on access to political arenas are two constraining factors for NGOs at the contextual level, although these factors could not be correlated with the various levels of NGO influence (see figure 6.1, page 265). Therefore these constraining factors are only *assumed* to be relevant. But relevant or not, it is unthinkable that NGOs can change the distribution of resources in the international system. Therefore this factor is ignored in this subsection. However, as far as the rules of the game are concerned, NGOs *can* make a difference. For example, they might try to expand the rules of the game in favour of their access to and participation in treaty formation and implementation. This can be done by lobbying governments and UN bodies for more liberal rules, either by demanding a general change of rules in the UN system or by urging officials and delegates to negotiation sessions to set specific precedents of NGO participation. Such precedents may, in turn, in the long run have unforeseen but positive institutional effects in the UN system as a whole.

However, a very substantial part of NGO input into international environmental policies is already dedicated to this subject. Almost any UN negotiating session is accompanied by struggles over NGO access among governments, officials and the NGOs themselves, for example on access to the closed, small, informal contact groups in which the most relevant intergovernmental negotiations take place and decisions are prepared. Therefore NGOs already engage in institutional struggles to enlarge their opportunities all the time. Given this quite well-developed NGO focus on the need for procedural change, any advise on increasing it may be superfluous, especially as this might happen at the cost of *substantive* inputs into the negotiations. Besides, the institutional question on NGO access in the UN system transcends the level of individual issue arenas and should therefore mainly be dealt with in other bodies (for example the NGO Commission of UN-ECOSOC). Nonetheless, climate and biodiversity NGOs could continue to try to enforce greater NGO participation in their own arenas.

> *Recommendation 14 — Continue your efforts to enforce precedents or new rules on better NGO access to intergovernmental negotiations and decision-making, but be careful that this does not overshadow your input on substance.*

7.2.4 Neutral factors

The following neutral factors were identified in table 6.9:

- specific (type of) NGO;
- specific (type of) NGO intervention;
- number of NGOs;
- direct impact/drafting;
- NGO co-operation/contacts;
- presence of BINGOs;
- nature of the environmental issue;
- nature of the regime element;
- environmental shock event.

Whether these factors change, increase or decrease, or are improved or not: given the analysis in the previous chapter, effects on the extent of NGO influence are quite unlikely. Therefore NGOs need not anticipate on these factors in order to increase their impact on policy outcomes. Given the fact that these organizations nonetheless tend to do so in a number of cases, the various factors will be briefly discussed below and additional recommendations will be made. The chronology is based on table 6.9 (page 263), although the sequence has been reshuffled slightly on the basis of the factors' nature and links.

Conclusion 5 in chapter 6 (page 237) revealed that all NGOs, all NGO interventions and all sizes or numbers of NGOs identified in this study may become influential *in principle.* There does not exist one typical NGO, one typical intervention or one specific size or number that guarantees success. However, this conclusion runs counter to some rather strong beliefs in the NGO community, as well as in NGO literature (Bramble and Porter, 1992; Doherty, 1994; Lowe, 1983; Mazey and Richardson, 1992; Van Noort *et al.*, 1987; Potter, 1996; Princen and Finger, 1994; Thompson-Feraru, 1974; Willets, 1982). First of all, it is believed that NGOs characterized by a more pragmatic ideology – such as WWF or IUCN – are generally more influential than those characterized by a more radical ideology – e.g. Greenpeace or Rafi. This belief does not appear to be confirmed by this study, although the nature of an NGO's *approach* does play a role and that nature may depend on the NGO's ideology (see recommendation 9, page 278). Nonetheless, the general ideology of a group and the approach of its representatives in a particular issue arena do not necessarily match. After all, so-called radical groups may still offer well-qualified experts with whom delegations can do business. It all depends on the quality of someone's specific input and not on someone's political background. Given the fact that representatives of various organizations such as WWF, IUCN, WRI, EDF, Greenpeace, TWN, GRAIN and Rafi – either individually or as group in CAN or Bionet – were influential in some cases and not in others, it seems that policy- and decision-makers mainly look at what individuals have to offer in particular cases and not so much at their organizational and ideological position in general. In other words, individuals seem more important than organizations as far as political influence is concerned (although the former need the latter in order to be able to function as effective pressurizers and lobbyists).

Recommendation 15 — Select, as an NGO community, your lobbyists on the basis of individual qualities, not on their organizational or ideological backgrounds.

Secondly, it is generally assumed that smaller NGOs or a smaller number of NGOs have less chance to become influential than big organizations or large groups. Consequently, NGOs invest a great deal of resources in increasing their staff and in mobilizing large groups of colleagues to pressurize or lobby governments on a policy topic. In addition, they try to get as many NGO representatives as possible to international negotiating sessions, because they believe that large numbers enable NGOs to impact outcomes. One may wonder, though, whether this focus on *quantity* is very useful from a political influence perspective. It seems not, as – to repeat – policy- and decision-makers are mainly interested in *quality* (expertise, ideas, innovative proposals, rational approaches etc.). True, there is of course a certain minimum number of NGOs required to be able to manage all the necessary activities in international arenas, but at least the focus should be shifted.

Recommendation 16 — Focus less on quantity and more on quality with regard to interventions in international political arenas.

This recommendation is also related to the substance of conclusion 22 in chapter 6, which states that political influence is not more likely through more NGO co-operation and contacts (page 259). Again, quantity does not seem decisive.

Thirdly, the type of NGO intervention does not matter much either, at least with regard to those types which were identified in the climate and biodiversity arenas. Oral statements, publications, text proposals, participation in drafting or parallel conferences: all these NGO initiatives may lead to political influence or – in other cases – may fail. In particular with regard to 'drafting' this is remarkable. One might expect that the participation of NGO representatives in the drafting of legal texts is a more than sufficient precondition for political influence. After all, if you impact legal language, you will impact the policy outcome concerned. However, your input may have been limited to technical advice or marginal notes alone, as for example case 12 revealed (access to genetic resources, Article 15 of the CBD, subsection 5.3.5). Given the strong belief among some NGOs that the 'drafter's role' is necessarily effective and should therefore be increased, as some respondents noted, the following recommendation seems useful:

Recommendation 17 — Consider a contribution to the drafting of legal texts equally (in)effective as any other NGO intervention.

The presence of a business lobby, next, is no obstacle to NGO influence, as was concluded earlier. Obviously, there is no such thing as a zero-sum game in this respect (influence of proponent X implies no influence of opponent Y and vice versa). Therefore NGOs need not to be too impressed by BINGOs. But there is no reason to believe that they *are.* NGOs even tend to *focus* on topics which are covered by a strong business lobby, in order to countervail the latter's influence as much as possible. In consequence, a recommendation which says that NGOs do not need to be 'frightened' by BINGOs is superfluous.

As was concluded in chapter 6, the different nature of the environmental issues of climate change and biodiversity loss cannot explain different outcomes in political influence (conclusion 10, page 241). Therefore NGOs need not anticipate on this factor in the sense that, given the chance of success, they should for example ignore the former issue and concentrate on the latter. But as there are no signs that NGOs tend to do so, no additional recommendation needs to be formulated here. The same applies to the factor 'regime element'. As principles, norms, regulations and procedures may all be as easy or as difficult to affect for NGOs, they should not focus on one element or another in order to increase their political influence (see conclusion 13, page 246). But they appear not to do this in practice. Finally, it has been assumed that the occurrence of an environmental shock event – for example an extremely hot summer which may point at climate change – does not increase NGO influence. This, however, does not imply that NGOs cannot use these occurrences as examples in their campaigns, as they simply can refer to them as manifestations of the environmental problems concerned. But they should not expect too much immediate effect from such events on their own influence.

— 7.3 The respondents' views —

The previous section produced a number of recommendations based on the analysis and hypotheses in chapter 6. Therefore these recommendations are also partly based on the respondents' explanations of NGO influence (see section 6.6). Nonetheless, additional views of respondents which relate to the subject matter of this chapter are available. After all, respondents were not only asked to *explain* political influence, but to express their views on how NGOs might *increase* political influence as well. True, these questions are highly interdependent. If a respondent for example believes that NGOs are influential because of factor X, he or she will definitely think that an improvement in the performance of NGOs can be reached by increasing X. It follows that the answers to these different questions will generally show a proportional pattern. Furthermore, it should be recognized – as was outlined before – that not all respondents were able to deal with all the questions. The number of respondents in each category of answer varies and the reasons for this variety were elucidated in chapter 3. Consequently, the analogy between tables 6.7 and 6.8 (pages 258 and 260) and table 7.1 (below) is not perfect. However, most expert respondents – 36 out of the 46 – were confronted with the interview question discussed here. This number is considerably higher than the numbers of respondents who reflected on the 'why' or 'why not' of political influence (26 and 18 respondents respectively). Therefore the table below is a better reflection of the totality of expert respondents than are those in the previous chapter.

The following arguments in table 7.1 confirm the substance of tables 6.7 and 6.8: expertise, legitimacy, NGO-NGO interaction (which is, again, only positively valued by NGO respondents themselves), innovative policy proposals (which is, again, mainly valued by government respondents), and watchdog. As these factors have already been covered by earlier recommendations, no additional recommendations on them will be formulated in this section. The opposite however applies to some of the factors which have been added in table 7.1. New factors which are regularly referred to are: involvement of Southern and Eastern NGOs, NGO-state interaction, 'be careful with NGO influence', as well as intersessional and national work of NGOs. These will be briefly discussed below.

The involvement of more NGOs from the Third World and Eastern Europe is considered a precondition for more impact. As one NGO respondent said:

> "One improvement would be the involvement of NGOs from more countries. I remember when I started after Rio it was mainly the Anglo-American NGOs. We should therefore bring in those NGOs, for example from the Third World, which are traditionally not involved. This will contribute a lot, because it will just enrich the whole spectrum."

Others add the argument that linkages with developing country delegations will be strengthened as soon as more Southern NGOs are involved. This might produce new possibilities for NGOs to impact decision-making. However, these arguments are brought forward by NGO respondents in the climate arena only. Obviously, it is not an issue for NGOs in the biodiversity arena. The question of why a similar phenomenon – the dominance of Northern NGOs – is perceived differently in those two policy arenas is hard to answer. It might be explained by, for example, different beliefs among the respondents involved, the nature of the mutual relationships between Western and non-Western NGO representatives, or the success – or lack of success – of the North-South co-operation in the NGO communities concerned. Besides, it should be recognized that government respondents – apart from one – do not believe that more effectiveness depends on the involvement of non-Western NGOs. Therefore it is not believed to be an important precondition for more NGO influence, although there may be other reasons to involve non-Western groups (justice, better representation of the population and diffusion of expertise, experience and skills among NGOs worldwide, etc.).

Whereas the extent of NGO-state interaction was not mentioned as an important factor to explain (the lack of) political influence in the previous chapter, something which has been considered a 'surprise', it is regularly referred to in the context of the interview question dealt with here. A number of respondents are convinced that these interactions should be improved to increase political influence. To quote one:

> "It has mostly to do with beginning a dialogue with country delegations and I cannot emphasize enough that this should be a constructive dialogue. I will give you the marine example, where NGOs took something that was wide open, did substantive work to provide a focus for it and worked closely with governments so that the governments and the NGOs were aligning themselves."

Both government and NGO respondents believe their interaction should be improved, although the numbers in table 7.1 are not impressive.

Recommendation 18 — Increase and improve your interactions with country delegations.

TABLE 7.1 *Factors to increase political influence mentioned by the respondents*

Respondents	*Government*		*NGO*		
Argument	*FCCC*	*CBD*	*FCCC*	*CBD*	*Total*
expertise	3	2	4	2	11
involvement Southern & Eastern NGOs	1	-	7	-	8
legitimacy	-	2	1	5	8
NGO-NGO interaction	-	1	3	4	8
NGO-state interaction	2	2	1	2	7
innovative policy proposals	1	4	1	1	7
'be careful with NGO influence'	1	-	2	3	6
intersessional work of NGOs	-	1	1	3	5
national work of NGOs	-	1	3	1	5
watchdog	1	3	1	-	5
resources (funding, time)	-	-	3	1	4
participation	-	3	-	-	3
NGO-others interaction	-	-	3	-	3
other factors (only mentioned once)	-	2	-	2	4
number of respondents	5	10	12	9	36

The question is still how far and under what conditions NGOs should interact with which governments. Certainly not with all governments in all situations. Firstly, they need to focus on the most appropriate delegation(s) in a specific case. These should definitely be the *key* country (or countries) involved. Secondly, they have to select parties which need to be approached: either like-minded countries or the opponents of the NGOs' preferences. It seems best to approach the former when there is a basic and general commitment among countries to deal with a certain policy topic. Then intergovernmental decision-making may be strengthened further by interacting with the most like-minded delegations. However, if there is a lot of disagreement among countries on dealing with a topic, with some laggards frustrating any progress, the NGOs should interact with these and try to soften their position.

For some in the NGO community, however, recommendation 19 will definitely be a contested one. Some fear that NGOs will loose their ability to critically monitor intergovernmental policies or that they will become co-opted if they are too close to governments. These possible dangers will be addressed in the next section.

It is also believed – although mainly by their own representatives – that NGOs could perform better if they improve their intersessional communicating and co-ordinating. This refers to a complaint regularly heard among the NGOs, which is that they only start to communicate, co-ordinate and co-operate with regard to an international meeting just before the session actually starts. And that is generally too late. NGOs should be better prepared to act effectively at those meetings. Therefore so-called intersessional work among NGOs should be strengthened. In the words of one NGO respondent:

> "I think we could improve our work by communicating even more regularly between conferences, for example just sit down and do a one page update of your region and put it up on our internet conference."

Others believe that the national work should be strengthened. After all, the policy positions of national states are set at home in advance of any international negotiating session. Moreover, these positions generally do not change – or only slightly – at international meetings, so that NGOs cannot achieve very much there any more. They should therefore put pressure on governments before such meetings, while they are formulating their national positions at home. Additionally, NGOs should gather and share information on the different positions of countries in advance of international sessions, otherwise they cannot produce an effective strategy.

Note that only NGOs respondents believe that intersessional and national work should be improved. This is no surprise, as these aspects refer to the preparatory work of NGOs outside global arenas. Government delegates are generally unaware of these NGO activities, except maybe for those at national level in their countries. As a result, these delegates probably have no insight in or an opinion on the effectiveness of such activities. The importance of the preparatory work is nonetheless taken up here, given the conclusions of this study on the need for *quality* and *co-ordinated* NGO interventions at global sessions (see next section). Therefore the following recommendation has been adopted:

Recommendation 19 — Increase and improve your national and intersessional work to prepare for your interventions at international sessions.

As far as this strengthening of the intersessional and national work is concerned, some biodiversity NGOs stress the importance of the first, whereas some climate

NGOs emphasize the second. This difference can be explained by the fact that the climate NGOs are relatively well organized at global level, so that there is already some intersessional co-ordination through the Climate Action Network (CAN), whereas the biodiversity NGOs are quite strong at national level which means that their work at home is already well organized.

Apart from the above views on how to increase the impact of NGOs on environmental negotiations, respondents also referred to the following aspects, but to some extent only:
- investment of more resources by NGOs (time and money);
- increase in participation of NGOs in international negotiating sessions;
- more emphasis by NGOs on interaction with other players such as scientists, international organizations and business.

To conclude this section, it should be noted that several NGO respondents issued a warning with regard to the wish to exert more political influence (see 'be careful with NGO influence' in table 7.1). According to them, NGOs should not and cannot sit on government chairs, nor share – or take over even – government responsibilities. This would definitely imply more NGO influence, but it also implies a number of drawbacks. Moreover, NGOs should not become too much involved in political games and formal procedures. One NGO representative put it as follows:

> "If we are going to run around after UNCED and say we are the NGOs, we are important, we know the answers, we are so clever ... maybe that gives us all a nice warm feeling around the heart, makes us all walk a little bit taller, but some clever guy from the government is going to turn to us one of these days and says: 'yes, you are right, you are the people who know how to do this, get on with it.' They lay off their responsibility and that is a real danger."

Another NGO respondent said:

> "We are not established as negotiators. That is not our primary goal. We should not fool ourselves. (...) One of the dangers is: when you get too close to the political process, you get absorbed by it yourself. Then you lose your personality, your ability to stand back and say: hey, where are we?"

In other words, these respondents see some possible drawbacks for an NGO community which becomes too much involved in negotiations and political decision-making. Their message is obvious: NGOs should remain independent, whereas governments should discharge their own responsibilities.

Although this warning should be taken seriously into consideration by all NGOs, it should not lead to a reluctance on their side to interact intensively with states, since political influence simply presupposes such interaction. Either one wishes to influence policy outcomes in favour of the environment, or one rejects such interaction and remains ineffective. It is one way or the other.

— 7.4 A strategic model —

The previous (sub)sections have yielded 19 Recommendations, of which some are more relevant than others given the various explanatory capabilities of the factors concerned. Taking into account this consideration, the prominent outstanding recommendations are:

1. the increase of expertise by NGOs (recommendation 1);
2. the marshalling of existing rules (recommendation 8);
3. the avoidance of 'wrong' approaches (recommendation 9);
4. the avoidance of policy topics already covered by other, like-minded players (recommendation 10);
5. the avoidance of policy topics characterized by North-South conflicts (recommendation 11);
6. the need to focus on quality instead of quantity with regard to NGO interventions (recommendation 16).

Further recommendations are: to organize more protest activities outside formal issue arenas; to introduce innovative proposals; to use victim-interests in arenas; to continue monitoring the policy performance of governments; to focus mainly on integrative negotiations; to address both treaty formation and implementation; not to ignore relevant topics or topics with much domestic policy effect irrespective of the fact that these are difficult to influence; to try to enforce more liberal rules on NGO access; to focus on individual qualities of lobbyists instead of organizational/ideological backgrounds; to avoid a single focus on the drafting of legal texts; to intensify the interaction with governments; and to improve intersessional and national work to prepare for international meetings.

It is obvious that this set of recommendations is both large and diverse. Therefore it does not offer a handy or integrated vision on how to improve the performance of NGOs, although all these individual recommendations may have their own value. In order to have such a comprehensive view, a strategic model will be presented below. This model *does* offer an integration of the recommendations into one strategic vision. An important aspect of this view is a specific division of labour between activists, advocates and lobbyists. It should be realized, though, that this strategic model cannot cover all the details of the above sections. Besides, it is based on a reinterpretation of the recommendations, whereby the author's own preferences on NGO strategies and his experiences at international negotiating sessions have definitely played a role.

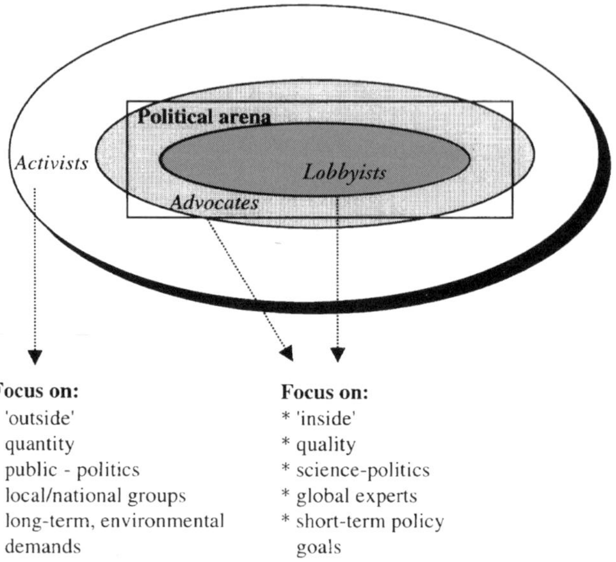

FIGURE 7.1 *Strategic model for global NGOs*

Figure 7.1 consists of an 'egg-like' model in which each ring represents a group of individuals focusing on different activities within the NGO community (respectively lobbying, advocacy and protest; see table 2.2, page 52). The distance between each ring and the formal political arena varies. This arena is represented by the rectangle in the middle of the figure. Lobbyists are inside, advocates are both in and outside, but mainly inside, whereas activists are outside, except for a small minority. Whether people are in or outside is determined by the preferences of the individuals themselves and by the rules of access to UN sessions. In general, it is not difficult for any NGO representative to get in through the official channels (accreditation as NGO observer or journalist). Most activists, though, will choose to stay out in order to organize their protest, or because they are not interested in being involved at all, although some will try to get in and disturb the routine of intergovernmental policy-making in order to 'wake up' delegates. Lobbyists and (most) advocates, however, will prefer to be inside. The latter may also choose to organize activities outside, such as parallel conferences. It is obvious that the focus of the groups is not the same and this has been added to the model. Although the distinction of the three categories of NGO activities and the extent of these categories' inclusion in the political arena are hardly innovative from a theoretical or strategic point of view, the contribution of this study may nonetheless lie in the specific interpretation of the three groups' orientation. This is elaborated on below and linked to the various recommendations.

In one of the previous sections, a strict spatial separation of protest on the one hand and advocacy/lobbying on the other has been argued for (see recommendation 2, page 272). This preference was based on the likely effectiveness of moderate protest outside formal issue arenas – for example around international negotiation sessions – but, at the same time, on the possibility that radical and/or 'inside' protest would annoy decision-makers too much. A separation and some restraint by activists would therefore be ways to improve political influence of NGOs. This view is also expressed in figure 7.1, as activists are situated outside. True, most will prefer to be out there, but it is considered necessary that *all* stay out, at least as 'activists-in-operation'. As individuals, though, they may participate, but only according to the rules inside. What is more, it is even *preferred* that some of them will be involved in issue arenas, to maintain contacts and substantive links with the pressurizers and lobbyists inside, but only as long as they leave the organization of protest to their colleagues outside. This preference is also expressed in figure 7.1, as the rectangle of the political arena to some extent overlaps with the ring of activists.

The organization of such outside and moderate protest is problematic for several reasons, as was already emphasized in subsection 7.2.2 (page 269). There the need for media coverage and public support, the danger of annoying policy- and decision-makers, and the difficult task of co-ordination were dealt with. Some additional arguments will follow here. Firstly, resources should be available or mobilized at local and national level to organize protest activities near negotiating sessions. This will not always be possible, in particular when UN meetings take place in Third World countries where strong NGOs and abundant resources may be lacking. However, Western and wealthier NGOs can then lend assistance. Of course, protest can also be organized in other places and at times other than those of the negotiating session involved, but then the activists should have the means at their disposal to effectively address the meeting concerned, for example through the media, the internet, satellite connections, reports, written demands etc. This will need the mobilization of a lot of resources as well. Secondly, activists need to be responsive to the 'prerequisites for political influence', which are separation and moderation of protest. They need to agree to limit their actions to the outside and refrain from protest strategies which are too radical and confrontational. It remains to be seen whether activists will indeed agree to this, given the fact that most of them are not very interested in political influence. Many dislike mainstream politics and just want to make their point: the need for dramatic social and political change in order to save the planet. They do not care much about what policy- and decision-makers may think of them. Nonetheless, advocates and lobbyists could try to link up with moderate activists – and vice versa – to purposely organize protest activities 'outside' issue arenas aimed at increasing political influence 'inside'. To do this, global NGOs should communicate with local and national activist groups in advance of any meeting and try to draw up a strategy. Thirdly, protest needs to be as massive as possible, otherwise it looks as if the demands in question are supported by some 'lonely' individuals only. In other words, activists need to focus on quantities. Again, massive events are quite difficult to organize, unless there are good (local) entrepreneurs and considerable (local) constituencies to support the initiatives. Given these considerations, however, one may wonder whether protest can be deliberately organized in such a way that it becomes effective and efficient in terms of political influence. Maybe not, and therefore NGOs should make a cost-benefit analysis in advance of each intergovernmental meeting. But if the circumstances are beneficial – e.g. many local groups are available as well as good press coverage – one should not hesitate to activate protest outside.

More important from the perspective of political influence, though, is the *quality* of the NGO interventions inside issue arenas, either of advocates or of lobbyists. Here the notion of 'quality' covers a number of items which have been addressed in the most relevant recommendations: the supply of solid knowledge, innovative proposals and information; the utilization of diplomatic skills; the utilization of similar rules in international environmental policies; the avoidance of a 'wrong' approach (e.g. too accusing, radical or confrontational); and refraining from 'wrong' topics (e.g. topics which are already covered by other, like-minded players or which are characterized by traditional block politics or negotiations which are too confrontational). Although quality presupposes some quantity – a number of experienced NGOs and individuals to execute all the tasks – that is considered of less importance here. A few experts may be as influential as a large NGO community. Therefore NGOs should no longer focus on getting as many NGO representatives to international meetings as possible, but instead invest those resources in – what one might also call – *professionalization.* The more NGOs become professionals in intergovernmental negotiations, the more chance they have to impact policy outcomes in accordance with (some of) their goals. This conclusion is, incidentally, not spectacular, as many will already be convinced of the need for professionalism, but it is expressed slightly more radically here.

Given the above elaborations, a specific division of labour can be distinguished – and promoted – among activists, advocates and lobbyists during intergovernmental negotiating sessions anywhere in the world. Activists should be mainly of local or national origin and capable of mobilizing their constituencies to address the meeting concerned through for example demonstrations, marches, campaigns, petitions etc. Of course, activists from other countries should be welcome as well, but for efficiency and cultural reasons, the first and main initiative should be at local/national level. Furthermore, activists should definitely stay on the outside of intergovernmental meetings and focus on mobilizing as spectacular a quantity of supporters and journalists as possible. However, their demands and strategies should be moderate, rather than too radical and confrontational. The aim is to indirectly insert public concern, grievances and demands for change into politics. The motto should be: "Let policy- and decision-makers feel that people outside do care!" Lobbyists, next, represent the opposite of all these aspects. They should be experts in global politics, remain inside issue arenas, focus on the quality of their interventions, feed science into politics, defend reasonable policy goals, employ rational and pragmatic approaches, build and maintain good contacts with delegates, and express a willingness to negotiate. Advocates should do the same, although they tend to focus not so much on individual delegates, which lobbyists do, but more on issue arenas and negotiating sessions as a whole. They should continue to formulate and disseminate position papers, publish journals and reports, organize workshops and round-table conferences, and make oral statements etc. inside arenas, in order to convince delegations of their cause. Again, the quality of such interventions is decisive.

A plea for absolute separation of protest and lobby/advocacy does, however, not imply that mutual contacts and links should be cut off. On the contrary. The separation is meant to be purely spatial and functional. Activists, advocates and lobbyist should ideally keep in touch, harmonize their activities, and mutually reflect on their premisses in order to avoid making protest too radical on the one hand and the co-opting of lobbyists by governments on the other. To do this, individuals and organizations should continuously be in touch. Nonetheless, they should respect the functional distinction of each domain as described above, at least if political influence is to be strived for.

— 7.5 Extrapolation —

To repeat, the criterion of extrapolation aims at judging the external validity of the previous recommendations and the strategic model. The following question is central: to what extent might these recommendations and this model be extrapolated in time and to other policy issues, respectively to the biodiversity and climate arenas of the near future and to other political arenas? Extrapolation in *time* implies a judgement on how 'date-sensitive' the recommendations and model are; in other words, how far they are deduced from a very specific situation in a certain era. *Thematic* extrapolation concerns a judgement on how 'issue-specific' the recommendations and the model are. Do they apply to the biodiversity and climate arenas only or to other policy issues as well? It should be recognized, though, that the judgements below are first thoughts rather than elaborations on a well-designed methodology. In addition, one should put into perspective any effort to extrapolate findings in a few cases to other eras and areas of global politics, which in themselves are so complex and dynamic.

As far as the possible extrapolation in time is concerned, the following remarks should be taken into consideration. The period which has been studied (1990-1995) was indeed a special time, as it was the 'peak period' of the UNCED process and of the so-called 'Rio Spirit'. This period seems over now, particularly since the failure of the recent special session of the UN General Assembly on Agenda 21 (June 1997). Remarkably enough, NGOs played, according to observers, a very active and prominent role in the negotiations of this special session, but never before have the outcomes of an UN meeting been so disappointing from an environmental point of view (Doc II.89). So the NGOs' performance in terms of participation and diplomatic skills seems to have improved since Rio, but the policy outcomes show an opposite trend, as probably does the NGOs' political influence. Whereas a qualitative improvement of the NGOs' performance might indeed have led to more political influence in the UNCED era, such an increase in impact may be frustrated today by a lack of commitment to the environmental cause by the international community. This likely proposition, therefore, urges one to be very careful with extrapolations in time. On the other hand, the recent success of the Kyoto Summit – the adoption of a binding protocol to the FCCC on greenhouse gas emission reductions after 2000 – puts a too pessimistic view on governments' commitments to the environment in perspective. Although the political climate has definitely changed since Rio, new agreements are obviously not out of the question either.

The context may have changed, but there are still phenomena which have remained relatively constant since 1992. For example, many key delegates and key officials who were active in the climate and biodiversity arenas in the period studied are still there in those same arenas today. The preferences which they expressed with regard to NGO input are probably still valid (and their emphasis on 'quality' is one). Another example of continuity is the rules on NGO access. Even today, several contact groups remain closed to NGO observers, so that for example recommendation 14 is still valid (page 283). In short, the recommendations and the strategic model will probably still hold in the biodiversity and climate arenas of today, *although the political environment has changed.* This implies that favourable policy outcomes are probably even harder to achieve than 5 years ago. Therefore better NGO interventions in accordance with the strategic model designed in the previous section will probably have more effect on current negotiation processes, but the final policy outcomes are – given contextual changes – not necessarily more favourable from an NGO standpoint. Finally, it should be recognized that for example NGO interventions and NGO-state interaction have probably continued to develop and – maybe – improve in both arenas in the period 1995-1998. Therefore some recommendations may already be out of date. It is up to the NGOs themselves to judge this, as their interventions, failures and successes since 1995 have not been intensively followed by the author.

As far as the possible thematic extrapolation is concerned a number of observations should be taken into account as well. Firstly, UNCED is also a relevant factor here. After all, the formation of both the CBD and the FCCC was formally linked to the UNCED process, in terms of substance, procedures and personnel. This may have positively impacted NGO-state interaction in the biodiversity and climate arenas, a development which other issue arenas may have lacked, for example the Convention on Long Range Transboundary Air Pollution (LRTAP), the Basel Convention on toxic waste, CITES on international trade in endangered species, the North Sea policies, ozone policies etc. One may therefore assume that NGO-state relations were – and are – closer in the climate and biodiversity arenas than elsewhere, leading to more opportunities for NGOs to impact policy outcomes. However, the literature on some of these other arenas does not seem to point to that (Bramble and Porter, 1992; Potter, 1996; Princen and Finger, 1994). At first sight, the NGOs' political influence in the biodiversity and climate arenas seems not to exceed that in other arenas, *although this is not thoroughly assessed here or by other studies.* On the other hand, UNCED as the institutional context for treaty formation did not necessarily guarantee positive effects either, given the failure of the forest negotiations and the lack of NGO influence in that arena (see chapter 1). Besides, UNCED may have had a similar impact on *all* environmental issue arenas, in the sense that the so-called major groups of Agenda 21 – and both NGOs and indigenous groups are examples of these – are taken (a bit) more seriously by policy- and decision-makers since then. In short, the fact that the climate and biodiversity arenas were formally linked to the UNCED does not seem a prominent factor when compared to other arenas which were not, at least from an NGOs' political influence perspective. Thematic extrapolation can therefore not be excluded because of the UNCED-linkage. It nonetheless *can* where biodiversity loss and climate change are specific and prominent environmental issues compared to others. However, the two issues *themselves* are totally different: a common resource (biodiversity) versus a common sink (the atmosphere); a national property (biodiversity) versus a common property (the atmosphere). Yet the outcomes with regard to the NGOs' political influence are more or less the same. Therefore this argument for excluding thematic extrapolation does not seem valid either.

One might also assume that the enabling, constraining and neutral factors on which the recommendations are based are so issue-specific that they do not have any explanatory capability with regard to other issues, such as toxic waste, oceans, desertification, ozone, etc. Nonetheless, most of the factors identified in table 6.9 (page 263) are quite general in nature. For example, the possibility that traditional block politics may hamper NGOs in successfully intervening seems plausible for other environmental arenas – or even international policy arenas in general – as well. The same may apply to: the utilization of existing rules, integrative negotiations, protest outside, rules of the game, domestic policy effects, etc. Exceptions may be some neutral factors – e.g. there may be arenas in which the number of NGOs *does* matter – and the preferences of NGO representatives and government key players on the nature of the NGO input. This last may be different in other arenas. Therefore one should be careful in extrapolating the respondents' views. To sum up, thematic extrapolation does not seem impossible, although every recommendation should be considered on its own merits, in particular those stemming from the respondents' views. This, incidentally, needs to be done by the NGOs themselves. Whether they feel challenged by the above analysis is up to them.

To conclude, temporal and thematic extrapolations of the recommendations cannot be (entirely) excluded, given 1. a certain degree of stability in the climate and biodiversity arenas; 2. the different nature of the two policy issues; 3. the general nature of several explanatory factors. At the same time, one should be cautious in extrapolating the recommendations too easily, as the international political climate as well as the governments' attitude towards the environment have changed since UNCED.

— 7.6 Conclusions —

This chapter has yielded 19 Recommendations and a strategic model to enhance the NGOs' political influence. These can be summarized on the basis of a few key notions. With regard to lobbying and pressure, the notion of *professionalization* is valid. In order to increase their impact on policy outcomes, NGOs should become more professional in their interventions in political arenas. There should be less focus on *quantities*, for example on the number of NGOs present at international negotiation sessions. As far as protest is concerned the following two notions are valid: *separation* and *moderation.* Protest activities may produce additional and effective pressure on policy- and decision-makers as long as these activities are spatially separated from formal political arenas (the negotiation sessions), and as long as the demands and strategies concerned are moderate. Here the emphasis should lie on *quantity.* The more supporters and journalists are mobilized outside arenas, the more effective on the inside.

The above views should be put into the right perspective. It is not believed that all the recommendations which stem from a certain era (1990-1995) and certain issue arenas (climate and biodiversity) can simply be extrapolated to international environmental policies in general, although the possibility that lessons may be learned by NGOs in current and different arenas is not excluded either. In addition, the above views may at best be considered as an (unrequested) input into the strategic debate of NGOs. As NGOs continuously reflect on their approaches in global politics through workshops, (internet) conferences and strategy papers, such an input can be fed into the debate at any time. This is not to say that NGOs should immediately follow the thoughts expressed here or that this would guarantee some easy success. On the contrary. The objective is to enrich the debate. Therefore NGO representatives can simply take note of the views presented here, judge them in the context of their own experiences, and adopt or reject them as they wish.

CHAPTER 8

Summary, General Conclusions and Discussion

— 8.1 Introduction —

This final chapter summarizes the study, draws general conclusions, confronts these with views from the literature – both academic and 'activist' – and reflects on the study as a whole. Section 8.2 deals with the main conclusions and discusses these. In doing so, the three research questions of this study are chronologically dealt with in subsections 8.2.1-8.2.3. To repeat, these questions as identified and elaborated on in chapter 1 are as follows (see subsection 1.5.2):

1 To what extent did global NGOs influence the formation and implementation of the UN Framework Convention on Climate Change (FCCC) and the UNEP Convention on Biological Diversity (CBD) in the period 1990-1995?

2 What factors explain the NGOs' political influence in the climate and biodiversity arenas?

3 How might global NGOs increase their political influence?

Section 8.3, next, presents some reflections on this study from the perspective of design, theory, methodology, validity and reliability. These aspects are highlighted in subsections 8.3.1-8.3.4. Finally, section 8.4 will summarize the chapter and formulate some recommendations on how to build on this study in new research programmes.

— 8.2 Summary and debate —

8.2.1 The political influence of global NGOs

This study aimed, firstly, at an assessment of the political influence of global NGOs on the Climate and Biodiversity Conventions (or treaties) in the period 1990-1995. The reasons for this were twofold:

1 To analyse the effectiveness of the NGO input into the United Nations Conference on Environment and Development (UNCED) to which the formation of the FCCC and CBD was formally linked;
2 To contribute to the body of knowledge on NGO influence at global level.

Political influence has been defined as the achievement of (a part of) one's policy goal with regard to an outcome in treaty formation and implementation, which is (at least partly) caused by one's own and intentional intervention in the political arena and process concerned. This implies that one's policy goal would not have been achieved – or would have been achieved to a lesser extent – if one had not intervened. Therefore the so-called *counterfactual* would have been different. But NGOs can also affect intergovernmental negotiations without becoming influential in the political sense (*process* influence). In such cases, NGOs affect debates and views, either temporarily or continuously, yet without achieving more of their goals in intergovernmental decision-making (counterfactual remains the same). Note that the domain of this study is limited to *intended* influence. Unintentional and unconscious influence are definitely existing phenomena, but they are not included or assessed here, purely for methodological reasons.

The notion of political influence should be distinguished from *political power*, which refers to a player's more or less permanent ability to influence policy outcomes. Therefore an influential player with regard to a specific outcome is not necessarily powerful in general, although political power is built on and sustained by political influence. At the structural level, a more or less permanent *distribution of resources* may be distinguished in the international system. This distribution co-determines the power positions of individual players in global politics and, in consequence, enables or constrains them in affecting policy outcomes. In general, NGOs are constrained by this distribution of resources at global level, whereas states – and in particular big ones such as the USA – are enabled by it. But this does not necessarily exclude NGO influence in specific cases. In addition, NGOs may be enabled or constrained by given regimes and rules of the game in world politics as well. Regimes are sets of substantive rules in specific issue-areas in international relations. As regimes frame negotiation and decision-making processes on related issues, they make some policy outcomes more likely than others. Formal rules of the game – and in particular those of the UN system – co-determine to what extent NGOs have access to treaty formation and implementation.

Political influence has been assessed by the so-called *EAC Method*, which combines elements of the well-known decision and reputation methods. *Ego-perception* covers the views of key representatives of NGOs on their own political influence. Ego-perception has also been used as a means to select relevant cases for further analysis. As the NGOs' views on their own political influence should be mistrusted – they will probably overestimate their own impact – other indicators have been designed. *Alter-perception* refers to the views of key government representatives on the NGOs' political influence, while the *Causal analysis* covers the reconstruction of negotiation and decision-making processes to detect possible causal links between NGO interventions on the one hand and effects on policy outcomes on the other. As such a reconstruction is difficult to achieve at the global level of policy-making, alter-perception has functioned as a tool to compensate for shortcomings in the causal analysis (and vice versa). But methodological shortcomings are hard to overcome in respect of *any* anatomy of political influence in global, complex negotiation and decision-making processes. Therefore the assessments have been labelled *informed guesses* in this study. True, the more informed the better, but some 'guessing' will remain in any case. The *extent* of political influence, next, has been determined by four dimensions. It may increase: 1. if more of the NGOs' goals are achieved with regard to a specific policy outcome; 2. if more of these are achieved due to their own and intended interventions; 3. the higher the political relevance of the policy outcome concerned is; and 4. the more direct the NGOs' impact is. These four dimensions have been transformed into one formula to determine the (relative) strength of NGO influence in each case. The following techniques to collect the necessary data were applied: interviews, document-analyses and personal observations.

Eight and ten cases respectively from the FCCC and the CBD have been assessed. The notion of 'case' as used here refers to a specific policy topic within the climate or biodiversity arena. Such a topic has been covered either by one or more articles of the treaty concerned (formation phase) or by one or more decisions of the Conferences of the Parties to the two conventions (implementation phase). Examples of such cases are the target of the FCCC (Article 4.2(a) and (b)), in-situ measures of the CBD (Article 8), the marine biodiversity topic (decision by the Conference of the Parties to the CBD), and the so-called Berlin Mandate (decision of the Conference of the Parties to the FCCC). NGOs claimed to have influenced all 8 FCCC cases and all 10 CBD cases (ego-perception). However, only half of these claims could be confirmed on the basis of alter-perception and causal analysis. Table 8.1 gives an overview of the cases concerned and the extent of political influence.

All in all, one has to conclude that NGOs *did* influence some policy outcomes of the FCCC and CBD, but only to a limited extent. Many topics remained unaffected by them, the level of goal-achievement was generally quite low, and most NGO impact was indirect in nature, although some very relevant policy outcomes were nonetheless impacted. This conclusion implies that other policy players were generally more influential than NGOs – e.g. governments, international organizations, scientific bodies, business groups -, although the political influence of the latter players was not thoroughly assessed in this study. It is remarkable that the figures on both conventions are more or less the same. Although climate change and biodiversity loss are quite different in nature and attracted quite different numbers and types of NGOs, this did not lead to very different outcomes on political influence.

TABLE 8.1 *FCCC and CBD cases and the extent of political influence of global NGOs*

	much political influence	*substantial political influence*	*some political influence*	*no political influence*
FCCC cases		- AOSIS Protocol	- Target - Joint implementation - Berlin Mandate	- Objective - Principles - Implementation - First review
CBD cases		- Preamble	- In-situ conservation - Indigenous people 2 - Marine biodiversity - Financial mechanism	- Indigenous people 1 - Genetic resources - Biosafety 1 - Forests - Biosafety 2

The above conclusions are hard to compare with those of other studies, as more or less similar analyses of the FCCC and CBD are simply lacking. However, several publications do nonetheless stress the importance of NGOs in the biodiversity and climate arenas (Bodansky, 1993; Burhenne-Guilmin and Casey-Lefkowitz, 1993; Glowka *et al.*, 1994; Heins, 1993; Mintzer and Leonard, 1994; O'Riordan and Jäger, 1996; Rowlands, 1995; Sanchez and Juma, 1994). Some of these even try to guess their impact. For example Rowlands (1995) believes that, despite increased NGO participation in the climate arena, their impact on the final outcome was negligible. He notes that: "When the chips were down, the NGO representatives were not in the room" (p. 240). But Rahman and Roncerel (in Mintzer and Leonard, 1994) are a bit more optimistic. They believe that NGOs affected negotiations on joint implementation, the principles of the FCCC, the target of the FCCC, and the prompt start of the convention. Given the fact, though, that these authors were themselves NGO representatives involved in the FCCC, their conclusions may be considered another example of 'ego-perception'. It is therefore no surprise that some of their conclusions are at odds with those here, e.g. on the FCCC principles. Examples of 'alter-perception' are to be found in the book of Mintzer and Leonard (1994) as well. For example the key Indian delegate Dasgupta also believes that NGOs impacted the pledge & review debate (see case 3, chapter 4). All in all, Mintzer and Leonard's book on the FCCC parallels the general conclusion of this study: NGOs have at best influenced some policy outcomes of the climate treaty. As far as the literature on the Biodiversity Convention is concerned, no clear statements on the NGOs' political influence have been made, although the entrepreneurial role of the World Conservation Union (IUCN) has been widely recognized.

Publications on NGOs, political influence, and environmental issues other than climate change and biodiversity loss – such as car emissions, wildlife, Antarctica, ozone and hazardous waste – have been quoted in chapter 1 (see subsection 1.4.1). These revealed that NGOs had been able to affect some policy outcomes in various arenas, although disagreement on the question of NGO influence among observers has continued. Whereas some for example believe that NGOs impacted elements of the Vienna Convention and Montreal Protocol on the ozone layer, others remained sceptical (Rowlands, 1995). Besides, examples of NGO failures were mentioned in chapter 1 as well (e.g. on forest policies). Therefore one may cautiously conclude that NGOs seem to matter in international environmental policy-making and implementation, but to a limited extent only. They fail to exert political influence in many cases, and when they are successful, one should not exaggerate the extent of their impact on policy outcomes. But small impacts may have huge implications in the long run, as case 3 has proved. This case covers the inclusion of the (diluted) stabilization target in the FCCC, which was partly due to NGO pressure. The adoption of this target has led to serious implementation problems for OECD countries and has legitimized the on-going debate on binding targets and timetables on greenhouse gas emission reductions up to now.

8.2.2 The explanation of political influence

In addition to an assessment of political influence, this study aimed, secondly, at an explanation of this phenomenon. To achieve this, the above cases were mutually compared on the basis of a number of possible *explanatory factors*. These factors were deduced from the *theoretical model* which was designed in chapter 2 (see figure 2.2, page 71). This model, which had been built on theories of NGOs, international relations and influence/power, distinguishes variables at the level of the political arena, at the level of the policy outcome and at the contextual level. In a *data-matrix*, the various cases which yielded different outcomes on political influence were linked to the possible explanatory factors. Subsequently, it was checked whether a *correlation* between each factor on the one hand and the extent of political influence on the other could be made. If so, it was hypothesized that this factor could explain NGO influence. At a later stage, the explanatory capability and possible overlap of each factor was assessed. Finally, a causal model on political influence was designed.

The following conclusions were formulated (see table 8.2). Firstly, the two most important factors which enable NGOs to exert political influence are expertise (in particular substantive knowledge and diplomatic skills on the NGOs' part) and the similarity between NGO demands on the one hand and existing regime rules on the other (such similarity increases the legitimacy of these demands). Less important, but still enabling, factors are: the integrative nature of negotiations (compared to confrontational negotiations), protest activities of (other) NGOs outside the political arena, and the implementation phase (compared to the formation phase of treaties).

TABLE 8.2 *The most prominent explanatory factors of the NGOs' political influence*

Enabling factors	*Constraining factors*	*Neutral factors*
- expertise - regime similarity	- 'wrong' approach of NGOs - like-minded and dominant states - traditional block politics (North versus South)	- type of NGO (intervention) - quantity of NGOs - environmental issue

Secondly, the main three important factors which seem to constrain NGOs in exerting political influence are: the use of a 'wrong' approach by NGOs (confrontational, radical, idealistic etc.), the presence of like-minded but dominant states (which means that these dominate a topic and are not dependent on any NGO input), and negotiations based on traditional block politics of North versus South. Less important constraining factors are: the formation phase (compared to the implementation phase), the political relevance of a policy topic, and the domestic policy effects of an intergovernmental decision.

Thirdly, a number of factors which seem relevant to explain NGO influence at first sight are found not to do so in the second instance (so-called neutral factors). Examples are: a specific type of NGO, a specific type of NGO intervention (including the co-drafting of legal texts), the number of NGOs which intervene, the extent of NGO co-operation, and the nature of an environmental issue. However, the validity of a number of contextual factors could not be verified, as they were relevant to all cases and remained more or less constant in the period which was studied (1990-1995). One can think here of the distribution of resources in world politics, the impact of the UNCED process on environmental negotiations in the early 1990s, and the UN rules on NGO access to intergovernmental negotiations. Although these factors could not be verified, they definitely have impacted the chances for NGOs to become influential in the climate and biodiversity arenas.

If one compares the above conclusions with explanations of NGO influence in (academic) literature, some similarities and differences will be striking.

1 Many publications stress the importance of the size and numbers of NGOs as well as of NGO coalitions for influence (Bramble and Porter, 1992; Doherty, 1994; Lowe, 1983; Mazey and Richardson, 1992; Potter, 1996; Thompson-Feraru, 1974).

"The more and bigger, the more influence", seems to be the message. This is not confirmed in this study. Here 'quality' is considered of more importance than 'quantity', for NGOs to exert political influence. This difference may, of course, have been caused by the different empirical domains which have been studied by the various authors. After all, it is quite possible that the factor 'quantity' is more important in some arenas than in others. Besides, the difference may be a *relative* one, since it has been observed in one of the previous chapters that some minimum level of NGO attendance and co-operation is a prerequisite for any effective lobbying. So numbers are important, but to some extent only.

2 It is regularly assumed in the literature that NGOs can more easily influence agenda-setting or policy implementation than policy-making (Hooghe, 1995; Liefferink *et al.*, 1993; Mazey and Richardson, 1992). This has been confirmed by this study, at least in as far as NGO influence on policy formation compared to policy implementation is concerned. The phase of agenda-setting has, however, not been addressed.

3 Some stress the importance of the nature of an environmental issue in explaining NGO influence. For example Potter (1996) believes that some issues are more easily impacted by NGOs than others. This belief has not been strengthened by this study, although one should acknowledge the fact that Potter refers to issues other than those dealt with here (e.g. forests and the ozone layer). Nonetheless, given the fact that climate change and biodiversity loss are environmental issues which are very different in nature, it is believed that some generalization is possible with regard to the conclusion that the nature of an issue does not matter. True, some policy topics *within* the domain of one environmental issue are more easily influenced than others, as this study has also revealed, but there is probably no issue out there which cannot be (partly) influenced by NGOs *in principle.*

4 Many publications confirm the importance of 'quality', both in terms of (substantive) knowledge and (organizational) skills of NGOs (Boulding, 1988; Breyman, 1993; Dawkins, 1991; Enge and Malkenes, 1993; Hooghe, 1995; Huberts, 1988; 1989; Lowe, 1983; Van Noort *et al.*, 1988; Princen and Finger, 1993; Potter, 1996; Rahman and Roncerel, 1994; Willets, 1982; 1996). The conclusion of this study on expertise as an explanatory variable is therefore widely reaffirmed in the literature.

5 Some also stress the importance of NGOs using existing rules in order to strengthen their demands, although this argument is heard less often (Bramble and Porter, 1992; Potter, 1996).

6 The possible drawbacks linked to radical or confrontational approaches have been emphasized by others as well (Mazey and Richardson, 1992; Van Noort *et al.*, 1987; Rowlands, 1995). On the other hand, Huberts (1988) concluded that the nature of NGO demands and strategies does not matter much (except for expertise and leadership).

7 Several authors also believe that protest activities by NGOs increase the effectiveness of their pressure and lobbying (Doherty, 1994; Hontelez, 1993; Potter, 1996). The same applies to the legitimacy of NGOs (Bramble and Porter, 1992; Princen and Finger, 1993; Willets, 1982). The more NGOs are perceived as legitimate representatives of socio-environmental interests, the more influential they will be, these authors claim. This has also been concluded in chapter 6, although the importance of this factor is considered quite low here, unlike in other sources.
8 Several publications draw attention to contextual factors, such as the rules of the game and the political structure, variables which constrain NGOs (Potter, 1996; Rowlands, 1995; Willets, 1982; 1996). These factors are considered important here as well, although this could not be properly verified for methodological reasons.
9 There are a number of factors which are missing in the literature and elaborated on in this study (and vice versa). An example is the nature of negotiations among states. As long as these are confrontational, in particular due to North-South clashes, the chances for NGOs to become influential seem negligible. This argument has not been put forward in other publications. The same applies to the role of like-minded states, scientific certainty and domestic policy effects.
10 On the other hand, the literature deals with the impact of factors such as 'time' and 'nature of target organization', all aspects which have been ignored in this book.

To conclude, this study reaffirms some views expressed in the literature on the background of the NGOs' political influence and rejects others. Reaffirmed are the impact of the following main factors on NGO influence: the quality of NGO interventions, the similarity between existing rules and NGO demands, the policy phase, and a 'wrong' approach by NGOs. Rejected are: the number of NGOs in a political arena and the nature of an environmental issue. Additionally, some factors are elaborated on in this study which are missing in the literature, among others the nature of intergovernmental negotiations. Finally, the capability of contextual factors – such as the unequal distribution of resources in world politics – to explain the (lack of) NGO influence has been assumed both in the literature and in this book, but could not be assessed on the basis of this study.

8.2.3 The increasing of political influence

Alongside the assessment and explanation of political influence, this study finally aimed at the formulation of recommendations for NGOs to increase their impact on treaty formation and implementation. The nature of these recommendations, though, should be put in the right perspective. They are at best considered an (unrequested) input into the strategic debate of NGOs, not a recipe which guarantees success. In addition, it is up to NGO representatives themselves to judge their value.

The explanatory conclusions were converted into recommendations on the basis of the criterion of *anticipation*. It was checked to what extent NGOs might anticipate on the neutral, constraining and enabling factors. It is obvious that some of these are more within their control – e.g. the development of expertise – than others – e.g. the changing of UN rules on NGO access to intergovernmental meetings. Therefore each factor should be judged in the context of possible anticipation by NGOs. This prescriptive analysis revealed that NGOs should above all: increase their expertise in both environmental issues and diplomatic skills; utilize existing rules on (other) environmental issues to strengthen their demands; avoid approaches which are disliked by policy- and decision-makers; avoid policy topics already covered by other (like-minded) players; avoid policy topics characterized by heavy North-South conflicts; and focus on quality instead of quantity with regard to NGO interventions. Further recommendations are (among others): to organize more protest activities outside formal issue arenas; to introduce innovative proposals; to mainly focus on integrative negotiations; to address both treaty formation and implementation; to try to enforce more liberal rules on NGO access; to avoid a single focus on the drafting of legal texts; to intensify the interaction with governments; and to enhance intersessional and national work to prepare for international meetings.

All these recommendations were fused into one strategic model which, may be summarized as follows (see figure 7.1, page 292). With regard to lobbying and advocacy *inside* political arenas, NGOs should focus on and improve the quality of their interventions (*professionalization*). The focus should be less on *quantity*, for example on the number of NGOs present at international meetings. As far as protest activities are concerned, different notions are valid (*separation* and *moderation*). Protest activities may produce additional and effective pressure on policy- and decision-makers, as long as these activities do not take place *inside* conference buildings where intergovernmental negotiating sessions are held and as long as the demands and strategies concerned are moderate. Here the emphasis should lie on *quantity*. The more supporters (and journalists) are mobilized outside arenas, the greater the effects will be on the inside. It was also assessed to what extent the strategic model may be extrapolated in time and to other issues. It was concluded that temporal and thematic extrapolations of the recommendations cannot be (entirely) excluded given: 1. a certain degree of stability in the climate and biodiversity arenas in the period 1990-1997; 2. the difference between the two policy issues (climate change versus biodiversity loss); 3. the general nature of several recommendations. At the same time, it was concluded that one should be cautious in extrapolating conclusions too easily, as the international political climate as well as the governments' general attitude towards the environment have changed in the last couple of years.

It should be recognized that most recommendations have already been reflected in the literature one way or the other. For example, the need for increased professionalization of NGO work, for a better organization of work, for a combination of lobbying and protest, as well as for improved NGO access to political arenas, have all been mentioned by academics, observers and activists (Bramble and Porter, 1992; Dawkins, 1991; Doherty, 1994; Hontelez, 1993; Princen and Finger, 1994; Van der Schot and Van de Veen, 1997; Willets, 1996). However, there is one important exception: the emphasis on the need for more NGOs as well as more NGO co-operation in political arenas, which is expressed in much of the literature but rejected here. The central message of this book is that NGOs in political arenas should focus more on quality than on quantity, at least if they wish to seriously strive for political influence. To repeat, decision-makers are not interested in numbers, they are interested primarily in innovative ideas and arguments.

It should be highlighted, finally, that not all (academic) observers and activists will share the above views, as the key words professionalization, separation and moderation are probably regarded with suspicion. For example Jamison (1996) believes that environmental groups may lose their critical position if they become too much engaged in processes of professionalization, which are in fact determined by mainstream science and politics. Instead, they should retain their independence and extend their critical role. As far as the interaction with governments is concerned, many fear that this could lead to the co-opting of NGOs by the establishment, either at national or at international level. This might lead to the legitimatization of bad politics and 'undemocratic' systems by NGOs. Therefore authors such as Chatterjee and Finger (1994) and McCoy and McCully (1993) believe that NGOs should stay out of mainstream politics and develop their own alternatives outside formal political arenas. In addition, the notions of 'separation' and 'moderation' with regard to protest will be mistrusted as well. After all, this might be interpreted as the external control of protest groups by mainstream lobbyists, in order to strengthen the latter's positions and please formal policy- and decision-makers. Thus the process of co-optation would even be extended to outside formal political arenas.

Although these 'dangers' attached to the professionalization of lobbying and advocacy and to the separation and moderation of protest are acknowledged, these possible drawbacks are judged differently in this book. Firstly, professionalization *of* NGOs and criticism of policy-making *by* NGOs are not mutually exclusive (Hontelez, 1993). The same individual may play several roles, for example being a diplomat in one case and a critic in another. Or NGO representatives can specialize. Some may, for example, lobby delegates, while others write critical comments on those same delegates' views. Secondly, co-optation is considered a less severe danger here, as it is believed that NGO representatives – just like other people – are knowledgeable and capable agents who cannot simply be controlled by others

(Giddens, 1984). Besides, a plea for a separation of protest and lobby/advocacy does not imply that mutual contacts and links should be cut off. On the contrary. The separation is meant to be purely spatial and functional. Activists, advocates and lobbyist should ideally keep in touch, harmonize their activities, and mutually reflect on their premisses. To do this, individuals and organizations should continuously cross into each other's territory, also at international negotiation sessions. Such movements within the NGO community diminish the chances of co-optation.

— 8.3 Reflections on this study —

8.3.1 Design

The design of this study was based on the *iterative* approach. Research questions, theory, methodology, field work and reporting were not chronologically designed nor applied, but were paralleled to a very large extent and – during the course of the project period – mutually adjusted on several details. This has definitely some advantages, as was explained in chapter 1. Verschuren and Doorewaard (1995), for example, refer to the enhancement of internal consistency, efficiency and creativity in research due to this approach. In addition, it urges the researcher to start field work early instead of focusing on theoretical and methodological dilemmas behind his desk for too long. This might have been a real threat to this study, had the iterative approach not been applied.

However, iteration has drawbacks as well, some of which have become obvious in this project. First of all: although the approach is meant to tune all the elements of a research project, the opposite may be the result. As for example theory building and (initial) field work are paralleled, these activities may develop in their own separate way, in particular if theoretical and methodological views have not yet matured. Then the researcher needs to pioneer in different areas at the same time and there is a chance he will make impulsive decisions in both areas which may be incompatible at a later stage. For example, interviews were already conducted at the first Conference of the Parties to the Biodiversity Convention in November 1994, although neither the methodology nor the theory of this study had been fully developed yet. The interview guide used there was, in consequence, not based entirely on thorough theoretical and methodological insight, but on empirical improvisation as well. This had repercussions at a later stage. Some interview questions turned out to be superfluous, while others which were needed were missing. In short, theory, methodology and (initial) field work for this study were not ideally harmonized.

Secondly, adjustments to a research design during its application may lead to omissions. For example, a change in the format of an interview guide due to an adjustment of the research questions during the period of field work, will leave the researcher with differences in the information and opinions from before and from after the intervention. The question then is how to combine or repair these differences with regard to the empirical data collected. This study faced this drawback as well, for example on the question regarding the *lack* of political influence (see section 6.6). Thirdly, the iterative approach may foster what (neo)positivists would probably label *subjectivist* and *eclecticist* products, because a researcher who follows the iterative approach is allowed to individually and continuously 'play' with theories, methods and empirical data. Although the labels of subjectivism and eclecticism are not necessarily negatives in the scientific scene today – Verschuren and Doorewaard even seem to consider them as sources of creativity – the iterative approach may indeed lead to combinations of theories and methodologies which are not entirely compatible (see subsections 8.3.2 and 8.3.3. below). Furthermore, accumulation of knowledge may be hampered, because every individual researcher creates his own scientific world-view. On the other hand, researchers who follow more classical approaches in science – for example (neo)positivists who defend a linear approach – are generally forced to employ iterative techniques as well, but hold these implicit. Therefore Verschuren and Doorewaard rightly defend the proposition that one should make such iterative techniques explicit.

A research design consists of a conceptual and a technical design. The former will be reflected on in the next subsection (theoretical model). As far as the technical design is concerned, some remarks should be made regarding the multiple-case design. Firstly, one may question the choices made regarding the (sub)cases. The selection of the FCCC and CBD was discussed in detail in chapter 1 and has not been regretted so far. The criteria to select these two conventions were both pragmatic and methodological in nature. The pragmatic criteria refer to the UNCED process as point of departure for this study and to the limited availability of resources which made an analysis of more than two cases unattainable. The methodological criteria, however, cover the procedures of 'literal reproduction' of cases within the UNCED domain (Yin, 1991). Although the selection of (more or less) similar cases is generally criticized in the literature – because of the phenomenon of case-contamination and a lack of external validity – it has some advantages as well, in particular for explorative studies (Rosenthal and 't Hart, 1994; Swanborn, 1994; 1996). After all, a new phenomenon has to be studied and therefore it is better to select identical cases. However, extrapolation to other domains and causal analyses is then difficult to employ, as the cases and their variables interact to a very large extent. These problems have also become obvious in this study, but were inevitable and have been dealt with in the best way possible.

The selection of *sub*cases may also attract criticism, as these were based on *ego-perception*, and therefore on the NGOs' own examples of political influence in the climate and biodiversity arenas in the period 1989-1995. Although the researcher initially intended to select subcases himself on the basis of theoretical or empirical criteria, it was decided to take the list of ego-perceptions as the point of departure, especially as: 1. it soon became obvious that these ego-perceptions covered most of the key topics of both treaties, and 2. several of these ego-perceptions proved to be wrong, at least according to the EAC Method, so that the list contained both successes and failures of NGOs. However, it would probably have been better if all respondents had not only been asked to mention successes of NGOs, but failures as well. Then we probably would also have dealt with 'real' failures and not with 'false' perceptions of successes alone.

Secondly, one may wonder whether a quantitative technique – for example a survey – would not have yielded better results on ego- and alter-perception. Hundreds of views on political influence of those involved in the climate and biodiversity arenas could have been collected worldwide and, consequently, a more reliable representation of these perceptions would have resulted. Besides, such a survey would definitely be compatible with qualitative analyses of a number of cases. Nonetheless, there were three good reasons to stick to a limited number of semi-structured interviews to assess ego- and alter-perceptions: 1. information collected from some *key* delegates and NGO representatives is as good as information gathered from hundreds of less important participants in the climate and biodiversity arenas; 2. *in-depth* knowledge and information was needed from the respondents, which is difficult to collect on the basis of a survey; 3. the response to a survey was expected to be *low*, given the large number of surveys which have been disseminated among participants in the UNCED process by several scientific institutes around the world since 1992 (personal observation).

Thirdly, one may criticize the delineation with respect to the period 1990-1995. This choice implied the exclusion of more of an historical analysis of both treaties as well as of the NGOs' impact on issue-raising and agenda-setting in the 1970s and 1980s. Although this exclusion was intended, given the study's focus on the role of NGOs in policy-formation and implementation, it turned out to be an omission at the same time. One cannot simply isolate the present from the past. Consequently, the historical role of NGOs in both arenas has probably not received enough attention in this book. Moreover, an extension of the time horizon would have made a contextual analysis of political influence more appropriate. As a change of macro-context – i.e. the international system – is generally a long-term process, its impact on daily politics can only be studied over a longer period. A period of 5 years is definitely not long enough to do so, as this study proves (see chapter 6). On the other hand, there are methodological reasons to limit oneself to the present or to the recent past if one wishes to assess political influence. Such an assessment presumes access to detailed and reliable information on policy processes and decision-making and the further in the past the politics concerned are, the more difficult it will be to have access to such information. Memories of those involved are restricted and policy documents only supply limited and biased information. Therefore it is even believed that a *reconstruction* of policy-making is a drawback for *any* study on political influence. It seems best to analyse *current* policy processes and try – as a researcher – to participate in and observe these processes as much as possible in order to collect sufficient data. Hence, it would probably have been better if the study had focused on the period 1993-1997 of the FCCC and CBD – the period in which the study was conducted – instead of the period 1990-1995. But then the formation phase (1990-1992) of both treaties would have been excluded and that would have been another disadvantage.

Fourthly, the delineation with respect to *global* NGOs is indeed problematic, as local, national and regional NGOs may have indirectly influenced the FCCC and CBD through their respective authorities as well. Consequently, national positions of country delegations may hide NGO preferences from the domestic level. The same may apply to the positions of regional economic integration organizations such as the EU. Besides, governments may act strategically at global meetings, because they have to 'sell' the outcome at home, including to their NGOs. But these aspects are barely covered by this study. It was simply impossible to analyse national policy-making on the FCCC and CBD in more than 150 countries around the world. Another drawback of this study's focus is the reduction of NGO diversity into one community or 'collective player'. In all cases, goal-achievement and political influence of *the* NGOs was assessed. This is, of course, a simplification, as NGOs differ in goals and interventions. Although references to specific (groups of) NGOs were made in the case-descriptions in chapters 4 and 5, there remained some reification. Again, there are only pragmatic arguments to justify this. There was simply not enough time to reconstruct the policy goals and interventions of *all* NGOs involved in each of the 18 cases individually, unless a number of cases were deleted.

Finally, the use of multiple sources of evidence – interviews, documents, observations – has given satisfaction. Confrontation of data from different sources probably made the reconstruction of treaty formation and implementation, as well as the assessment and explanation of political influence, more successful. Omissions from one source could be compensated for by others, confirmation of data from different sources strengthened tentative conclusions, and contradictions urged the researcher to look for additional data. One aspect, though, should be highlighted: the value of direct observation of policy- and decision-making. This proved to be very useful. However, only part of the policy processes of the FCCC and CBD in the period 1990-1995 was observed, namely the Conference of the Parties to both treaties in 1994 and 1995. The main global sessions before 1994 were *not*, and the reconstruction of policy- and decision-making proved more difficult for that period. Therefore it would be preferable to include observations of policy- and decision-making in any assessment of political influence, and this – again – implies a focus on *current* policy processes.

8.3.2 Theory

As the academic literature consulted did not offer an appropriate theory or model on NGOs in treaty formation and implementation at global level, a new model was designed. In order to do so, a specific procedure was followed: 1. an elaboration of the main concepts of the study; 2. a nesting of these concepts in relevant theories; 3. a linking of these concepts and theories in a comprehensive model. Identified as the main concepts of the study were 'NGO', 'political influence', 'convention' and 'context'. These concepts were subsequently related to theories on: 1. non-state players in international political arenas; 2. influence and power, both at the agent and the structural level; 3. conventions and regimes; 4. the nature of the international system. Finally, a political arena model was designed on the basis of these elaborations (see figure 2.2, page 71). It has functioned as a heuristic lens through which the researcher observed and explained his empirical domain. As such, the model played a valuable role. It helped to choose related methodologies and indicators to assess the political influence of NGOs (see figure 3.1, page 82). It also helped to focus on relevant variables to explain the political influence of NGOs (see table 3.3, page 90).

The model positions global NGOs in treaty formation and implementation in a specific way. The basic assumption is that NGOs are in principle able to impact policy outcomes, but that they, in doing so, are generally constrained by other players (e.g. states, firms, international bodies) as well as by contextual factors (e.g. rules of the game, regimes, distribution of resources). At the same time, NGOs might in other cases be enabled by these (f)actors as well. This depends on the specific circumstances under which treaties are formulated or implemented. Although these premisses seem more than obvious at first sight, they are nonetheless the opposite to some of the mainstream views in the political sciences. First of all, the model rejects the *structuralist* and *realist* idea that NGOs are at best background variables in international politics. According to the model, NGOs are in principle influential players. Secondly, the *co-optation thesis* is also rejected. This says that NGOs which participate in mainstream politics are of necessity controlled by (inter)governmental elites. Here, NGOs are – instead – considered knowledgeable and capable players who are able to critically reflect and anticipate on day-to-day political practices and, thus, prevent co-optation. At the same time the model considers states as the most dominant players in current international politics. After all, governments are the ones to negotiate and decide on international treaties, at least formally, and not NGOs. Therefore the concept of *policy network*, which generally points to (more) horizontal relationships between states and non-state players is rejected as well. Instead, the notion of *political arena* was chosen. Thirdly, contrary to the realist view on international relations, the theoretical model accepts the relevance of rules (*regimes*) in global politics. It is assumed that these rules (co-)frame the actions of states and other policy players, including the formation and implementation of treaties. Finally, contextual factors are not only considered a constraint on the conduct of NGOs, but enabling as well. This runs counter to classical structuralist theories on power (see chapter 2). All in all, the above theoretical model is an effort to position NGOs as relevant policy players in global politics without defending either too voluntarist or too determinist a view.

Although this study did not aim at testing the theoretical model, one may nonetheless reflect on the question how the model relates to the empirical findings (see for an overview: section 8.2). First of all, the conclusions concerning the political influence of global NGOs on the FCCC and CBD seem to confirm the basic premisse of the model: NGOs make a difference in treaty formation and implementation at global level. However, the extent of NGO influence should not be exaggerated (see table 8.1, page 303). To repeat, two cases were substantially impacted by NGOs, seven to some extent, and nine not at all. Hence, the model may express a too optimistic view on the possible impact of NGOs on treaty formation and implementation (see arrow *(b)* in figure 2.2, page 71). As far as the explanation of political influence is concerned, secondly, most factors identified in the model seem to have either enabled or constrained NGOs in impacting policy outcomes in the climate and biodiversity arenas, although not all factors could be fully put to the test or verified on the basis of this study. Therefore some variables and their mutual relations in the model have remained mere assumptions so far. For example, the precise impact of the distribution of resources and the rules of the game on the chances for NGOs to become influential players have remained unclear. A more historical analysis of the NGOs' political influence on international environmental policies – for example since the 1960s – might have shed some light on this question. The recommendations for NGOs, finally, do only indirectly relate to the theoretical model. They are based on more specific factors than those identified in the model, especially as (specific) arguments from the respondents were added to the explanatory and prescriptive analyses in chapters 6 and 7. It is obvious that most recommendations refer more to a desired 'change' in the NGOs themselves and less to a desired 'change' in other players or contextual factors (see the overview of recommendations in section 7.4, page 291). This implies that the recommendations only cover *some* parts of the model. Obviously, several factors at the arena, outcome and contextual levels seem to have been pushed into the background by the analyses in chapters 6 and 7. Again, this was due to a lack of testing and verification of the more structural variables of the model in subsequent analyses.

As was noted above, the development of the theoretical framework alongside the design of the research questions, methodology and field work was not without problems. The various elements had, due to the iterative approach, partly developed their own dynamic and the mutual fit was not always ideal. In particular the fit between the theoretical model as developed in chapter 2 and the EAC Method from chapter 3 could have been better, although there are many links, for example through the pluralist tradition of political analysis (decision-making method). Nonetheless, a focus entirely on one comprehensive theory and methodology – as for example Huberts (1988) did – would probably have yielded more internal consistency. But given the fact that such a comprehensive framework was absent with regard to *global* politics, some improvisation proved to be inevitable.

In addition, the theoretical framework looks quite eclectic, as elements of transnational, pluralist, structuration and regime theories were used. This was due to the decision to elaborate on individual concepts in chapter 2, to subsequently nest these in theoretical traditions and to fuse these into one framework. Again, the fit was not always ideal, for example between the 'world-view' of pluralism and structuration theory. Therefore it is probably better to take one theory as the point of departure and amend this in accordance with the empirical domain to be studied. *However, this approach was originally intended, but an attempt to put Giddens' scheme into operation with respect to this study had failed.* In particular a consistency of structuration theory with current methodologies on political influence was absent. In consequence, it was decided to build a theoretical framework around the central concepts. Besides, eclecticism is not necessarily a bad thing, as the same Giddens (1984, p. xxii) notes:

> "To some [structuration theory] may appear an unacceptable eclecticism, but I have never been able to see the force of this type of objection. (..) If ideas are important and illuminating, what matters much more than their origin is to be able to sharpen them so as to demonstrate their usefulness."

The usefulness of the 'eclectic' model as designed in this study consisted of: 1. offering a comprehensive and explanatory 'world-view' to the researcher; 2. positioning NGOs between the agent and the structural level of international politics; 3. bridging the gap between (pluralist) methodologies on power/influence on the one hand and theories on international relations on the other.

8.3.3 Methodology

The 'pros' and 'cons' of the EAC Method, which is based on ego-perception, alter-perception and causal analysis, were extensively dealt with in chapter 3 (see also subsection 8.2.1). One positive aspect is definitely the in-built (mutual) confrontation of perceptions of key individuals on the one hand and information from all kinds of government and non-government policy documents on the other. It has become obvious that such a confrontation is definitely needed, given the fact that perceptions may be wrong (at least 'wrong' in the context of this study). Only 50% of the NGOs' claims could be confirmed, while some perceptions of government representatives turned out to be incorrect as well. For example, they did not credit NGOs with any influence in case 18 (financial mechanism/CBD), whereas the causal analysis produced a positive outcome on political influence. The opposite applies to case 10 (indigenous people/CBD). Here many government delegates did credit NGOs with some influence, but the final conclusion was different. On the other hand, policy documents do not produce decisive answers in many cases either, as this information tends to be as selective and biased as that of respondents (although the latter are also bound by their memories). Therefore the combination of interviews with key respondents and analyses of policy documents has proven to be satisfactory.

Another positive aspect of the EAC Method is the fact that it offers tools to get at least some 'control' over very complex policy processes. The interviews with key participants make a quick introduction to the field of inquiry possible and, moreover, help the researcher to focus on certain aspects of the process concerned and to leave others alone. The study of documents, subsequently, allows an in-depth analysis of these aspects and, moreover, confronts the perceptions of the key participants with new information. Omissions in the documents may, next, be compensated for by further interviews with (other) key participants. Hence, the method is also quite flexible and it is up to the researcher to repeat the rounds of interviews and document-analysis again and again and to improve the depth and quality of his conclusions.

A negative aspect of the EAC Method is definitely the fact that it does not exceed the level of an informed guess and – in some cases – remains more of a guess than an informed conclusion (but this is probably a drawback of any method to assess political influence or power). It appeared that there are no exact criteria to draw a positive or negative conclusion on players' impacts on policy outcomes and, hence, on the precise counterfactuals, if the information concerned is incomplete or unclear. Then the conclusion remains very much a *guess* and the outcome depends to a large extent on the researcher's personal insight (control-effect). An example of this is case 10 on indigenous people. Doubts on its final conclusion have remained until the final days of finishing the text of this study. In other cases, though, the NGOs' impact was quite obvious and confirmed by all sides (government and non-government respondents and documents). Then the informed guess appeared indeed to be *informed*. An example in this regard is case 15 on marine biodiversity. In addition, the *extent* of political influence was also difficult to determine, especially as the influence of players other than NGOs was not thoroughly assessed. Nonetheless, the formula to determine the *relative* extent of political influence on the basis of the dimensions 'goal-achievement', 'own intervention', 'political relevance' and 'directness', helped to discriminate among the various cases in this respect (see subsection 3.2.5).

A final drawback of the EAC Method as applied in this study is its focus on only one player (NGOs) instead of *all* relevant policy players (NGOs, states, business lobby, science, intergovernmental organizations). As a consequence, the political influence of the others has remained unknown. Given these considerations, the EAC Method could be improved along four lines: 1. the inclusion of perceived *failures* as an additional item to be covered by ego- and alter-perception; 2. the study of more *recent* or even *current* policy processes (although there should be 'policy products' available to assess goal-achievement); 3. the development of additional criteria to determine the *counterfactual* in cases where there is only unclear or incomplete information available; 4. the inclusion of all – or at least more – relevant policy players.

As far as the method to explain the NGOs' political influence is concerned, some 'pros' and 'cons' are obvious as well. The data-matrix certainly helped to structure and streamline the comparative case-analysis. However, it was not always easy to insert the correct data into the various cells and the correlations which were found between the extent of political influence and the (possible) explanatory factors were generally quite weak. Moreover, many of the factors which were investigated overlapped or interacted, whereas the issue of 'correlation versus causation' remained unresolved (see chapter 3). Besides, it was not totally obvious how to combine outcomes of the comparative case-analysis with perceptions of respondents, as these outcomes are of a rather different nature. But given the fact that the pretentions of the researcher were modest – no statistical generalizations were aimed at – one may conclude that the method worked well. The same applies to the method used to convert explanations into recommendations. Again, the pretentions were modest, as the recommendations at best have been considered as another input into the debate on NGO strategies.

8.3.4 Validity and reliability

Yin (1991, p. 40) distinguishes four criteria to judge the quality of a case-study: 1. construct validity; 2. internal validity; 3. external validity; 4. reliability. The first criterion covers the correct putting into operation of concepts. In other words: did one collect those data which one needed to according to his theory and methodology? It is believed this was done in this study, at least with regard to the central concept of political influence. A solid operational scheme was designed in chapter 3 for the causal analysis and this was applied to all cases. Moreover, it was explained to the respondents how political influence should be interpreted before any interview started. When views of respondents nonetheless remained unclear, additional questions were asked in order to find out whether he or she referred to political influence or not. In addition, the four main groups of respondents – governments/FCCC, NGOs/FCCC, governments/CBD, NGOs/CBD – were all interviewed over a short period, in order to prevent them mutually deliberating on and co-ordinating their conduct vis-à-vis the researcher. Additionally, several sources of evidence were used and some procedures were done again, for example the classification of respondents' arguments in chapters 6 and 7 (techniques which Yin advises to enhance construct validity). However, 'subjective' elements are present, as was outlined before, e.g. in cases where information remained incomplete or unclear and when in the causal analysis it was difficult to conclude one way or the other. As far as theoretical notions other than political influence are concerned, though, the putting into operation of factors was less 'deep'. This sometimes led to problems of interpretation, as chapter 6 revealed.

The second of Yin's criteria – internal validity – refers to the establishing of correct causal relationships in a study (as distinguished from spurious relationships). In particular the explanation of political influence is an important aspect in this respect, as such an explanation implies the establishing of causal relationships between explanatory factors on the one hand and the NGOs' political influence on the other. The question is whether the relationships which have been established are – according to Yin – correct, complete and airtight. As far as 'correctness' is concerned, one should recognize the fact that this criterion is problematic, as *the* correct explanation of a phenomenon does not exist. Within the context of *this* study and its theoretical model, though, the explanations of political influence *are* correct, but then the question arises whether one accepts the premisses of this study. The explanations are definitely complete, as variables of different theories as well as the views of key respondents were included. However, they are probably not airtight, given the weak correlations in the data-matrix, the issue of 'correlation versus causation' which has remained unresolved, the incomplete elaboration on several (structural) variables in the analyses, and the mutual interaction and overlap of explanatory factors. Consequently, the explanations may be slightly spurious in a number of cases.

According to Yin (and others such as Swanborn, 1996), the external validity of case studies is limited by its nature, at least in *statistical* terms. In other words, the findings of this study are hard to extrapolate to other issue arenas in international environmental policies, although some generalizations were cautiously formulated in chapter 7 on the basis of NGO literature on the one hand and the different nature of the climate change and biodiversity loss issues on the other (see section 7.5). Such extrapolations may, of course, be disputed, but they were nonetheless considered legitimate in the context of this study. However, *theoretical* generalizations may definitely be formulated on the basis of case studies, according to Yin (although Swanborn believes he is too optimistic about that). This was also done in this study, in terms of generalizing the causal model on the explanation of political influence in chapter 6 (see section 6.7). As a consequence, this model might become the input of new research, for example to test it in other empirical domains. In short, one should be cautious in pretending to have a case study which is characterized by strong external validity, but to reject any such validity – as some (neo)positivists would do – is not appropriate either.

The last of Yin's quality criteria is 'reliability'. The essential point here is to demonstrate that the same case study can be repeated with the same results. This implies that a secondary analysis and interpretation of the researcher's data and documentation will yield similar conclusions. All interviews were taped, except for one where the respondent objected and some minor interviews of which records were nonetheless made. In addition, all documents consulted were stored. Therefore anyone who wishes to do so can do a secondary analysis. Of course, whether another interpreter will produce similar outcomes remains to be seen, but the researcher is convinced that it will be so with regard to (nearly) all cases. To conclude, this study is considered strongest on construct validity and reliability, moderate strong on internal validity, and least strong on external validity.

— 8.4 Conclusions —

To summarize the conclusions of this study: 1. NGOs make *some* difference in global treaty formation and implementation (although certainly not in all cases); 2. their political influence is mainly dependent on their own expertise, on the attitude and conduct of (like-minded) states, on the substance of current environmental regimes, and on the nature of intergovernmental negotiations; 3. NGOs may increase their political influence, mainly through professionalizing their lobbying and advocacy in political arenas and through organizing moderate protest outside.

Although the design, theory, methodology, validity and reliability of the study have proven to be satisfactory, some drawbacks were highlighted in this chapter as well. The most important were: some lack of internal consistency in the design of the study due to the iterative approach; the exclusion of some categories of NGO (national ones and NGOs involved before 1990); the eclectic nature and partial putting into operation of the theoretical model; the one-sided focus of the EAC Method on 'successes' (ignoring 'failures'); and the limits of case studies in terms of statistical generalizations.

Given these conclusions, one might build on this study in future research programmes as follows. First of all, it would be useful to extend this research to other policy issues in international environmental politics, for example the Convention on International Trade in Endangered Species (CITES) or the Basel Convention on hazardous waste, in order to verify whether the above conclusions are valid for other empirical domains as well. If they are, then one might start to think about 'real' statistical generalizations of findings. Furthermore, one might – at some point – even draw conclusions on the political *power* of NGOs, if enough empirical domains and eras are studied (given the fact that political power refers to a more or less permanent ability to exert political influence). Secondly, the theoretical model should be refined and more fully put into operation in subsequent research, in particular the structural variables. To do this, different research strategies need to be followed, for example a more historical analysis of international environmental policies. Thirdly, the EAC Method definitely requires both application and improvement. To repeat, it turned out to be an appropriate and flexible method to assess international and complex policy processes, but proposals to improve it were formulated above as well. In the future, it should also focus on failures (or *lack* of political influence), more on *current* policy processes (to increase validity), on an improvement of counterfactual reasoning (development of criteria to decide under conditions of uncertainty), and on more policy players (to reconstruct the *division* of political influence). In addition, a focus on current policy processes would also yield information on the possibility of extrapolating the findings of this study in time (temporal extrapolation). Finally, the *national* level of international treaty formation and implementation and the role of NGOs in these processes should be addressed as well. One way to do this is to study the role of some key countries and their NGOs at both levels of treaty formation and implementation. Then one might gain a more complete insight into the political influence of NGOs on this 'two-level game'.

It should be realized that the above recommendations are sometimes contradictory. For example, a better understanding of the theoretical model requires more *historical* research, whereas a refinement of the EAC Method urges one to study more *current* policy processes. Therefore more options are available. Ideally, three projects should be conducted on each new policy issue: one historical, one contemporary focusing on the global level only, and one contemporary focusing on the global-national links. Then one will gain insight into: 1. the historical role of NGOs and the impact of structural variables on their performance; 2. their political influence on current policy processes compared to the political influence of other policy players; and 3. the NGOs' role and impact at different levels of policy-making.

Samenvatting

Internationale milieuproblemen, zoals zure regen, ontbossing, klimaatverandering en aantasting van de ozonlaag, zijn al jaren in het nieuws. Die aandacht heeft regeringen ertoe aangezet internationaal milieubeleid te formuleren, meestal in Europees verband of in het kader van de Verenigde Naties (VN). Maar ook niet-gouvernementele organisaties (NGO's), zoals het Wereld Natuur Fonds, de World Conservation Union en Greenpeace, hebben aan het opstellen van internationale milieumaatregelen bijgedragen.

De inbreng van NGO's kwam vooral tot uitdrukking in het zogenoemde UNCED-proces. De *United Nations Conference on Environment and Development* werd in juni 1992 in Rio de Janeiro, Brazilië, gehouden. Deze VN-conferentie had als doel een wereldwijd actieplan op het gebied van duurzame ontwikkeling op te stellen. Agenda 21, zoals dit (juridisch niet-bindende) document wordt genoemd, behandelt de meest uiteenlopende zaken: de armoedeproblematiek in de Derde Wereld, de consumptiepatronen in de rijke landen, het beheer van natuurlijke hulpbronnen, gevaarlijk afval, menselijke gezondheid en natuurbescherming. Prachtige voornemens, maar tegelijkertijd moeilijk uitvoerbaar, omdat voldoende financiële middelen ontbreken om ze daadwerkelijk te implementeren.

Naast Agenda 21 heeft de UNCED nog andere producten opgeleverd: de Rio Declaratie, de Bossenverklaring en twee juridisch bindende verdragen: het Klimaatverdrag en het Biodiversiteitsverdrag. De Rio Declaratie bevat een aantal beginselen op het gebied van milieubeleid, zoals het 'recht op ontwikkeling' en het 'voorzorgprincipe'. De Bossenverklaring omschrijft eveneens een aantal principes, maar dan op het gebied van bescherming en duurzaam gebruik van bossen. Het Klimaatverdrag heeft als doel voor de mens gevaarlijke klimaatverandering tegen te gaan, terwijl het Biodiversiteitsverdrag beoogt de bescherming en het duurzaam gebruik van de biologische diversiteit te bevorderen.

Opvallend in het UNCED-proces was de inbreng van NGO's. Tijdens de conferentie zelf waren zo'n 9.000 NGO's aanwezig, waaronder vele milieu-organisaties. Ook de voorbereidende bijeenkomsten werden vaak door meer dan duizend NGO's bezocht. Ze probeerden de agenda en de uitkomst van de UNCED te beïnvloeden door gebruik te maken van hun formele rechten – NGO's kunnen VN-bijeenkomsten bijwonen en mondelinge interventies plegen – en door middel van lobby, pressie en actie. In de wandelgangen probeerden sommige NGO-vertegenwoordigers delegatieleden van (belangrijke) landen te beïnvloeden, terwijl andere NGO's publieksmanifestaties in de buurt van de congresgebouwen organiseerden.

Ondanks al hun inspanningen waren veel NGO's teleurgesteld over de uitkomsten van de UNCED. De overeengekomen documenten bleken vaag en bevatten weinig nieuwe maatregelen. Bovendien bleek een aantal thema's buiten de UNCED gehouden – zoals de (vermeende) tegenstelling tussen vrijhandel en milieubescherming – en kwamen de rijke landen nauwelijks met financiële toezeggingen, hetzij om de afspraken te implementeren, hetzij om de ontwikkelingslanden op milieugebied te assisteren. Sommige NGO's waren dan ook erg negatief over de effectiviteit van hun inbreng. Zij meenden geen enkele invloed te hebben gehad en kwamen tot de conclusie dat NGO's dergelijke grote VN-conferenties maar beter zouden mijden. Andere organisaties waren minder negatief. Zij meenden dat NGO's wel degelijk invloed hadden gehad op de inhoud en het verloop van UNCED. Zij wezen op sommige tekstgedeelten in Agenda 21 – bijvoorbeeld het hoofdstuk over armoedebestrijding – waarin tekstvoorstellen van NGO's door regeringen zouden zijn overgenomen.

— Onderzoeksvragen —

De vraag of NGO's nu wel of geen invloed op de uitkomsten van de UNCED hebben gehad, vormde het startpunt van dit onderzoek. Uiteindelijk zijn het Klimaatverdrag en het Biodiversiteitsverdrag onderwerp van studie geweest. Deze twee verdragen zijn parallel aan de UNCED opgesteld en in Rio door meer dan 150 landen ondertekend. De volgende onderzoeksvragen stonden centraal:

1 In hoeverre hebben NGO's de formulering en implementatie van het Klimaat- en het Biodiversiteitsverdrag beïnvloed in de periode 1990-1995?
2 Welke factoren kunnen de politieke invloed van NGO's verklaren?
3 Hoe kunnen NGO's hun politieke invloed vergroten?

De onderzochte NGO's zijn lobby- en pressiegroepen die zich op het mondiale milieubeleid richten. Wat betreft het Klimaat- en het Biodiversiteitsverdrag zijn dat milieu-, natuurbeschermings- en ontwikkelingsorganisaties alsmede organisaties van inheemse volkeren. Voorbeelden zijn: Friends of the Earth International, World Wide Fund for Nature, Environmental Defense Fund, World Resources Institute, Cultural Survival, Greenpeace International, Climate Action Network, Biodiversity Action Network, Centre for Science and Development, World Conservation Union, Environment Liaison Centre International, Third World Network en Genetic Resources Action International.

De onderzoeksperiode 1990-1995 is gebaseerd enerzijds op de start van de formele onderhandelingen over beide verdragen in 1990 en anderzijds op het tijdstip van de eerste Conferentie van Verdragspartijen (respectievelijk november 1994 voor het Biodiversiteitsverdrag en april 1995 voor het Klimaatverdrag). Met deze periode van onderzoek zijn twee beleidsfasen gegeven: de formulering en vaststelling van beide verdragen in de periode 1990-1992 en de (eerste aanzetten tot) implementatie in de periode 1992-1995.

— De politieke invloed van NGO's —

Om de eerste onderzoeksvraag te kunnen beantwoorden, is een aantal stappen gezet. Allereerst is het concept *politieke invloed* afgebakend. In deze studie wordt politieke invloed gedefinieerd als het door een actor (gedeeltelijk) bereiken van zijn doel inzake een specifieke beleidsuitkomst, hetgeen (mede) is veroorzaakt door zijn eigen, doelbewuste interventie in het betreffende politieke proces. Dit begrip dient te worden onderscheiden van het concept *politieke macht*, dat in dit onderzoek wordt opgevat als een min of meer duurzaam vermogen van een actor om politieke invloed uit te oefenen. Iemand is dus pas machtig als hij bewezen heeft vaak en veel invloed te kunnen uitoefenen op politieke besluitvorming. Omgekeerd geldt dat iemand die één enkele keer invloed uitoefent nog geen machtige politieke actor is. Zo zijn de Verenigde Staten en de Europese Unie machtige actoren in het internationale milieubeleid, terwijl kleine landen op zijn best wel eens invloed hebben.

Vervolgens is er een methode ontworpen om de politieke invloed van NGO's in te schatten. Met opzet wordt hier het woord 'meten' vermeden, omdat het onmogelijk is – zeker in complexe politieke processen op mondiale schaal – de invloed van een actor exact vast te stellen. De ontworpen methode combineert drie elementen:

1 *ego-perceptie:* dit is de visie van NGO-vertegenwoordigers op hun eigen politieke invloed;

2 *alter-perceptie:* dit is de visie van regeringsvertegenwoordigers op de politieke invloed van NGO's;

3 *causale analyse:* dit is een (selectieve) reconstructie van de onderhandelingen en van het besluitvormingsproces door de onderzoeker om oorzakelijke verbanden tussen NGO-interventies en beleidsuitkomsten vast te stellen.

Deze zogenaamde EAC-methode (Ego, Alter, Causaal) is als volgt toegepast: Ten eerste zijn op internationale bijeenkomsten over het Klimaat- en het Biodiversiteitsverdrag sleutelfiguren uit de NGO-gemeenschap geïnterviewd. Hen werd gevraagd hun eigen invloed in te schatten en te onderbouwen met voorbeelden uit de praktijk (ego-perceptie). Zo kon voor beide onderwerpen – klimaat en biodiversiteit – een lijst van beleidsthema's worden opgesteld, die *mogelijk* door NGO's zijn beïnvloed. 'Mogelijk', omdat men ervan moet uitgaan dat mensen de neiging hebben hun eigen invloed te overschatten. Deze exercitie werd herhaald aan de kant van regeringen. Aan vertegenwoordigers van belangrijke landen en aan sleutelonderhandelaars werd eveneens gevraagd de invloed van NGO's in te schatten en te onderbouwen met praktijkvoorbeelden (alter-perceptie). Deze laatste interviews zijn – in tegenstelling tot die met de NGO's – telefonisch gehouden. Ten derde is een aantal casus – 18 in totaal – geselecteerd en aan een causale analyse onderworpen. Dat waren die beleidsthema's die het meest door NGO-respondenten als voorbeelden van politieke invloed werden genoemd en die bovendien uitgebreid door hen werden behandeld.

Wat betreft het Klimaatverdrag zijn uiteindelijk acht casus onderzocht: 1. de doelstelling van het verdrag; 2. de principes van het verdrag; 3. de norm inzake de stabilisatie van de uitstoot van broeikasgassen; 4. de artikelen over implementatie in het verdrag; 5. het voorstel van de kleine eilandstaten voor een nieuw protocol bij het verdrag (AOSIS Protocol); 6. het proefproject inzake 'joint implementation'; 7. de eerste beoordeling van de implementatie van de stabilisatiedoelstelling door de rijke landen; en 8. het onderhandelinsmandaat dat voortkwam uit de Klimaat Top in Berlijn (1995) om het verdrag aan te scherpen ('Berlin Mandate'). NGO-vertegenwoordigers claimen dat zij invloed hebben gehad op beleidsuitkomsten van al deze acht casus.

Eenzelfde NGO-claim geldt voor tien kwesties van het Biodiversiteitsverdrag: 1. de preambule (inleiding) van het verdrag; 2. de paragrafen over inheemse bevolkingsgroepen in het verdrag; 3. de maatregelen voor de *in situ* bescherming van biologische diversiteit; 4. het artikel over de toegang tot genetische hulpbronnen van andere landen; 5. de paragrafen over veiligheidseisen omtrent de import en export van genetisch gemodificeerde organismen ('biosafety'); 6. de (eerste) implementatie van de paragrafen over inheemse bevolkingsgroepen vanaf 1992; 7. de agendering van het onderwerp 'mariene biodiversiteit' door de Conferentie van Verdragspartijen in 1994; 8. de door die Conferentie genomen beslissingen over het onderwerp 'bossen'; 9. de strijd om het opstellen van een 'biosafety' protocol; en 10. het debat over het financiële mechanisme bij het verdrag, het *Global Environment Facility* (GEF).

Op al deze casus is een causale analyse toegepast. Gekeken is in hoeverre kon worden vastgesteld of interventies van NGO's daadwerkelijk tot *andere* beleidsuitkomsten hebben geleid; 'anders' in die zin dat zij meer van hun beleidsdoelen wisten te realiseren. Om dat te achterhalen is gebruik gemaakt van beleidsdocumenten, informatie uit de ego- en alter-perceptie en aanvullende interviews met de hoofdrolspelers. Met die informatie zijn de onderhandelingen en het besluitvormingsproces van iedere casus gereconstrueerd, met nadruk op de doelbereiking en de interventies van NGO's. Uiteindelijk is per casus een eindoordeel over hun politieke invloed geveld.

TABEL 1 *De invloed van NGO's op het Klimaat- en het Biodiversiteitsverdrag*

	veel invloed	*substantiële invloed*	*enige invloed*	*geen invloed*
Klimaat-verdrag		- AOSIS Protocol	- Stabilisatie-norm - 'Joint implementation' - 'Berlin Mandate'	- Doelstelling - Principes - Implementatie - Eerste beoordeling rijke landen
Biodiversiteits-verdrag		- Preambule	- In situ bescherming - Inheemse bevolkingsgroepen (implementatie) - Mariene biodiversiteit - Financieel mechanisme	- Inheemse bevolkingsgroepen (verdragstekst) - Toegang genetische hulpbronnen - 'Biosafety' (verdragstekst) - Bossen - 'Biosafety' (protocol)

Onderdeel van die analyse was eveneens een inschatting van de *mate* van politieke invloed. Die inschatting is gebaseerd op vier dimensies: doelbereiking, eigen interventie, politieke relevantie en directheid. NGO's oefenen meer invloed uit, indien: 1. zij meer van hun doelen hebben bereikt; 2. zij dat meer aan hun *eigen* interventies hebben te danken; 3. de politieke relevantie van het te beïnvloeden beleidsthema groter is; en 4. hun invloed directer is. Op basis van deze vier dimensies is een kwalitatieve schaal van politieke invloed ontwikkeld: geen, enige, substantiële en veel invloed.

Voor beide verdragen geldt dat de helft van de casus door NGO's is beïnvloed (zie tabel 1). In de meeste van deze gevallen is sprake van *enige* politieke invloed. Slechts voor twee casus – één behorende bij het Klimaatverdrag en één bij het Biodiversiteitsverdrag – geldt dat er sprake is van *substantiële* invloed. In de andere helft van de casus kan echter *geen* invloed van NGO's worden aangetoond, ondanks het feit dat zij zelf het tegendeel beweren. Kortom, NGO's hebben beide verdragen beïnvloed, al moet de effecten niet worden overschat.

— De verklaring van politieke invloed —

De tweede onderzoeksvraag van deze studie betreft de verklaring van NGO-invloed. Om deze vraag te kunnen beantwoorden zijn de bovengenoemde casus onderling vergeleken aan de hand van een aantal mogelijk verklarende factoren. Die factoren zijn afgeleid uit het theoretische model dat voor deze studie is opgesteld. Dit *politieke-arenamodel* onderscheidt factoren op het niveau van de politieke context (het internationale systeem), op het niveau van de politieke arena (beleidsactoren en hun handelingen) en op het niveau van de beleidsuitkomst (de verdragen). Deze factoren zijn, evenals de verschillende casus, in een data-matrix ondergebracht. Vervolgens is geanalyseerd in hoeverre (kwalitatieve) correlaties voorkomen tussen de mate van NGO-invloed aan de ene kant en de mate waarin de factoren een rol spelen aan de andere. Indien inderdaad van een correlatie sprake was, werd geconcludeerd dat de betreffende factor een *verklarende* factor is. Daarnaast is tijdens de interviews aan alle respondenten gevraagd de oorzaken van NGO-invloed te duiden.

Sommige factoren blijken de kansen voor NGO's om invloed uit te oefenen te vergroten. In de eerste plaats is hun eigen expertise doorslaggevend, zowel in de zin van beleidsinhoudelijke kennis als in de zin van diplomatieke vaardigheden. Hoe meer zij die in huis hebben, des te groter hun kans op invloed. Ten tweede blijken internationale regels, of *regimes*, op het gebied van het milieu van belang. Wanneer de wensen en eisen van NGO's inhoudelijk overeenkomen met reeds bestaande regelgeving, dan is de kans groter dat zij invloed uitoefenen op nieuw beleid. Zo hebben NGO-vertegenwoordigers regelmatig verwezen naar de relatief strenge regelgeving op het gebied van de bescherming van de ozonlaag, die een precedent voor het Klimaatverdrag zou moeten zijn. Een dergelijke argumentatie heeft (enigszins) gewerkt. Naast 'expertise' en 'regime correspondentie' zijn andere, kansenverruimende, factoren van belang. Een constructieve sfeer onder de landen tijdens onderhandelingen verruimt – in tegenstelling tot confrontaties – de mogelijkheden voor NGO's om beleidsuitkomsten te beïnvloeden. En de kansen worden ook groter wanneer NGO's (mild) protest rondom internationale onderhandelingen organiseren of zich op de implementatie van beleid richten, omdat die fase makkelijker te beïnvloeden is dan beleidsformulering.

Eveneens is een aantal factoren onderscheiden die afnemende invloed kan verklaren. Ten eerste verkleinen NGO's hun kans op invloed door een 'verkeerde' benadering toe te passen, althans verkeerd in de ogen van de beleidsmakers en de besluitvormers.

Dat geldt bijvoorbeeld voor een te confronterende, radicale of idealistische benadering. Ten tweede kunnen (min of meer) gelijkgezinde landen NGO's voor de voeten lopen. Zo heeft een aantal ontwikkelingslanden het thema 'biosafety' voor zich opgeëist, waardoor NGO's nauwelijks de kans kregen zich op dit thema te profileren. Ten derde maken NGO's minder kans invloed uit te oefenen op thema's die sterk beheerst worden door de tegenstelling tussen de rijke en arme landen, zoals het thema 'bossen'. De rijke landen willen immers vergaande afspraken maken over de bescherming van (tropische) bossen, terwijl de arme landen – vanwege hun economische belangen in de bosbouw – daar tegen zijn. Het gevolg is dat het onderwerp sterk is gepolitiseerd en nauwelijks ruimte biedt aan NGO's om te interveniëren. Verder is nog een aantal andere beperkende factoren van belang. Voorbeelden zijn de politieke relevantie van een onderwerp (hoe relevanter, hoe kleiner de kans op NGO-invloed) en de te verwachten effecten van internationale besluiten op het beleid binnen landen (hoe meer effecten, hoe minder politieke invloed).

Ten slotte onderscheidt deze studie factoren, waarvan op grond van de literatuur verwacht mag worden dat zij NGO-invloed verklaren, terwijl zij dat uiteindelijk *niet* doen. De twee belangrijkste zijn de aard van het milieu-thema en het aantal NGO's. Sommige auteurs geloven dat sommige thema's makkelijker door NGO's te beïnvloeden zijn dan andere. Voor klimaatverandering en biodiversiteit geldt die observatie echter niet. Wat NGO-invloed betreft ontlopen deze twee thema's elkaar nauwelijks, terwijl ze inhoudelijk zeer van elkaar verschillen. Voorts doet in de literatuur nog steeds het idee opgang dat 'hoe meer NGO's aanwezig zijn, hoe meer invloed ze hebben'. Het blijkt echter dat een paar goed opererende NGO's meer invloed kunnen hebben dan een grote groep collega's. Kortom, meer dan de *kwantiteit* van NGO's en hun interventies, is de *kwaliteit* van belang.

— Aanbevelingen —

De derde onderzoeksvraag betreft de mogelijkheden voor NGO's om hun invloed te vergroten en heeft als doel aanbevelingen te formuleren. Overigens zijn deze met de nodige bescheidenheid opgesteld. De aanbevelingen moeten worden beschouwd als bijdragen aan de strategiediscussie van NGO's, niet als wetmatigheden die succes garanderen. Bovendien is het aan de NGO's zelf om de aanbevelingen op al hun merites, inclusief hun geldigheid, te toetsen.

De verklaringen voor de politieke invloed van NGO's zijn omgezet in aanbevelingen op basis van het criterium *anticipatie*. Er is nagegaan in hoeverre NGO's kunnen anticiperen op die factoren die hun kansen verruimen dan wel verkleinen. Sommige van die factoren zijn immers meer door hen te controleren dan andere. Het is bijvoorbeeld 'makkelijker' om je eigen expertise te vergroten, dan om de aard van internationale milieu-onderhandelingen te veranderen. Vandaar dat iedere factor is bekeken in het licht van mogelijk anticipatie door NGO's. Daarnaast is alle respondenten naar hun mening gevraagd over hoe NGO's hun invloed zouden kunnen vergroten.

In totaal zijn 19 aanbevelingen opgesteld, die als volgt kunnen worden samengevat. Wat lobby en pressie binnen formele politieke arena's betreft, zouden NGO's hun inbreng verder moeten *professionaliseren*. Dat houdt onder meer in: het vergroten van inhoudelijke kennis, het verbeteren van diplomatieke vaardigheden, het mijden van 'verkeerde' benaderingen, het gebruikmaken van bestaande regelgeving om argumenten kracht bij te zetten, het mijden van onderwerpen die te zeer door gelijkgezinde landen of door Noord-Zuid-conflicten worden beheerst en meer letten op de kwaliteit van de interventies dan op hun kwantiteit.

Wat betreft protestactiviteiten van NGO's, zoals demonstraties: deze kunnen lobby en pressie van NGO's binnen arena's versterken, indien zij gematigd van toon zijn en bovendien ruimtelijk strikt zijn gescheiden van de plaats van formele onderhandelingen en besluitvorming. Radicale acties en acties in het hol van de leeuw – daar waar de beleidsmakers en besluitvormers aan het werk zijn – roepen namelijk irritaties op. Dan is protest in de regel contraproductief, althans vanuit het perspectief van politieke invloed. Milde en omvangrijke acties rondom congresgebouwen leiden echter *wel* tot een aanvullende druk op onderhandelaars, zoals bijvoorbeeld op de Klimaat Top in Berlijn (1995) is gebleken.

Niet iedereen in de milieubeweging zal bovenstaande aanbevelingen delen. Ten eerste zijn sommigen bevreesd dat een te ver doorgeschoten professionalisering in het diplomatieke circuit de onafhankelijke positie van NGO's zal aantasten. Zo worden NGO's mogelijk door regeringen gecoöpteerd. Ten tweede zien veel actievoerders weinig heil in een matiging en een ruimtelijke scheiding van hun protest. Immers, activisten zijn niet alleen in politieke invloed geïnteresseerd, maar ook in (media-)aandacht voor hun eisen, of in radicale veranderingen van de samenleving. Marginale 'winst' in de dagelijkse politiek kan hen in de regel gestolen worden.

— Conclusie —

De bevindingen van deze studie kunnen als volgt worden samengevat:

1 NGO's zijn in staat om de formulering en implementatie van internationale milieuverdragen te beïnvloeden, al moeten de effecten niet worden overschat. De mate van hun politieke invloed is veelal beperkt en betreft in de regel slechts enkele beleidskwesties.

2 De politieke invloed van NGO's kan met name verklaard worden door de omvang van hun eigen expertise en de aard van hun benaderingen, door de inhoud van bestaande milieuregelgeving, door de concurrentie met gelijkgestemde landen en door de mate waarin internationale milieu-onderhandelingen door de Noord-Zuid-tegenstelling worden beheerst.

3 NGO's zouden hun invloed kunnen vergroten door hun inbreng binnen formele politieke arena's verder te professionaliseren en door mild protest buiten die arena's te organiseren.

Bibliography

Adams, W. (1991), *Sustainable Development and the 'Greening' of Development Theory.* Paper, Niccos-seminar 'Development Theories in the Nineties', KU Nijmegen.

Agarwal, A. and S. Narain (1991), 'Global Warming in an Unequal World. A Case of Environmental Colonialism', *Development and Change,* No. 5, 1991, pp. 12-15.

Aggarwal, V. (1985), *Liberal Protectionism. The Internationale Politics of Organized Textile Trade.* University of California Press, Berkeley.

Albinski, M., ed. (1981), *Onderzoekstypen in de sociologie.* Van Gorcum, Assen.

Archer, C. (1983), *International Organizations.* Allen & Unwin, London.

Arts, B. (1993), *Duurzame Ontwikkeling, Duurzame Samenwerking? Allianties tussen sociale bewegingen.* Cahier 54, Studiecentrum voor Vredesvraagstukken, Nijmegen.

Arts, B. and W. Rüdig (1995), 'Negotiating the "Berlin Mandate": Reflections on the First Conference of the Parties to the UN Framework Convention on Climate Change.' *Environmental Politics,* Vol. 4, No. 3, Autumn 1995, pp. 481-487.

Arts, B. and E. Mulleneers (1995), "Vijftig jaar Verenigde Naties, vijfentwintig jaar mondiaal milieubeleid." In: B. Bomert and W. van Genugten (eds.), *Vijftig Jaar Verenigde Naties: Grote Problemen, Bescheiden Middelen.* Cahier 66, Studiecentrum voor Vredesvraagstukken, Nijmegen, 1995, pp. 125-156.

Bader, V. (1991), *Collectief handelen: sociale ongelijkheid en collectief handelen.* Deel 2. Wolters-Noordhoff, Groningen.

Bandyopadhyay, J. and V. Shiva (1989), 'Development, Poverty and the Green Movement in India', *The Ecologist,* Vol. 19, No. 3, 1989, pp. 11-17.

Bachrach, P. and M. Baratz (1962), 'Two Faces of Power', *American Political Science Review,* Vol. 56, 1962, pp. 947-952.

Bergesen, H. and G. Parmann, eds. (1993), *Green Globe Yearbook 1993. An independent publication on environment and development from the Fridtjof Nansen Institute, Norway.* Oxford University Press, Oxford.

Blowers, A. and P. Glasbergen, eds. (1996), *Environmental Policy in an International Context III. Prospects for Environmental Change.* Arnold, London.

Bodansky, D. (1993), 'The United Nations Framework Convention on Climate Change: A Commentary.' *Yale Journal of International Law,* Vol. 18, No. 451, 1993, pp. 453-559.

Bodansky, D. (1994), 'Prologue to the Climate Change Convention'. In: I.M. Mintzer and J.A. Leonard (eds.): *Negotiating Climate Change. The Inside Story of the Rio Convention.* Cambridge University Press, Cambridge, 1994, pp. 45-76.

Boehmer-Christiansen, S. (1994), 'Scientific Uncertainty and Power Politics: The Framework Convention on Climate Change and the Role of Scientific Advice'. In: B. Spector,

G. Sjöstedt and I. Zartman: *Negotiating International Regimes. Lessons Learned from the United Nations Conference on Environment and Development (UNCED).* Graham & Trotham/Martinus Nijhoff, London, 1994, pp. 171-180.

Boersema, J.J., J.W. Copius Peereboom and W.T. de Groot, eds., (1991), *Basisboek Milieukunde*, 4th ed. Boom, Meppel.

Bos, J.M. and P.A.T.M. Geurts (1994), 'Procesmethode.' In: L.W.J.C. Huberts and J. Kleinnijenhuis, *Methoden van invloedsanalyse.* Boom, Amsterdam/Meppel, 1994, pp. 61-82.

Boulding, E. (1988), *Building a Global Civic Culture. Education for an Interdependent World.* Syracuse University Press, Syracuse.

Bouwer, K. (1985), 'Grensoverschrijdende Milieuproblemen in een Geografisch Kader', *Geografisch Tijdschrift*, Vol. 19, No. 15, pp. 385-398.

Bouwer, K. (1990), 'Interdependentie van economie, milieuproblemen en milieubeleid'. In: B. Arts (ed.), *Ontwikkeling, Milieu en Veiligheid – Mondiale Interdependentie.* Studiecentrum voor Vredesvraagstukken, Nijmegen, 1990, pp. 33-44.

Bouwer, K. and P. Leroy, eds. (1995), *Milieu en ruimte – analyse en beleid.* Boom, Meppel.

Braam, G. (1973), *Invloed van bedrijven op de overheid. Een empirische studie over de verdeling van maatschappelijke invloed.* Boom, Meppel.

Bramble, B. J. and G. Porter (1992), 'Non-Governmental Organizations and the Making of US International Environmental Policy.' In: A. Hurrell and B. Kingsbury (eds.), *The International Politics of the Environment. Actors, Interests and Institutions.* Clarendon Press, Oxford, 1992, pp. 313-353.

Bray, A. J. (1991), 'Remembering the Scare of Global Cooling', *Policy Review*, Fall 1991, pp. 82-84.

Breyman, S. (1993), 'Knowledge as Power: Ecology Movements and Global Environmental problems'. In: R. Lipschutz and K. Conca (eds.), *The State and Social Power in Global Environmental Politics.* Columbia University Press, New York, 1993, pp. 124-157.

Brouns, M. (1993), *De homo economicus als winkeldochter. Theorieën over arbeid, macht en sekse.* SUA, Amsterdam.

Bruinsma, G. and M. Zwanenburg, eds. (1992), *Methodologie voor bestuurskundigen. Stromingen en methoden.* Coutinho, Muiderberg.

Burhenne-Guilmin, F. and S. Casey-Lefkowitz (1992), 'The New Law of Biodiversity.' In: G. Handl (ed.), *Yearbook of International Environmental Law.* Graham & Trotman/ Martinus Nijhof, London, 1992, pp. 43-59.

Burke, T. (1982), 'Friends of the Earth and the Conservation of Resources'. In: P. Willets (ed.), *Pressure Groups in the Global System.* Frances Pinter, London, 1982, pp. 105-124.

Buzan, B., C. Jones and R. Little (1993), *The Logic of Anarchy. From Neorealism to Structural Realism.* Columbia University Press, New York.

Caldwell, L.K. (1988), 'Beyond Environmental Diplomacy: The Changing Institutional Structure of International Cooperation'. In: Caroll J.E. (ed.) *International Environmental Diplomacy.* Cambridge University Press, Cambridge, 1988, pp. 13-28.

Caldwell, L.K. (1990), *International Environmental Policy. Emergence and Dimensions.* Duke University Press, Durham and London.

Carroll, J.E., ed. (1988), *International Environmental Diplomacy.* Cambridge University Press, Cambridge.

Chatterjee, P. and M. Finger (1994), *The Earth Brokers. Power, Politics and World Development.* Routledge, London.

Clark, M. L. (1994), 'The Antarctic Environmental Protocol: NGOs in the protection of Antarctica'. In: T. Princen and M. Finger (eds.), *Environmental NGOs in World Politics. Linking the global and the local.* Routledge, London, 1994, pp. 160-185.

Clegg, S. R. (1989), *Frameworks of power.* SAGE, London.

Conca, K. (1993), 'Environmental Change and the Deep Structure of World Politics'. In: R. Lipschutz and K. Conca (eds.): *The State and Social Power in Global Environmental Politics.* Columbia University Press, New York, 1993, pp. 306-326.

Cox, R. and H. Jacobson, eds. (1973), *The Anatomy of Influence.* Yale University Press, New Haven.

Cox, R. (1989), 'Social Forces, States and World Orders: Beyond International Theory', *Millennium*, Vol. 10, No. 2, 1981, pp. 126-155.

Cramer, J. (1989), *De groene golf: geschiedenis en toekomst van de milieubeweging.* Jan van Arkel, Utrecht.

Crucible Group (1994), *People, Plants, and Patents. The impact of intellectual property rights on biodiversity, conservation, trade, and rural society.* International Development Research Centre, Ottawa.

Dahl, R. A. (1961), *Who Governs? Democracy and Power in an American City.* Yale University Press, New Haven.

Dankelman, I., P. Nijhof and J. Westerman (1981), *Bewaar de Aarde. In het Perspectief van de World Conservation Strategy.* Meulenhoff, Amsterdam.

Dawkins, K. (1991), *Sharing Rights and Responsibilities for the Environment: Assessing potential roles for non-governmental organizations in international decision-making.* Massachusetts Institute of Technology, Boston.

Dessler, D. (1989), 'What's at stake in the agent-structure debate?', *International Organization*, Vol. 43, No. 3, Summer 1989, pp. 441-473.

De Wilde, J. (1991), *Saved from Oblivion: Interdependence Theory in the First Half of the 20th Century.* Dartmouth, Aldershot.

Doherthy, A. (1994), 'The Role of Non-Governmental Organizations in UNCED'. In: B. Spector, G. Sjöstedt and I. Zartman (eds.), *Negotiating International Regimes. Lessons learned from the United Nations Conference on Environment and Development (UNCED).* Graham & Trotham/Martinus Nijhoff, London, 1994, pp. 199-220.

Dryzek, J.S., M.L. Clark, and G. McKenzie (1989), 'Subject and system in international interaction', *International Organization*, Vol. 43, No. 3, Summer 1989, pp. 475-504.

Duyvendak, J., H. van der Heijden, R. Koopmans and L. Wijmans (1992), *Tussen verbeelding en macht: 25 jaar nieuwe sociale bewegingen in Nederland.* SUA, Amsterdam.

Enge, E. and R.I. Malkenes (1993), 'Non-governmental Organizations at UNCED: Another Successful Failure?'. In: H. Bergesen and G. Parmann (eds.), *Green Globe Yearbook 1993.* Oxford University Press, Oxford, 1993, pp. 25-34.

Everts, P.P. (1985), *Controversies at Home: Domestic factors in the Foreign Policy of the Netherlands.* Martinus Nijhoff, Dordrecht.

Feld, W. and R. Jordan (1983), *International Organizations. A Comparative Approach.* Praeger, New York.

Felling, A.J.A. and H.J.M. Hüttner (1981), 'De analyse van sociale netwerken'. In: M. Albinski (ed.), *Onderzoekstypen in de sociologie.* Van Gorcum, Assen, 1981, pp. 244-261.

Finger, M. (1994), 'Environmental NGOs in the UNCED process.' In: T. Princen and M. Finger, *Environmental NGOs in world politics. Linking the local to the global.* Routledge, London, 1994, pp. 48-68.

Fowler, C. (1993), 'Biological Diversity in a North-South Context'. In: H. Bergesen and G. Parmann (eds), *Green Globe Yearbook 1993*. Oxford University Press, Oxford, 1993, pp. 35-44.

French, H.F. (1994), 'Making Environmental Treaties Work', *Scientific American*, December 1994, pp. 94-97.

Friberg, M. and B. Hettne (1988), 'Local Mobilization and World System Politics', *International Social Science Journal*, 1988, No. 117, pp. 341-360.

Frouws, J. (1993), *Mest en Macht. Een politiek-sociologische studie naar belangenbehartiging en beleidsvorming inzake de mestproblematiek in Nederland vanaf 1970*. PhD Thesis, Landbouw Universiteit Wageningen.

Giddens, A. (1984), *The Constitution of Society. Outline of the theory of structuration*. Polity Press, Cambridge.

Glasbergen, P. and A. Blowers, eds. (1995), *Environmental Policy in an International Context I. Perspectives on Environmental Problems*. Arnold, London.

Glowka, L., F. Burhenne-Guilmin and H. Synge (1994), *A Guide to the Convention on Biological Diversity*. IUCN, Gland.

Godfroy, A. (1981), *Netwerken van Organisaties. Strategieën, spelen, structuren*. VUGA, Amsterdam.

Godfroy, A. and N. Nelissen, eds. (1993), *Verschuivingen in de besturing van de samenleving*. Coutinho, Bussum.

Goldberg, D.M. (1993), 'The Montreal Protocol Multilateral Fund. A Model for the Framework Convention on Climate Change', *International Environment Reporter*, 24 March 1993, pp. 233-237.

Grieco, J.M. (1995), 'Anarchy and the limits of Cooperation: A Realist Critique of the Newest Liberal Institutionalism'. In: C.W. Kegley, *Controversies in International Relations Theory. Realism and the Neoliberal Challenge*. St. Martin's Press, New York, 1995, pp. 151-171.

Gupta, J., G. Junne and R. van der Wurff (1993), *Determinants of Regime Formation*. Institute for Environmental Studies, Free University of Amsterdam.

Guzzini, S. (1993), 'Structural power: the limits of neorealist power analysis', *International Organization*, Vol. 47, No. 3, Summer 1993, pp. 443-478.

Haas, P. M. (1993), 'Epistemic Communities and the Dynamics of International Environmental Co-Operation'. In: V. Rittberger (ed.), *Regime Theory and International Relations*. Clarendon Press, Oxford, 1993, pp. 168-201.

Haggard, A. and B.A. Simmons (1987), 'Theories of international regimes', *International Organization*, Vol. 41, No. 3, 1987, pp. 491-517.

Haufler, V. (1993), 'Crossing the Boundary between Public and Private: International Regimes and Non-State Actors'. In: V. Rittberger (ed.), *Regime Theory and International Relations*. Clarendon Press, Oxford, pp. 94-111.

Hayes, P. and K. Smith (1993), *The global greenhouse regime. Who pays?* Earthscan, London.

Hecht, L. (1993)., 'The Coming (or Present) Ice Age', *21st Century*, Vol. 6, No. 4, Winter 1993-1994, pp. 22-35., Heijnsbergen, P. (1991), 'Biodiversity and international law', *Internationale Spectator*, Vol. 45, No. 11, November 1991, p. 681-685.

Heins, V. (1993), 'Survival of the Fattest? Genetische Resourcen und globale Biopolitik', *Peripherie*, Vol. 13, No. 51/52, December 1993, pp. 69-85.

Hocking, B. and M. Smith (1990), *World Politics. An introduction to international relations*. Harvester/Wheatsheaf, New York.

Hoering, U. (1985), 'Bhopal und keine Ende, oder: der Second-hand-Kapitalismus und die Ökologie'. *Peripherie,* Vol. 5, No. 21, pp. 53-59.

Hontelez, J.J.A.M. (1993), 'Internationaal milieubeleid en de milieubeweging'. In: G. Spaargaren *et al.* (eds.), *Internationaal Milieubeleid.* SDU, Den Haag, 1993, pp. 171-193..

Hooghe, M. (1995), 'De invloed van de milieubeweging op het milieubeleid. Neo-corporatisme of culturele vernieuwing?' *Res Publica,* 1995, Vol. 37, No. 1, pp. 93-108.

Huberts, L. and W. van Noort (1986), 'Invloedsmeting: een oplosbaar probleem?', *Sociologische Gids,* No. 1, 1986, pp. 45-53.

Huberts, L. (1988), *De politieke invloed van protest en pressie. Besluitvormingsprocessen over rijkswegen.* DSWO Press, Leiden.

Huberts, L. and W. van Noort (eds.) (1989), *Sociale bewegingen in de jaren negentig. Stand van zaken and vooruitblik.* DSWO Press, Leiden.

Huberts, L. (1989), 'The influence of social movements on government policy', *International Social Movement Research,* Vol. 2, 1989, pp. 395-426.

Huberts, L.W.J.C. (1994), 'Intensieve procesanalyse'. In: L.W.J.C. Huberts and J. Kleinnijenhuis, *Methoden van invloedsanalyse.* Boom, Amsterdam/Meppel, 1994, pp. 38-60.

Huberts, L.W.J.C. and J. Kleinnijenhuis (1994), *Methoden van invloedsanalyse.* Boom, Amsterdam/Meppel.

Hufen, J. and A. Ringeling (1990), *Beleidsnetwerken. Overheids-, semi-overheids- en particuliere organisaties in wisselwerking.* VUGA, 's-Gravenhage.

Huizinga, F. and B.J. Verbeek (1995), *Reconstructing the Logic of Appropriateness. Reflections on Regime Theory.* Unpublished Paper, Free University Amsterdam.

Humphreys, D. (1994), *The Global Politics of Forest Conservation, 1983-1994.* PhD Thesis, City University London.

Hurrell, A. (1993), 'International Society and the Study of Regimes'. In: V. Rittberger (ed.), *Regime Theory and International Relations.* Clarendon Press, Oxford, 1993, pp. 49-72.

Hurrell, A. and B. Kingsbury, eds. (1992), *The International Politics of the Environment. Actors, Interests, and Institutions.* Clarendon Press, Oxford.

Hutjes, J.M. and J.A. van Buuren (1992), *De gevalstudie. Strategie van kwalitatief onderzoek.* Boom, Meppel.

Jacobson, H. (1984), *Networks of Interdependence. International organizations and the global political system.* Alfred A. Knopf, New York.

Jamison, A. (1996), 'The Shaping of the Global Environmental Agenda: The Role of Non-Governmental Organisations'. In: S. Lash, B. Szerszynski and B. Wynne (eds.), *Risk, Environment & Modernity.* Sage, London, 1996, pp. 224-245.

Johnson, S. (1993), *The Earth Summit: The United Nations Conference on Environment and Development.* Graham & Trotman/Martinus Nijhof, London.

Johnson, M. and B. Maiguashca, eds. (1994), 'Special Issue: Social Movements and World Politics', *Millennium,* Vol. 23, No. 3, Winter 1994.

Jönsson, C., A. Young Kronsell and P. Söderholm (1995), *Prelude to Regimes: International Organizations and Agenda Setting.* Paper to the Second Pan-European Conference in International Relations, Paris.

Juma, C. (1989), *The Gene Hunters. Biotechnology and the Scramble for Seeds.* Zed Books, London.

Junne, G. (1992), 'Beyond Regime Theory', *Acta Politica,* Vol. 27, No. 1, 1992, pp. 9-28.

Junne, G. (1993), 'Transnational Environmental Policies'. In: Bob Reinalda (ed.), *Internationalisering van Beleid. Grenzen en Perspectieven.* KU Nijmegen, 1993, pp. 20-32.

Kaelberer, M. (1992), *State Power and International Environmental Cooperation: A Comparison of Collaborative Efforts on Global Warming and Ozone Layer Depletion.* Princeton University, Department of Politics, Paper ISA Convention, Atlanta.

Käkönen, J., ed. (1994), *Green Security or Militarized Environment.* Dartmouth, Aldershot.

Kamieniecki, S., ed. (1993), *Environmental Politics in the International Arena. Movements, Parties, Organizations, and Policy.* University of New York Press, Albany.

Kamminga, M (1995), 'Principles of International Law'. In: P. Glasbergen and A. Blowers (eds.), *Environmental Policy in an International Context I. Perspectives on Environmental Problems.* Arnold, London, 1995, pp. 111-132.

Karlen, W., E. Friis-Christensen and B. Dahlström (1993), *The Earth's Climate – Natural variations and human influence.* Elforsk, Stockholm.

Kaufmann, J. (1980), *United Nations Decision Making.* Sijthof & Noordhof, Alphen aan den Rijn.

Keohane, R.O. (1982), 'The demand for international regimes', *International Organization,* Vol. 36, No. 2, Spring 1982, pp. 325-356., Keohane, R.O. (1984), *After Hegemony. Cooperation and Discord in the World Political Economy.* Princeton University Press, Princeton.

Keohane, R.O., ed. (1986), *Neorealism and its critics.* Columbia University Press, New York.

Keohane, R.O. (1993), 'The Analysis of International Regimes: Towards an European-American Research Programme'. In: V. Rittberger (ed.), *Regime Theory and International Relations.* Clarendon Press, Oxford, 1993, pp. 23-48.

Keohane, R.O. and J.S Nye, eds. (1971), *Transnational Relations and World Politics.* Harvard University Press, Cambridge.

Keohane, R.O. and J.S. Nye (1989), *Power and Interdependence. World Politics in Transition,* 2nd ed. Scott, Foresman & Company, Glenview.

Klotz, A. (1995), 'Norms reconstituting interests: global racial equality and U.S. sanctions against South Africa', *International Organization,* Vol. 49, No. 3, Summer 1995, pp. 451-478.

Koch, K., R. Soetenendorp and A. van Staden, eds. (1994), *Internationale Betrekkingen. Theorieën en Benaderingen.* Het Spectrum, Utrecht.

Kolk, A. (1996), *Forests in International Environmental Politics. International Organisations, NGOs and the Brazilian Amazon.* Van Arkel/International Books, Utrecht.

Koppenjan, J. (1993), *Management van de beleidsvorming. Een studie naar de totstandkoming van beleid op het terrein van het binnenlands bestuur.* VUGA, 's-Gravenhage.

Krasner, S. D. (1982), 'Structural causes and regime consequences: regimes as intervening variables', *International Organization,* Vol. 36, No. 2, Spring 1982, pp. 185-206.

Krasner, S.D. (1985), *Structural Conflict. The Third World Against Global Liberalism.* University of California Press, Berkeley.

Kratochwil, F. and J.G. Ruggie (1986), 'International organization: a state of the art on an art of the state', *International Organization,* Vol. 40, No. 4, Autumn 1986, pp. 762-775.

Kuypers, G. (1973):, *Grondbegrippen van politiek.* Het Spectrum, Utrecht/Antwerpen.

Lafferière, E. (1994), 'Environmentalism and the Global Divide', *Environmental Politics,* Vol. 3, No. 1, Spring 1994, pp. 91-113.

Liberatore, A. (1995), 'The social construction of environmental problems'. In: P. Glasbergen and A. Blowers (eds.), *Environmental Policy in an International Context I. Perspectives on Environmental Problems.* Arnold, London, 1995, pp. 59-84.

Liefferink, J.D. (1995), *Environmental Policy on the Way to Brussels.* PhD Thesis Landbouw Universiteit Wageningen.

Liefferink, J.D., P.D. Lowe and A.P.J. Mol (1993), *European Integration and Environmental Policy.* Belhaven Press, London.

Lieshout, R. (1993), *Anarchie en Hierarchie. Een theorie van internationale betrekkingen en buitenlandse politiek.* Coutinho, Bussum.

Lipschutz, R. and K. Conca, eds. (1993), *The State and Social Power in Global Environmental Politics.* Columbia University Press, New York.

List, M. and V. Rittberger (1992), 'Regime Theory and International Environmental Management'. In: A. Hurrell and B. Kingsbury (eds.), *The International Politics of the Environment. Actors, Interests, and Institutions.* Clarendon Press, Oxford, 1992, pp. 85-110.

Lowe, P. (1983), *Environmental Groups in Politics.* Allen & Unwin, London.

Lukes, S. (1974), *Power. A Radical View.* Macmillan, London.

Manno, J.P. (1994), 'Advocacy and diplomacy: NGOs and the Great Lakes Water Quality Agreement'. In: T. Princen and M. Finger (eds.), *Environmental NGOs in World Politics. Linking the global and the local.* Routledge, London, 1994, pp. 69-120.

Martens, J. (1993), 'Dabeisein ist noch nicht alles. Die NGOs in den Vereinten Nationen: Akteure, Kritiker, Nutzniesser', *Vereinte Nationen,* No. 5, 1993, pp. 168-171.

Mazey, S. and J. Richardson (1992), 'Environmental Groups and the European Community: Challenges and Opportunities', *Environmental Politics,* Vol. 1, No. 4, Winter 1992, pp. 109-128.

McCormick, J. (1989), *The Global Environmental Movement.* Belhaven Press, London.

McCoy, M. and P. McCully (1993), *The Road from Rio. An NGO Guide to Environment and Development.* WISE/International Books, Amsterdam.

Melucci, A. (1988), 'Social Movements and the Democratization of Everyday Life'. In: J. Keane (ed.), *Civil Society and the State.* Verso, London/New York, 1988, pp. 245-260.

Metz, B. and P. Vellinga (1992), 'Rio: flop of investering?', *Internationale Spectator,* Vol. 45, No. 10, 1992, pp. 585-589.

Miller, M.A.L. (1995), *The Third World in Global Environmental Politics.* Open University Press, Buckingham.

Mintzer, I.M. and J.A. Leonard, eds. (1994), *Negotiating Climate Change. The inside story of the Rio convention.* Cambridge University Press, Cambridge.

Morgenthau, H. J. (1993), *Politics among Nations. The struggle for power and peace,* 6th ed. McGraw-Hill, New York.

Morss, E. (1991), 'The New Global Players: How They Compete and Collaborate', *World Development,* Vol. 19, No. 1, 1991, pp. 55-64.

Munters, Q., E. Meijer, H. Mommaas, H. van der Poel and G. Spaargaren (1991), *Anthony Giddens. Een kennismaking met de structuratietheorie.* Wageningse Sociologische Studies, Wageningen.

Nayar, B.R. (1995), 'Regimes, Power, and International Aviation', *International Organization,* Vol. 49, No. 1, Winter 1995, pp. 139-170.

Nelissen, N.J.M. (1992), *Besturen binnen verschuivende grenzen.* Kerckebosch, Zeist.

Nelissen, N., J. van der Straaten and L. Klinkers, eds. (1997), *Classics in Environmental Studies. An Overview of Classic Texts in Environmental Studies.* Van Arkel/International Books, Utrecht.

Niou, E. and P. Ordeshook (1994), 'Less Filling, Tastes Great. The Realist-Neoliberal Debate', *World Politics,* Vol. 46, No. 2, January 1994, pp. 209-234.

Nollkeamper, A. (1992), 'On the Effectiveness of International Rules', *Acta Politica,* Vol. 27, No. 1, 1992, pp. 49-70.

Nowicki, N. (1994), 'The Role of Non-Governmental Organizations'. In: Graham Bennett (ed.), *Conserving Europe's Natural heritage. Towards an European Ecological Network.* Graham & Trotman/Martinus Nijhoff, London, 1994, pp. 211-215.

Odum, E.P. (1971), *Fundamentals of Ecology,* 3rd ed. Saunders, Philadelphia.

Opschoor, H. (1989a), *Na ons geen zondvloed. Voorwaarden voor duurzaam milieugebruik.* Kok Agora, Kampen.

Opschoor, H. (1989b), 'Het nieuwe denken in het Kremlin en milieuveiligheid', *Transaktie,* Vol. 18, No. 4, 1989, pp. 347-356.

O'Riordan, T. and J. Jäger (1996), *Politics of Climate Change. A European Perspective.* Routledge, London.

Paterson, M. (1996), 'IR theory: neorealism, neoinstitutionalism and the Climate Change Convention'. In: J. Vogler and M. Imber (eds.), *The Environment and International Relations.* Routledge, London, 1996, pp. 59-76.

Pereira, W. (1993), *Tending the Earth. Traditional, Sustainable Agriculture in India.* Earth Care Books, Bombay.

Pistorius, R. and J. van Wijk (1997), 'Kleine boeren en de internationale strijd om genetisch materiaal', *Derde Wereld,* Vol. 15, No. 3-4, 1997, pp. 241-246.

Posey, D. (1994), 'International agreements for protecting indigenous knowledge'. In: V. Sanchez and C. Juma (eds.), *Biodiplomacy. Genetic Resources and International Relations.* ACTS Press, Nairobi, 1994, pp. 119-138.

Potter, D. (1996), 'Non-Governmental Organizations and Environmental Policies'. In: A. Blowers and P. Glasbergen (eds.), *Environmental Policy in an International Context III. Prospects for Environmental Change.* Arnold, London, 1996, pp. 25-50.

Powell, R. (1994), 'Anarchy in international relations theory: the neorealist-neoliberal debate', *International Organization,* Vol. 48, No. 2, Spring 1994, pp. 313-44.

Princen, T. (1994), 'The ivory trade ban: NGOs and international conservation'. In: T. Princen and M. Finger, *Environmental NGOs in world politics. Linking the local to the global.* Routledge, London, 1994, pp. 121-159.

Princen, T. and M. Finger (1994), *Environmental NGOs in world politics. Linking the local to the global.* Routledge, London.

Puchala, D.J. and R.F. Hopkins (1982), 'International regimes: lessons from inductive analysis', *International Organization,* Vol. 36, No. 2, Spring 1982, 245-276.

Rahman, A. and A. Roncerel (1994), 'A View from the Ground Up'. In: I.M. Mintzer and J.A. Leonard (eds.), *Negotiating Climate Change. The inside story of the Rio convention.* Cambridge University Press, Cambridge, 1994, pp. 239-276.

Ray, L.R. (1987), *Global Politics,* 3th ed. Houghton Mifflin, Boston.

Redclift, M. (1984), *Development and the Environmental Crisis: Red or Green Alternatives.* Methuen, London.

Redclift, M. (1987), *Sustainable Development: Exploring the Contradictions.* Methuen, London.

Regal, P.J. (1994), 'Scientific principles for ecologically based risk assessment of transgenic organisms', *Molecular Ecology,* No. 3, 1994, pp. 5-13.

Reinalda, B. (1997), *Private in Form and Public in Purpose: (I)NGOs as Political Actors in World Politics.* Paper for the ECPR 25th Joint Sessions of Workshops, Bern, Switzerland, March 1997.

Risse-Kappen, T., ed. (1995), *Bringing Transnational Relations Back In. Non-State Actors, Domestic Structures and International Institutions.* Cambridge University Press, Cambridge.

Rittberger, V., ed. (1993), *Regime Theory and International Relations.* Clarendon Press, Oxford.

Rood, J. (1987), 'Internationale regimes', in: R. Soetendorp and A. van Staden (eds.), *Internationale Betrekkingen in Perspectief.* Het Spectrum, Utrecht, pp. 213-242.

Rood, J. (1994), 'Neoliberaal institutionalisme: internationale samenwerking en regimes'. In: K. Koch, R. Soetenendorp and A. van Staden (eds.), *Internationale betrekkingen. Theorieën en benaderingen.* Het Spectrum, Utrecht, 1994, pp. 177-202.

Rosenau, J. N. (1993), 'Environmental Challenges in a Turbulent World'. In: R. Lipschutz and K. Conca (eds.), *The State and Social Power in Global Environmental Politics.* Columbia University Press, New York, pp. 71-92.

Rosenthal, U. and P. 't Hart (1994), 'Het één en het ander: case-contaminatie en andere methodologische implicaties in beleidswetenschappelijk onderzoek', *Beleidswetenschap,* No. 8, 1994, pp. 141-163.

Rowlands, I.H. (1995), *The politics of global atmospheric change.* Manchester University Press, Manchester.

Rowlands, I. and M. Greene, eds. (1992), *Global Environmental Change and International Relations.* Macmillan, London.

Rozendaal, S. (1997), 'De Broeikas-Mythe', *Elsevier,* Vol. 53, No. 13, 29 March 1997, pp. 96-101.

Sabatier, P.A. (1987), 'Knowledge, Policy-Oriented Learning, and Policy Change', *Knowledge: Creation, Diffusion, Utilization,* Vol. 8, No. 4, June 1987, 649-692.

Sanchez, V. and C. Juma, eds. (1994), *Biodiplomacy. Genetic Resources and International Relations.* ACTS Press, Nairobi.

Schermers, H.G. (1985a), *Internationaal Publiekrecht voor de Rechtspraktijk,* 3th ed. Kluwer, Deventer.

Schermers, H.G. (1985b), *Inleiding tot het Internationale Institutionele Recht.* 2nd ed. Kluwer, Deventer.

Schmitter, P.C. and W. Streeck (1991), 'Belangenorganisaties en het Europa van 1992', *Beleid & Maatschappij,* No. 6, 1991, pp. 261-278.

Schneider, S. H. (1989), 'The Changing Climate', *Scientific America,* September 1989, pp. 38-47.

Segers, J. and J. Hagenaars, eds. (1980), *Sociologische onderzoeksmethoden. Deel* II – Technieken van causale analyse. Van Gorcum, Assen.

Self, P. (1985), *Political Theories of Modern Government.* Allen & Unwin, London.

Simms, A. (1993), 'If Not Then, When? Non-Governmental Organizations and the Earth Summit Process', *Environmental Politics,* No. 1, 1993, pp. 94-100.

Soetendorp, R. and A. van Staden, eds. (1987), *Internationale betrekkingen in perspectief.* Het Spectrum, Utrecht.

Spaargaren, G., J.D. Liefferink, A.P.J. Mol, W. Brussaard and W.J. Kakebeeke, eds. (1993), *Internationaal Milieubeleid,* 2de druk. SDU, Den Haag.

Spector, B., G. Sjöstedt and I. Zartman (1994), *Negotiating International Regimes. Lessons Learned from the United Nations Conference on Environment and Development (UNCED).* Graham & Trotham/Martinus Nijhoff, London.

Sprinz, D. and T. Vaahtoranto (1994), 'The interest-based explanation of international environmental policy', *International Organization,* Vol. 48, No. 1, Winter 1994, pp. 77-105.

Stairs, K. and P. Taylor (1992), 'Non-Governmental organizations and the Legal protection of the Oceans: A Case Study'. In: A. Hurrell and B. Kingsbury (eds.), *The International*

Politics of the Environment. Actors, Interests, and Institutions. Clarendon Press, Oxford, 1992, pp. 110-131.

Stokman, F.N. (1994), 'Besluitvormingsmodellen binnen beleidsnetwerken.' In: L.W.J.C. Huberts and J. Kleinnijenhuis, *Methoden van invloedsanalyse.* Boom, Amsterdam/ Meppel, 1994, pp. 165-187.

Strange, S. (1982), 'Cave! Hic Dragons. A critique of regime theory', *International Organization,* Vol. 36, No. 2, Spring 1982, pp. 479-496.

Strange, S. (1988), *States and Markets,* 2nd ed. Pinter, London.

Strong, M. (1992), 'The Passage from Rio', *Development Forum,* Vol. 20, No. 4, 1992, p. 3.

Svarstadt, H. (1994), 'National sovereignty and genetic resources'. In: V. Sanchez and C. Juma (eds.), *Biodiplomacy. Genetic Resources and International Relations.* ACTS Press, Nairobi, 1994, pp. 45-66.

Swanborn, P.G. (1992), 'Verschillende opvattingen over theoriegebruik'. In: G. Bruinsma and M. Zwanenburg (eds.), *Methodologie voor bestuurskundigen. Stromingen en methoden.* Coutinho, Muiderberg, 1992, pp. 157-175.

Swanborn, P.G. (1994), 'Het ontwerpen van case-studies: enkele keuzen', *Mens en Maatschappij,* No. 69, 1994, pp. 322-335.

Swanborn, P.G. (1996), *Case-study's. Wat, wanneer en hoe?* Boom, Amsterdam.

Tacq, J. (1982), 'Causaliteit in sociologisch onderzoek', *Sociologische Gids,* No. 1, 1982, pp. 4-39.

Taylor, A. (1993), *Setting Environmental Agendas: NGOs, Democracy and Global Politics. A Framework and Methodology for the Research.* GECOU, Open University, Milton Keynes.

Thomas, C. (1992), *The Environment in International Relations.* Royal Institute of International Affairs, London.

Thompson-Feraru, A. (1974), 'Transnational Political Interests and the Global Environment', *International Organization,* Vol. 28, No.1, 1974, pp. 31-60.

Tolba, M.K. and O.A. El-Kholy (1992), *The World Environment 1972-1992. Two decades of challenge.* Chapman and Hall, London.

Vakgroep Methoden van Onderzoek (1994), *Interviewen met een interview guide.* Reader, KU Nijmegen.

Van de Graaf, H. and R. Hoppe (1992), *Beleid en Politiek. Een inleiding tot de beleidswetenschap en beleidskunde.* Coutinho, Muiderberg.

Van den Abbeele, E., ed. (1985), *Ontmanteling van de groei. Leesboek over een andere economie.* Markant, Nijmegen.

Van der Loo, H., E. Snel and B. van Steenbergen (1984), *Een wenkend perspectief? Nieuwe sociale bewegingen en culturele veranderingen.* De Horstink, Amersfoort.

Van der Pijl, K. (1992), *Wereldorde en Machtspolitiek. Visies op de internationale betrekkingen van Dante tot Fukuyama.* Het Spinhuis, Amsterdam.

Van der Pijl, K. (1995), *Globalisation as a strategic concept.* Paper for the 'Nederlandse Politicologenetmaal', Soesterberg, 1-2 June 1995.

Van der Schot, J. and H. van de Veen (1997), 'De nieuwe macht van NGO's'. In: K. Waagmeester (ed.), *Houdbare Economie. Kroniek van Duurzaam Nederland.* NCDO/Kok Agora, Amsterdam, 1997, pp. 197-209.

Van der Wurff, R. (1997), *International Climate Politics – Interests and Perceptions. A comparative study on climate change politics in Germany, the United Kingdom and the United States.* PhD Thesis, University of Amsterdam.

Van Goor, H., J. Ellemers and G. Braam (1985), 'Een nieuwe methode om invloed te meten: positiewisselingen en invloedsprocessen', *Sociologische Gids*, No. 4, 1985, pp. 250-270.

Van Goor, H., J. Ellemers and G. Braam (1986), 'Strategieën in het onderzoek naar macht en invloed: een repliek', *Sociologische Gids*, No. 1, 1986, pp. 54-60.

Van Noort, W., L. Huberts and L. Rademakers (1987), *Protest en Pressie. Een systematische analyse van collectieve actie.* Van Gorcum, Assen.

Van Noort, W. (1988), *Bevlogen bewegingen. Een vergelijking van de anti-kernenergie-, kraak- en milieubeweging.* SUA, Amsterdam.

Van Schendelen, M.P.C.M. (1981), 'Politieke invloed'. In: M.P.C.M. van Schendelen (ed.), *Kernthema's van de politicologie.* Boom, Meppel, 1981, pp. 11-138.

Van Tatenhove, J. (1993), *Milieubeleid onder dak? Beleidsvoeringsprocessen in het Nederlandse milieubeleid in de periode 1970-1990; nader uitgewerkt voor de Gelderse Vallei.* PhD Thesis, Landbouwuniversiteit Wageningen.

Van Tatenhove, J. and P. Leroy (1995), 'Beleidsnetwerken: Een kritische analyse', *Beleidswetenschap*, Vol. 9, No. 2, 1995, pp. 128-145.

Van Waarden, F. (1992), 'Dimensions and types of policy networks', *European Journal of Political Research*, Vol. 21, No. 1-2, February 1992, pp. 29-52.

Verschuren, P. (1993), *Research designing.* KU Nijmegen.

Verschuren, P. (1994), *Toegepast Evaluatie-Onderzoek. Mogelijkheden en kansen voor een kwalitatieve benadering.* KU Nijmegen.

Verschuren, P. and H. Doorewaard (1995), *Het ontwerpen van een onderzoek.* Lemma, Utrecht.

Vogler, J. and M. Imber, eds. (1996), *The Environment and International Relations.* Routledge, London.

Volkert, E.H., ed. (1951), *Social Behavior and Personality. Contributions of W.I. Thomas to Theory and Social Research.* Social Science Research Council, New York.

Von Moltke, K. (1995), 'Book Review of: B. Spector, G. Sjöstedt and I. Zartman, "Negotiating International Regimes. Lessons Learned from the United Nations Conference on Environment and Development (UNCED)." Graham & Trotham/Martinus Nijhoff, London, 1994', *International Environmental Affairs*, Vol. 2, No. 3, Summer 1995, pp. 289-290.

VROM (1993), *VN Conferentie inzake Milieu en Ontwikkeling. Verklaring van Rio, Agenda 21, Bossenverklaring, Biodiversiteitsverdrag en Klimaatverdrag.* VROM/DGIS, Den Haag.

Waltz, K.N. (1979), *Theory of International Politics.* McCraw-Hill, New York.

Wapner, P. (1995), 'Environmental Activism and World Civic Politics', *World Politics*, Vol. 47, 1995, pp. 311-340.

Waters, M. (1995), *Globalization.* Routledge, London.

WCED/World Commission on Environment and Development (1987), *Our Common future.* Oxford University Press, Oxford.

Weale, A. (1992), *The new politics of pollution.* Manchester University Press, Manchester., Weiss, T.G., D.P. Fortsythe and R.A. Coate (1994), *The United Nations and Changing World Politics.* Westview Press, Boulder.

Wendt, A.E. (1987), 'The agent-structure problem in international relation theory', *International Organization*, Vol. 41, No. 3, Summer 1987, pp. 335-370.

Wendt, A.E. (1992), 'Anarchy is what states make of it: the social construction of power politics', *International Organization*, Vol. 46, No. 2, Spring 1992, pp. 391-425.

Wester, F. (1991), *Strategieën van kwalitatief onderzoek.* Coutinho, Muidenberg.

Westerheijden, D. (1987), 'The substance of a shadow. A critique of power measurement methods', *Acta Politica*, No. 1, 1987, pp. 39-59.

Willets, P., ed. (1982), *Pressure Groups in the Global System.* Frances Pinter, London.

Willets, P. (1989), 'Interdependence: new wines in old bottles'. In: J. Rosenau and H. Tromp (eds.), *Interdependence and Conflict in World Politics.* Gower, Aldershot, 1989, pp. 195-210.

Willets, P. (1996), 'From Stockholm to Rio and beyond: the impact of the environmental movement on the United Nations consultative arrangements for NGOs', *Review of International Studies*, Vol. 22, 1996, pp. 57-80.

Williams, M. (1993), 'Re-articulating the Third World Coalition: the role of the environmental agenda', *Third World Quarterly*, Vol. 14, No. 1, 1993, pp. 7-29.

Williamson, P.J. (1989), *Corporatism in Perspective. An Introductionary Guide to Corporatist Theory.* SAGE, London.

Wilson, E.O. (1989), 'Threats to Biodiversity', *Scientific American*, September 1989, pp. 60-66.

Wilson, G. (1981), *Interest Groups in the United States.* Clarendon Press, Oxford.

Witmeyer, D. (1994), 'The Convention on Biological Diversity Changes Rules on the Game for International Plant Genetic Resources Regime', *Diversity*, Vol. 10, No. 3, 1994, pp. 28-31.

Woods, L. (1993), 'Nongovernmental Organizations and the United Nations System: Reflection upon the Earth Summit', *International Studies Notes*, No. 1, 1993, pp. 9-15.

Yin, R. (1991), *Case Study Research. Design and Methods,* 7th ed. SAGE, London.

Young, O.R. (1975), *Bargaining. Formal Theories of Negotiation.* University of Illinois Press, Urbana.

Young, O.R. (1980), 'International Regimes: Problems of Concept Formation', *World Politics*, Vol. 32, No. 3, April 1980, pp. 331-356.

Young, O.R. (1982), 'Regime Dynamics: the rise and fall of international regimes', *International Organization*, Vol. 36, No. 2, Spring 1982, pp. 277-298.

Young, O.R. (1993), 'International Organizations and International Institutions: Lessons Learned from Environmental Regimes'. In: S. Kamieniecki (ed.), *Environmental Politics in the International Arena. Movements, Parties, Organizations, and Policy.* University of New York Press, Albany, 1993, pp. 145-165.

Young, O.R. and G. Osherenko (1993), 'Testing Theories of Regime Formation: Finding from a Large Collaborative Research Project'. In: V. Rittberger (ed.), *Regime Theory and International Relations.* Clarendon Press, Oxford, 1993, pp. 223-251.

Zald, M. and J. McCarthy (1980), 'Social Movement Industries: Competition and Cooperation Among Movement Organizations'. In: L. Kriesberg (ed.), *Research in Social Movements, Conflicts and Change. A Research Annual.* Vol. 3, 1980, pp. 1-20.

APPENDIX I

FCCC Documents

Most relevant government and non-government policy documents and editions of journals

— Government documents —

1 EU/General Secretariat of the Council, *Community Strategy on Climate Change: Preparation of the First Conference of the Parties to the UN Framework Convention on Climate Change.* 10 March 1995.

2 FCCC/CP/1995/Misc.1, *Matters relating to Commitments. Review of the Adequacy of Article 4, Paragraph 2(a) and (b)*, 9 March 1995.

3 FCCC/CP/1995/1, *Provisional Agenda and Annotations, including suggestions for the organization of work*, 20 March 1995.

4 FCCC/CP/1995/Inf.3, *Matters relating to Commitments, Review of Information Communicated by Each Party Included in Annex I to the Convention, Status of submissions of first communications from Annex I Parties, Note by the interim secretariat.* 27 March, 1995.

5 FCCC/CP/1995/L.6, *Adoption of the Report of the Conference of the Parties. Draft Report of the Conference of the Parties on its First Session.* 5 April 1995.

6 FCCC/1995/Inf.5/Rev.2, *Directory of Participants.* 6 April 1995.

7 FCCC/CP/1995/L.13, *Conclusion of Outstanding Issues and Adoption of Decisions. Activities Implemented Jointly under the Pilot Phase.* 6 April 1995.

8 FCCC/CP/1995/L.14, *Conclusion of Outstanding Issues and Adoption of Decisions. Review of the Adequacy of Article 4, Paragraph 2(a) and (b) of the Convention, including proposals related to a protocol and decisions on follow-up.* 7 April 1995.

9 G-77 and China, *Position on Proposals from the Co-chairs of Working Group I. Matters relating to Commitments. Criteria for Joint Implementation. Draft Decision of the Committee.* February 15, 1995.

10 Germany, *Elements for a Comprehensive Protocol to the FCCC.* 26 September 1994.

11 Green Group, *Proposed Elements of a Mandate for Consultation on Commitments in Articles 4.2 (a) and (b).* 2 April 1995.

12 INC, A/AC.237/6, *Report of the Intergovernmental Negotiation Committee for a Framework Convention on Climate Change on the Work of the First Session, held at Washington,* DC. March 1991.

13 INC, A/AC.237/Misc.2, *Compilation of Possible Elements for a Framework Convention on Climate Change.* 13 June 1991.

14 INC, A/AC.237/9. *Report of the Intergovernmental Negotiation Committee for a Framework Convention on Climate Change on the Work of its Second Session, held at Geneva.* 19 August 1991.

15 INC, A/AC.237/WG.1/CRP.1/Rev.1, *Principles.* 18 September 1991.

16 INC, A/AC.237/WG.1/CRP.2 /Rev.1, *Commitments on Sources and Sink.* 19 September 1991.

17 INC, A/AC.237/15, *Report of the Intergovernmental Negotiation Committee for a Framework Convention on Climate Change on the Work of its Third Session, held at Geneva.* September 1991.

18 INC, A/AC.237/18 (Part 1), *Report of the Intergovernmental Negotiation Committee for a Framework Convention on Climate Change on the Work of the First Part of its Fifth Session, held at New York.* 10 March, 1992.

19 INC, A/AC.237/CRP.1/Add.2, *Completion of a Framework Convention on Climate Change. Working Papers by the Chairman.* 30 April 1992.

20 INC, A/AC.237/49, *Matters relating to Commitments. Criteria for Joint Implementation.* 17 January 1994.

21 INC, A/AC.237/81, *Matters relating to Commitments, First Review of Information Communicated by Each Party Included in Annex I to the Convention, Compilation and Synthesis of national Communications from Annex I Parties.* 7 December 1994.

22 INC, A/AC.237/91, *Report of the Intergovernmental Negotiation Committee for a Framework Convention on Climate Change on the Work of its Eleventh Session.* 2 March 1995.

23 INC, A/AC.237/91/Add. 1, *Part two: Recommendations to the Conference of the Parties and other decisions and conclusions of the Committee.* 8 March 1995.

24 IPCC, *Policymakers Summary of the Scientific Assessment of Climate Change. Report to IPCC from Working Group I.* June 1990.

25 IPCC, *Policymakers Summary of the Potential Impacts of Climate Change. Report from Working Group II to IPCC.* June 1990.

26 IPCC, *Policymakers Summary of the Formulation of Response Strategies. Report prepared for IPCC by Working Group III.* June 1990.

27 IPCC, *Report to the Eleventh Session of the Intergovernmental Negotiating Committee for a Framework Convention on Climate Change (INC/FCCC).* 6 February 1995.

28 IPCC, Statement to the First Session of the Conference of the Parties to the *UN Framework Convention on Climate Change.* 28 March 28 1995.

29 *Ministerial Declaration of the Second World Climate Conference.* 7 November 1990.

30 Ministry of Environment, Physical Planning and Housing, *Reports of the Dutch Delegation to: INC1, INC2, INC3, INC4, INC5, INC6, INC7, INC8, INC9, INC10, INC11, and COP1.* 1991-1995.

31 *Noordwijk Declaration on Climate Change.* 7 November 1989.

32 Norway, *Proposal on a Clearing House Mechanism.* Undated.

33 OECD, *Positions Taken by Delegations from Various Countries at the first INC Meeting.* 1991.

34 OECD, *Comments on the Paper Produced by a Group of Developing Countries.* 3 April 1995.

35 *Second World Climate Conference. Conference Statement.* Geneva, 29 October - 7 November 1990.

36 The White House, *Remarks by the Vice President.* George Washington University, 17 March 1995.

37 Trinidad and Tobago to the United Nations, *Draft Protocol to the United Nations Framework Convention on Climate Change on Greenhouse Gas Emission Reduction.* 20 September 1994.

38 UNEP/WMO Information Unit on Climate Change, *United Nations Framework Convention on Climate Change. Text.* 1992.

39 USA, *Statement on Joint Implementation at INC11.* 6 February 1995.

— Non-government documents —

40 A SEED, *The Greenhouse Gathering. Berlin,* 2-8 April 1995.

41 Australian Conservation Foundation/B. Hare, *United Nations Framework Convention on Climate Change.* 10 May 1992.

42 *Business and Industry Statement to the First Conference of the Parties to the FCCC.* 30 March 1995.

43 Centre for Science and Environment, *Global Warming in a Unequal World. A case of environmental colonialism. Update 1.* 15 February 1991.

44 Climate Institute, *Report of the international workshop on a framework convention and associated protocols: a nongovernmental perspective.* 2-4 February 1990.

45 CAN (Climate Action Network), *Position Statement, Second INC Meeting.* 20 June 1991.

46 CAN-US, *Comments on working group texts. Executive summary and detailed notes.* December 1991.

47 CAN-US, *Technical comments on negotiating documents for INC 5 New York, February 18 through 28.* 2 February 1992.

48 CAN, *No Joint Implementation and Emissions Trading.* February 1992.

49 CAN-Europe, *Joint Implementation from a European NGO Perspective.* July 1994.

50 CAN, *Statement before INC-9.* Undated.

51 CAN-US and CAN-Europe, *Independent NGO Evaluations of National Plans for Climate Change Mitigation, OECD Countries.* January 1995.

52 CAN – Central and Eastern Europe, *Independent NGO Evaluations of National Plans for Climate Change Mitigation, Central and Eastern Europe.* January 1995.

53 CAN, *Rio is not enough. Climate Action Network conference.* 27 March, 1995.

54 Environmental Defense Fund, *Joint Implementation Discussion Paper.* 21 February 1992.

55 Foundation for International Environmental Law and Development, *Field in Brief.* Vol. 1, No. 1, 1995.

56 German NGO Forum on Environment & Development, *Positions on the Climate Summit '95.* 1995.

57 Greenpeace International, *Preventing a Climate Holocaust: a Global Defence Agenda.* June 1991.

58 Greenpeace International, *Preventing A Climate Holocaust: A National Legislative Agenda, Proposals Submitted by Greenpeace International to the Second Session of the Intergovernmental Negotiating Committee on a Draft Framework Climate Convention.* 19-28 June 1991.

59 Greenpeace International/P. Hohnen, *Draft Climate Convention.* 27 March 1992.

60 Greenpeace International, *Greenpeace Briefing on Joint Implementation.* August 1993.

61 Greenpeace International, *Climate Time Bomb. Signs of Climate Change from the Greenpeace database.* June 1994.
62 Greenpeace International/J. Leggett, *Why it is imperative to agree a 20% cut in carbon dioxide emissions in the First Protocol to the Climate Convention.* August 1994.
63 Greenpeace International/J. Leggett, *The Threats of Climate Change: A Dozen Reasons to Fear for the Future.* Climate Action Network NGO Conference 'Rio is not enough'. 27 March 1995.
64 Greenpeace International, *The OECD's Response to Climate Change: Undermining the Climate Convention, Prepared for the First Session of the Conference of the Parties to the Framework Convention on Climate Change (COP 1).* 28 March - 7 April 1995.
65 Greenpeace International, *Joint Implementation.* Undated.
66 Greenpeace International, *Key-Fact Briefings. Arguments for AOSIS Negotiators.* Undated.
67 International NGO Forum, *The NGO Alternative Treaties and the NGO Treaty Process. A complete edition of the alternative treaties from the International NGO Forum, Rio de Janeiro, Brazil.* February 1994.
68 Natural Resources Defense Council, *A Global Greenhouse Gas Convention: Recommendations on Structure and Substance.* 2 October 1989.
69 Natural Resources Defense Council/D.A. Lashof, *Concern about Global Warming: Panic or Prudence?* 10 January 1990.
70 Natural Resources Defense Council/B. Gebers & D.A. Lashof, *NGO Proposal for a "Berlin Protocol" to the Framework Convention on Climate Change.* Institute for Applied Ecology for Climate Network Europe. August 1994.
71 NGOS, *NGO Position for an International Climate Change Convention. 30 January 1991.*
72 NGOS, *The NGO Statement to the Plenary – COP 1, Berlin.* 30 March 1995.
73 Öko-Institut/G. Betty & U. Fritsche, *NGO Working Paper on Protocols to the Convention on Climate Change.* January 1994.
74 Öko-Institut/G. Betty & U. Fritsche, *Proposal for a 'Berlin Protocol' to the FCCC.* August 1994.
75 Vereniging Milieudefensie, *Persconferentie gezamenlijke milieubeweging en AOSIS-delegatie,* Den Haag. 1 March 1995.
76 World Wide Fund for Nature, *WWF Position Statement. Joint Implementation.* March 1995.
77 WWF/M.D. Davidson and G. de Wit, *Saving the Climate, That's My Job!, Case Study: Nederland.* Centre for Energy Conservation and Environmental Technology, WWF Netherlands. March 1995.

— Journals —

78 *Berliner Zeitung,* Editions of 28 March to 7 April, 1995.
79 *Earth Negotiations Bulletin,* Vol. 12, No. 21, 10 April 1995 (Summary of the First Conference of the Parties to the FCCC).
80 *Earth Negotiations Bulletin,* Vol. 12, No. 38, 22 July 1996 (Summary of the Second Conference of the Parties to the FCCC)'
81 *Earth Negotiations Bulletin,* Vol. 12, No. 68, 1 December 1997 (Report of the First Day of the Third Conference of the Parties to the FCCC).

82 *Eco*, INC2, 26 June 1991.
83 *Eco*, INC33, 13 September 1991.
84 *Eco*, INC5, 1 May 1992.
85 *Eco*, INC8, 26 August 1993.
86 *Eco*, INC8, August 1993 (Special Joint Implementation Insert).
87 *Joint Implementation Quarterly*, Vol. 1, No. 0, Spring 1995.

APPENDIX II

CBD Documents

Most relevant government and non-government policy documents and editions of journals

— Government documents —

1 FAO, *Outline Prepared by FAO for a Draft International Convention on the Conservation and Sustainable Use of Biological Diversity.* 9 November 1990.
2 Global Environment Facility, *Quarterly Operational Report.* August 1994.
3 Global Environment Facility, *GEF and Biodiversity. Undated.*
4 Interim Secretariat of the Convention on Biological Diversity, *Convention on Biological Diversity. Text and Annexes.* 1992.
5 Ministry of Environment, Physical Planning and Housing, *Reports of the Dutch Delegation to: UNEP Expert meeting 1, UNEP Expert meeting 2, INC1, INC2, INC3, INC4, INC5, UNEP Panels, Trondheim Conference, ICCBD1*, Mexico meeting, *ICCBD2, COP1.* 1990-1994.
6 United Kingdom/the Netherlands, International Technical Guidelines for Safety in Biotechnology. Draft. 21 May 1994.
7 UNEP, A Background Paper on Conventions including Lists of Special Areas. Undated.
8 UNEP/Bio.Div/WG.2/1/3, *Elements for Possible Inclusion in a Global Framework Legal Instrument on Biological Diversity.* November 1990.
9 UNEP/Bio.Div/SWGB.1/4. *Biotechnology and Biodiversity (executive summary). Ad hoc Working Group of Experts on Biological Diversity. Sub-working Group on Biotechnology, Nairobi.* November 1990.
10 UNEP/Bio.Div./WG.2/ 2/2, *Draft Convention on Biological Diversity.* 22 January 1991.
11 UNEP/Bio.Div/WG.2/1/4/add.1, *Report of the ad hoc Working Group of Legal and Technical Experts on Biological Diversity on the Work of its First Session.* 5 February 1991.
12 UNEP/Bio.Div/N4-INC.2/5, *Draft Articles adopted by the Plenary as the Basis for Future Negotiations.* October 1991.
13 UNEP/Bio.Div/N5-INC.3/4, *Draft Articles on which there is a General Understanding in Working Group II.* December 1991.
14 UNEP/Bio.Div/N5-INC.3/11, *Articles of the Draft Convention on Biological Diversity Revised by Working Groups I and II.* December 1991.

15 UNEP/Bio.Div/N6-INC.4/2, *Fourth revised draft Convention on Biological Diversity.* 16 December 1991.

16 UNEP/Bio.Div/N6-INC/WGI/CRP.6, *Working Group I, Proposal by the Netherlands for a new paragraph under the preamble.* 10 February 1992.

17 UNEP/CBD/COP/1/1, *Provisional agenda.* 21 September 1994.

18 UNEP/CBD/COP/1/3, *Report of the Intergovernmental Committee on the Convention on Biological Diversity on the Work of its First Session.* 21 September 1994.

19 UNEP/CBD/COP/1/4, *Report of the Intergovernmental Committee on the Convention on Biological Diversity on the Work of its Second Session.* 21 September 1994.

20 UNEP/CBD/COP/1/2, *Draft Rules of Procedure for Meetings of the Conference of the Parties to the Convention on Biological Diversity.* 4 October 1994.

21 UNEP/CBD/COP/1/6, *Institutional Structure to operate the Financial Mechanism under the Convention,* 6 October 1994.

22 UNEP/CBD/COP/1/16, *Report of the Open-ended Intergovernmental Meeting of Scientific Experts on Biological Diversity, including the agenda for scientific and technological research.* 7 November 1994.

23 UNEP/CBD/COP/1/12, *Preparation of the Participation of the Convention on Biological Diversity in the third session of the Commission on Sustainable Development (CSD),* 28 November-9 December 1994.

24 UNEP/CBD/COP/1/13, *Medium-Term Programme of Work of the Conference of the Parties. Note by the Interim Secretariat.* 28 November-9 December 1994.

25 UNEP/CBD/COP/1/CW/L.1, *Financial Resources and Mechanism,* 1 December 1994.

26 UNEP/CBD/COP/1/CW/L.2, *Medium-Term Programme of Work of the Conference of the Parties.* 2 December 1994.

27 UNEP/CBD/COP/1/CW/L.6, *Medium-Term Programme of Work of the Conference of the Parties. Safety in biotechnology.* 3 december 1994.

28 UNEP/CBD/COP/1/L.10, *Preparation of the Participation of the Convention on Biological Diversity in the Third Session of the Commission on Sustainable Development.* 3 December 1994.

29 UNEP/CBD/COP/1/Inf. 11, *List of Participants.* 8 December 1994.

30 UNEP/CBD/COP/1/Inf. 12, *Declaration by the Ministers of the Alliance of Small Island States (AOSIS) at the First Conference of the Parties to the Convention on Biological Diversity.* 8 December 1994.

31 UNEP/CBD/COP/1/L.12, *Rules of Procedure for the Conference of the Parties.* 8 December 1994.

32 UNEP/CBD/COP/1/17, *Report of the First Meeting of the Conference of the Parties to the Convention on Biological Diversity.* 28 February 1995.

33 UNEP/CBD/COP/2/19, *Report of the Second Meeting of the Conference of the Parties to the Convention on Biological Diversity.* 30 November 1995.

34 UNEP/CBD/BSWG/1/3, *Elaboration of the Terms of Reference for the open-ended ad hoc Working Group on Biosafety.* 23 May 1996.

— Non-government documents —

35 Accion Ecologica, FOE, TWN, CNI, *Biosafety protocol: is the OECD promoting double standards?* 6 December 1994.

36 Biodiversity Action Network (Bionet), *Proposals for action by the first Conference of the Parties to the Convention on Biological Diversity (final version).* November 1994.

37 CNI/Richard Schweiger, *NGO Call for a legally binding protocol on biosafety.* 24 February 1995.

38 CNI (USA), Greenpeace, Institute for Agriculture and Trade Policy (USA), TWN, *NGO/Government Roundtable on the Urgent Need for a Biosafety Protocol.* 1 December 1994.

39 Friends of the Earth, *Intellectual Property Rights and the Biodiversity Convention: the Impact of GATT.* November 1994.

40 Friends of the Earth, Seven suggestions for the consolidation of the Convention on Biological Diversity. Undated.

41 Genetic Resources Action International (GRAIN), Genes for Sustainable Development. Briefing on Biodiversity Nr. 1. October 1990.

42 Genetic Resources Action International (GRAIN), *Genes for Sustainable Development. Briefing on Biodiversity Nr. 2.* February 1991.

43 Genetic Resources Action International (GRAIN), *Briefing on Biodiversity Nr. 3.* August 1992.

44 Genetic Resources Action International (GRAIN), *Biobriefing Nr. 4.* June 1994.

45 German NGO Working Group on Indigenous People and the Protection of Nature, *The protection of nature and the preservation of biological diversity with special reference to the role of indigenous peoples. Paper submitted to the First Conference of the Parties to the Convention on Biological Diversity.* November 1994.

46 German NGOs of the Forum Umwelt & Entwicklung, *Position paper on a new international political instrument for the global protection of forests.* Undated

47 Greenpeace, *Letter to delegates at the UNEP Biodiversity Conference in Trondheim, 24-28 May 1993, concerning Norway to allow a resumption of commercial whaling.* 24 May 1993.

48 Greenpeace International/Jesper Grolin, *Statement on the Institutional Structure to operate the financial mechanism.* Undated.

49 Guaymi General Congress of Panama, *Letter to protest against US patent claim on a Guaymi cell line.* 5 October 1993.

50 IIPA/Ashish Kothari, *The Biodiversity Convention and the Commission on Sustainable Development: missing elements in the proposed statement.* Undated.

51 International NGO Forum (INGOF), *The NGO Alternative Treaties and the NGO Treaty Process. A complete edition of the alternative treaties from INGOF, Rio de Janeiro, Brazil.* February 1994.

52 IUCN, *Draft articles prepared by IUCN for inclusion in a proposed convention on the conservation of biological diversity and for the establishment of a fund for that purpose.* June 1989.

53 IUCN/TNC/WWF, *Report of the First Global Forum on Environmental Funds.* Santa Cruz, Bolivia. 30 May-2 June 1994.

54 IUCN/WRI/UNEP, *Global Biodiversity Forum, Nassau, Bahamas,* 26-27 November 1994.

55 IUCN, *Implementing the Convention on Biological Diversity: IUCN's capacities and possible contribution.* Undated.

56 IUCN/WWF, *Discussion paper. Establishing priorities for the conservation and sustainable use of biological diversity in the context of the Convention on Biological Diversity.* Undated.

57 NGOS, *NGO concerns on biodiversity discussions in WGI.* 24 March 1992.

58 NGOS, *Draft discussion paper for the First Conference of the Parties to the Biodiversity Convention. Moving from words to action. A 10-Point NGO Action Plan for consideration by Governments.* November 1994.

59 NGOS, *NGO Statement on forests requested to be attached as an Annex to the Paper transmitted from the COP to the Commission for Sustainable Development.* Undated.

60 NGOS-Nederland, *Vragenlijst Biodiversiteitsconventie.* Undated.

61 NGO-Task Group on Biodiversity, *Draft report of the latest negotiations on the Biological Diversity Convention.* January 1992.

62 NGO-Task Group on Biodiversity, *Verslag consultatiebijeenkomst Nederlandse NGO's voor de zesde sessie van de Intergovernmental Negotiating Committee for a Convention on Biodiversity.* 22 January 1992

63 NGO-Task Group on Biodiversity, *Report of the sixth negotiating session on the Convention on Biological Diversity.* 6-15 February 1992.

64 NGO-Task Group on Biodiversity, *Draft recommendations of the ad hoc working group on biotechnology and biodiversity.* 10 March 1992.

65 NGO-Task Group on Biodiversity, *Report of the NGO-Task Group on the Biodiversity Convention. Overview of the viewpoints of a number of NGOs.* Undated.

66 NGO-Task Group on Biodiversity, *Draft report of the latest negotiations on the Biological Diversity Convention.* Undated.

67 NGO-Task Group on Biodiversity, *Recommendations on access to genetic resources and rights of indigenous people and local communities.* Undated.

68 NGO-Task Group on Biodiversity/Simone Bilderbeek, *Proposal for a provision including liability for genetically modified organisms.* Undated.

69 NGO-Task Group on Financial Matters under the CBD, *A statement to the first Conference of the Parties to the Convention on Biological Diversity.* Undated.

70 Sierra Club, *Policy on Biotechnology.* September 1993.

71 Sierra Club, *The Extinction Crisis.* Undated.

72 TWN/Gurdial Singh Nijar, *A conceptual framework and essential elements of a rights regime for the protection of indigenous rights and biodiversity.* Undated.

73 TWN, *Biodiversity Convention briefings.* June 1994.

74 TWN, *Biodiversity Convention briefings: Reports on the extraordinary session of the FAO commission on plant genetic resources. Germplasm collections brought under UN umbrella.* November 1994.

75 TWN, Greenpeace, CNI, FOE, *Is UNEP Sidetracking work on a biosafety protocol?* 5 December 1994.

76 TWN/G.S. Nijar, Why the ex-situ collections are so crucial. Undated.

77 WRI/IUCN/UNEP, *Presentation by the Global Biodiversity Forum to ICCBDI,* 11 October 1993.

78 WWF/Simon Lyster, Proposed Biological Diversity Convention. 5 December 1990.

79 WWF, *A WWF International Project Portfolio: Biodiversity.* April 1993.

80 WWF, *A WWF International Position Paper: The Conference of the Parties to the Convention on Biological Diversity, First Meeting.* June 1994.

81 WWF, *The need for a biosafety protocol as part of the Convention on Biological Diversity. A background paper produced by WWF for the first Conference of the Parties to the Convention on Biological Diversity. Nassau, the Bahama's.* 28 November-9 December 1994.

— Journals —

82 *Earth Negotiations Bulletin,* Vol. 9, No. 6, 22 October 1993 (Summary of the First Session of the Intergovernmental Committee on the Convention on Biological Diversity, 11-15 October 1993).
83 *Earth Negotiations Bulletin,* Vol. 9, No. 17, 6 July 1994 (Summary of the Second Session of the Intergovernmental Committee on the Convention on Biological Diversity, 20 June - 1 July 1994).
84 *Earth Negotiations Bulletin,* Vol. 9, No. 18, 28 November 1994 (First Meeting of the Conference of the Parties to the Convention on Biological Diversity, 28 November - 9 December 1994).
85 *Earth Negotiations Bulletin,* Vol. 9, No. 21, 1 December 1994 (First Meeting of the Conference of the Parties to the Convention on Biological Diversity, 28 November - 9 December 1994).
86 *Earth Negotiations Bulletin,* Vol. 9, No. 39, 19 November 1995 (A Summary Report on the Second Session of the Conference of the Parties to the Convention on Biological Diversity).
87 *Earth Negotiations Bulletin,* Vol. 9, No. 65, 18 November 1996 (A Summary Report on the Third Session of the Conference of the Parties to the Convention on Biological Diversity).
88 *Earth Negotiations Bulletin,* Vol. 13, No. 34, 24 February 1997 (Summary of the Fourth Session of the Intergovernmental Panel on Forests).
89 *Earth Negotiations Bulletin,* Vol. 5, No. 88, 30 June 1997 (Summary of the Nineteenth United Nations General Assembly Special Session to Review Implementation of Agenda 21).
90 *Eco,* ICCBD1, 1 October 1993.
91 *Eco,* ICCBD1, 4 October 1993.
92 *Eco,* ICCBD2, 22 June 1994.
93 *Eco,* ICCBD2, 24 June 1994.
94 *Eco,* ICCBD2, 26 June 1994.
95 *Eco,* ICCBD2, 1 July 1994.
96 *Eco,* COP1, 29 November 1994.
97 *Eco,* COP1, 1 December 1994.
98 *Eco,* COP1, 5 December 1994.
99 *Eco,* COP1, 7 December 1994.
100 *Environment Bulletin,* Vol. 6, No. 2, Spring 1994.
101 *The Biodiversity Coalition,* Newsletter No. 9, September 1994.

APPENDIX III

Background of Respondents

One or more representatives of the following countries, UN bodies and NGOs were interviewed:

Countries	*UN Bodies*	*NGOs*
- Argentina	- FAO	- Bangladesh Centre for Advanced Studies
- Brazil	- Interim Secretariat CBD	- Bionet
- Canada	- Interim Secretariat FCCC	- Birdlife International
- Denmark		- Climate Action Network – Africa
- Chile		- Climate Action Network – Central and Eastern Europe
- France		- Climate Action Network – Europe
- Ghana		- Climate Action Network – Latin America
- Japan		- Climate Action Network – Southeast Asia
- Kenya		- Climate Action Network – South Asia
- Malaysia		- Climate Action Network – USA
- Nauru		- Environment Liaison Centre International
- Netherlands		- Foundation for International Environmental Law and Development
- Norway		- Friends of the Earth International
- Sweden		- Genetic Resources Action International
- United Kingdom		- Greenpeace International
- United States		- Indian Institute of Public Administration
		- Natural Resources Defense Council
		- Stichting Natuur en Milieu
		- Third World Network
		- Vereniging Milieudefensie
		- World Conservation Union (IUCN)
		- World Resources Institute
		- World Wide Fund for Nature (WWF)

APPENDIX IV

Curriculum Vitae

Bas Arts was born in Langenboom (municipal of Mill & St. Hubert, the Netherlands) in 1961. He passed secondary school VWO-b at Elzendaal College in Boxmeer in 1980. In the period 1980-1988 he studied biology at the University of Nijmegen, specializing in the ecology of forests and the ecology of mammals, as well as in the issue of environment and development. The study included a 9 month stay in Nepal (field study on the exploitation and management of fir forests) as well as a 1 year course in development studies (specialization in environment and development).

Bas Arts was affiliated to the Peace Research Centre of the University of Nijmegen in the period 1989-1993. He co-ordinated a scientific project on the substance of the famous United Nations' Brundtland report *Our Common Future*, during which an academic course on the subject of sustainable development was designed and a study on the role of Dutch NGOs in the UNCED process was conducted. The output of this study was published in the book *Duurzame Ontwikkeling, Duurzame Samenwerking? Allianties tussen sociale bewegingen* (Sustainable Development, Sustainable Co-operation? Alliances between social movements).

Since 1993, Bas Arts has been attached to the Department of Environmental Policy Sciences, Faculty of Policy Sciences, University of Nijmegen. The PhD study *The Political Influence of Global NGOs* was conducted and finalized in the period 1993-1997. Since 1 September 1997, Bas Arts is a lecturer at the same institution.